T0231118

The Wadsworth & Brooks/Cole Statistics/Probability Series

R. Becker, J. Chambers, A. Wilks, *The New S Language: A Programming Environment for Data Analysis and Graphics*

P. Bickel, K. Doksum, J. Hodges, Jr., *A Festschrift for Erich L. Lehmann*

G. Box, *The Collected Works of George E. P. Box*, Volumes I and II, G. Tiao, editor-in-chief

L. Breiman, J. Friedman, R. Olshen, C. Stone, *Classification and Regression Trees*

J. Chambers, W. S. Cleveland, B. Kleiner, P. Tukey, *Graphical Methods for Data Analysis*

W. S. Cleveland, M. McGill, *Dynamic Graphics for Statistics*

K. Dehnad, *Quality Control, Robust Design, and the Taguchi Method*

R. Durrett, *Lecture Notes on Particle Systems and Percolation*

F. Graybill, *Matrices with Applications in Statistics*, Second Edition

L. Le Cam, R. Olshen, *Proceedings of the Berkeley Conference in Honor of Jerzy Neyman and Jack Kiefer*, Volumes I and II

P. Lewis, E. Orav, *Simulation Methodology for Statisticians, Operations Analysts, and Engineers*

H. J. Newton, *TIMESLAB*

J. Rawlings, *Applied Regression Analysis: A Research Tool*

J. Rice, *Mathematical Statistics and Data Analysis*

J. Romano, A. Siegel, *Counterexamples in Probability and Statistics*

J. Tanur, F. Mosteller, W. Kruskal, E. Lehmann, R. Link, R. Pieters, G. Rising, *Statistics: A Guide to the Unknown*, Third Edition

J. Tukey, *The Collected Works of J. W. Tukey*, W. S. Cleveland, editor-in-chief
Volume I: *Time Series: 1949–1964*, edited by D. Brillinger
Volume II: *Time Series: 1965–1984*, edited by D. Brillinger
Volume III: *Philosophy and Principles of Data Analysis: 1949–1964*, edited by L. Jones
Volume IV: *Philosophy and Principles of Data Analysis: 1965–1986*, edited by L. Jones
Volume V: *Graphics: 1965–1985*, edited by W. S. Cleveland

Simulation Methodology for Statisticians, Operations Analysts, and Engineers

Volume I

P. A. W. Lewis
Naval Postgraduate School

E. J. Orav
Harvard School of Public Health

CRC Press
Taylor & Francis Group
Boca Raton London New York

CRC Press is an imprint of the
Taylor & Francis Group, an **informa** business

First published 1989 by Wadsworth & Brooks/Cole Advanced Books & Software
Taylor & Francis Group
6000 Broken Sound Parkway NW, Suite 300
Boca Raton, FL 33487-2742

Reissued 2018 by CRC Press

© 1989 by Taylor & Francis
CRC Press is an imprint of Taylor & Francis Group, an Informa business

No claim to original U.S. Government works

A Library of Congress record exists under LC control number: 88005536

Publisher's Note
The publisher has gone to great lengths to ensure the quality of this reprint but points out that some imperfections in the original copies may be apparent.

Disclaimer
The publisher has made every effort to trace copyright holders and welcomes correspondence from those they have been unable to contact.

ISBN 13: 978-1-138-10537-9 (hbk)
ISBN 13: 978-1-138-56187-8 (pbk)
ISBN 13: 978-0-203-71031-9 (ebk)

Visit the Taylor & Francis Web site at http://www.taylorandfrancis.com and the CRC Press Web site at http://www.crcpress.com

Interior and Cover Design: *Flora Pomeroy*
Art Coordinator: *Sue C. Howard*
Interior Illustration: *Graphic Arts*
Typesetting: *Asco Trade Typesetting Limited*
Cover Printing: *Phillips Offset*

Preface

Simulation is essentially a controlled statistical sampling technique that, with a model, is used to obtain approximate answers for questions about complex, multi-factor probabilistic problems. It is most useful when analytical and numerical techniques are unable to supply exact answers. In attempting to provide these approximate answers by repeated sampling on a digital computer, a host of techniques and disciplines are called into play:

- Probability concepts in the modeling of systems
- Human experience and intuition in the laying out of the assumptions and specifications of the model
- Number theory in the generation of uniform pseudo-random numbers
- Probability theory in the generation of nonuniform pseudo-random numbers
- Stochastic process theory for estimation techniques based on sample paths (ergodic theory)
- Computing science in the implementation of random number generators and for the organization and graphical display of data sets
- Statistical theory for the all-important design of the simulation and the analysis of the data obtained as output

It is this interaction of experience, applied mathematics, statistics, and computing science that makes simulation such a stimulating subject, but at the same time a subject that is difficult to teach and write about. The sophisticated practice of simulation requires a broad background on the part of the user and healthy interdisciplinary interaction.

In fact, we have only to examine the contents of introductory courses on simulation to see the complexity and diversity of the subject. Some courses deal primarily with random number generation, some concentrate on modeling, and some teach only simulation programming languages. Seldom do we find a course with a basic emphasis on the statistical and modeling techniques, which is where we feel the emphasis should be. In writing this simulation text, we have attempted to provide in one place many of the prerequisite statistical and graphical techniques that are unavailable or unknown to engineers, scientists, and computing specialists.

We have also tried to stress the strong interaction of applied mathematics, statistics, and computing in simulation. Finally, we emphasize, mostly by example, the probabilistic and human sides of modeling operational systems. Thus, our goal is to train a practitioner who can detect the shortcomings of reported simulation results and who can personally carry out a valid simulation.

The work has been divided into two volumes, more for the sake of manageability and readability than because of any deep logical division between the contents of the two volumes. In fact, the main division of the complete work is between crude simulation techniques, which are given in Part I of Volume I, and sophisticated simulation techniques, which take up the rest of Volume I and all of Volume II. This first volume is a prerequisite for the second volume and for the running of any simulation, in that these five chapters contain the basic ideas for modeling and random number generation, while the remaining six chapters discuss more advanced but fundamental topics such as variance reduction and comparison of simulations. Moreover, if the simulation produces replicates of independent output, either univariate or bivariate, then the statistical and graphical output analysis techniques described in this first volume should be applied. Such independent output is almost always the result of statistical simulations and, when performed crudely, of systems simulations as well. Volume II goes on to consider more sophisticated methods for analyzing the dependent output that can often result from systems simulations. It also covers the sophisticated generation of random variables and the generation of random stochastic sequences as input to simulations. These are very advanced topics not essential to most beginning and intermediate simulation practitioners.

Three features of the book make it unique.

First, we deal with simulation methodology for problems in both mathematical statistics and systems simulation. The problems in these two areas vary somewhat but also have a large common stratum, which is laid out in Volume I.

Second, we have included statistical methods based on graphical techniques and exploratory data analysis, thereby emphasizing the importance of the analysis of simulation output data.

Third, the book is organized to present the simplest ideas of simulation (crude simulation) before proceeding to ideas of sophisticated simulation. Thus, the first five chapters of Volume I should serve as an introduction to simulation for engineers, operations analysts, scientists, statisticians, and computing specialists, who may have only a minimal background in statistics and probability theory. These five chapters are also written at a lower level of sophistication in probability and statistics than the remainder of the book.

This complete first volume is most appropriate for an introductory simulation course of one or more semesters in the second half of a master's level program in operations research, industrial engineering, civil engineering, electrical engineering, computer science, or statistics. Preferred prerequisites are one quarter of probability theory, one quarter of stochastic processes, one quarter of statistical theory (inference) at a master's level, and one quarter's introduction to computing and programming. Although not necessary, it is helpful to have (again at a graduate level) a one quarter course in regression analysis and analysis of variance, a one quarter course in probabilistic models, and an introductory course in computer science dealing with machine organization and languages. The book, however, does develop the material that is particularly germane to simulation and that normally would be covered in the latter courses. Also, references are given to fill in many of the details presumed already known.

The second volume serves as the basis for an advanced course in simulation with an emphasis on discrete-event and systems simulation. Prerequisites would include a course based on the first volume, as well as additional background in stochastic processes and applied probability.

We believe that the purpose of simulation is, generally, to facilitate choice among several different and competing schemes for a practical operational or statistical situation, such as which of several queue disciplines or which of several inventory policies is better. Therefore, it is with reluctance that we have limited much of the book to techniques for analyzing one sample in detail and for comparing two samples. Extensive graphical methods, however, are given for comparing multiple samples, as well as an introduction in Chapter 8 to the literature available on experimental design and multifactor experiment analysis. We have done this primarily because the comparison of two samples is fairly straightforward, whereas the comparison or ranking of more than two samples creates a quantum jump in the amount and sophistication of statistical methodology needed.

Finally, we point out that one reason many practical operational problems seem intractable is that they overtax available or foreseeable analytical techniques (exact and approximate); moreover, the problems often are too complex to be solved with sufficient precision by simulation on presently available digital computers. The limitations imposed on simulation by the size and speed of computers are likely to become less important because of new developments in computer hardware and software. The remaining problem will then be the available statistical methodology that is applicable in these simulations.

We hope that by presenting the available and newly developed statistical and modeling methodology, we will stimulate the development of new ideas to extend the use of digital simulation techniques.

Acknowledgments

Many people have contributed to this book, mostly by reading it, correcting it, and commenting on it. Special thanks are due to Professor Arie Hordijk, University of Leiden, whose invitation to lecture on simulation to students at the Free University in Amsterdam, Amsterdam University, and the Mathematics Center in Amsterdam provoked the first draft of this book. We also thank Professor Bruce Schmeiser, whose initial reading of the chapter on variance reduction made us rethink the whole topic.

The following people have read and commented on parts of the book: David Burns, Paul Fishbeck, Tony Lawrence, Thomas Lewis, Eddie McKenzie, Steve Pilnich, Richard Ressler, Edward Rockower, Lee Schruben, and Al Washburn.

Many students in simulation classes at the Naval Postgraduate School helped to correct the manuscript and to test its suitability as a textbook. Typing was performed principally by Susie "Turbo" Pickens; her good cheer and incredible competence were a great source of inspiration. We also thank Barbara Potkay, Ellen Saunders, and Sabrina O'Jack for typing and editorial assistance.

Luis Uribe provided valuable insights into computing problems and is the coauthor, with us, of the software supplement to the book. Many figures in the book, particularly the figures from the SMTBPC program, were created with this package. This software supplement has gone through three editions, all published by Wadsworth & Brooks/Cole under the titles *Introductory Simulation and Statistics Package* (1985), *Advanced Simulation and Statistics Package* (1986), and the current version, *Enhanced Simulation and Statistics Package* (1989).

Reviewers of this book, William Biles, Louisiana State University; Larry George, Lawrence Livermore National Laboratory; Donald Haber, University of Idaho; and Lee Schruben, Cornell University, also provided valuable comments.

Some of the figures in this book were created with the APL-based graphics program GRAFSTAT from IBM Research. We are indebted to Dr. Peter D. Welch for making this program available on a test-bed basis at the Naval Postgraduate School. Other members of the IBM research staff who helped with our use of GRAFSTAT are P. Heidelberger, D. Stein, T. Lane, A. Blum, and G. Berkland. Other figures were created using the STATGRAPHICS program distributed by STSC, Inc.

Finally, the research of P. A. W. Lewis was supported, for much of the time during which this book was being written, by the Office of Naval Research under various grants. We are grateful for their continued support.

P. A. W. Lewis
E. J. Orav

Brief Contents of Volume I

Contents of Volume I

Chapter 11 Variance Reduction   334

List of Examples

Introduction

Simulation, as defined in Chapter 1 of this book, can be divided into three major, usually overlapping, areas:

1. *Setting up the simulation problem.* This includes identification of the question being asked, the assumptions that need to be made, and the model that is to be used.

2. *Running the simulation on a digital computer.* Here we consider the techniques and methodologies, simple or advanced, that are used to produce the simulation output. Issues include random variable generation, precision of estimates, and variance reduction techniques. Advanced topics such as discrete event simulation methodology, sophisticated generation of random variables, and the generation of stochastic sequences, are taken up in Volume II of this text.

3. *Analyzing the output.* Given a question and the proper simulation output, we can go on to statistical estimates and graphical displays in order to interpret the output and make decisions.

In addressing the first of these points, Chapters 1, 2, and 3 of this book define "models" and "simulation," give the purposes of simulation, and discuss the basic principles and golden rules of simulation. Also included is the all-important question of modeling and five typical examples of systems or problems in which simulation might be required to obtain answers to questions that have been posed. We say "might be required" because in some cases it will suffice to use simple approximations to the problem for which analytic solutions are available.

Chapter 4 goes on to give some basic elements needed to run what might be called a crude (straightforward) Monte Carlo or simulation; there are five of these elements, all of which have straightforward but generally inefficient techniques associated with them. There are also many efficient and advanced techniques for performing and analyzing a simulation, and these are described under the heading of "Sophisticated Monte Carlo," starting with Chapter 6. The use of sophisticated simulation techniques is almost essential if one is going to achieve sufficient precision in a simulation to answer probabilistic questions in a definite way.

Generation of pseudo-random numbers is discussed in Chapter 5. All simulations discussed in this book assume that a source of such numbers, which essentially simulate sequences of independent uniformly distributed random variables, is available. Fortunately, today, with reasonable care, this is a valid assumption, whereas 10 or 15 years ago pseudo-random number generation was problematical and was the subject of intense research and discussion.

Chapters 1 through 4, with the possible addition of Chapter 5, can be used as a complete text for an introduction to simulation methodology and modeling for students who have only a minimal preparation in probability and statistics.

Presentation of some of the different aspects of sophisticated simulation (Monte Carlo) starts in Chapters 6 & 7 with methodology for analyzing and displaying single batches of data. Readers with sophisticated backgrounds in statistics will want to examine only the simulation examples in these chapters and use them as a reference. Design and comparison techniques for multiple outputs from a multi-factor simulation are discussed in Chapter 8. Examples such as the analysis of a multiple-server queue, which were initially addressed in Chapters 6 and 7, are completed here. Techniques such as sectioning, jackknifing, and bootstrapping for assessing the variability of the output are considered in Chapter 9, bivariate problems in Chapter 10, and variance reduction in Chapter 11. Although these topics are not comprehensive of all techniques available in simulation, they provide the reader with all the basic, and some of the advanced tools, necessary to design, run, and analyze a simulation. Moreover, the topics reflect our chosen emphasis on the use of statistics and graphics in simulation, and the use of simulation in statistics. The other broad area of simulation application, systems simulation, is dealt with in Volume II of this work, where we look at the techniques for generating stochastic sequences and analyzing dependent output. Also included in the second volume are advanced techniques for generating random variables.

References and Literature

There are many books on the various aspects of simulation, and we describe briefly some of the more important works. We also discuss the differences between these books and the present volume.

Five recent books on simulation are those by Fishman (1978), Law and Kelton (1982), Bratley, Fox, and Schrage (1983, 1987), Morgan (1984) and Ripley (1987). The first three are distinguished by an emphasis on *systems simulation* and, in particular, discrete event simulation. Discrete event simulation refers to systems like queues, whose state changes only at distinct event times so that their sample paths are easily simulated. In this respect, Fishman (1973) is still a valuable reference. The very particular discrete event simulation called the regenerative method is detailed in Crane and Lemoine (1977) and Iglehart and Shedler (1980). Recently, there has been interest in so-called continuous simulation—for instance, simulation of runs of homing torpedos (see Korn and Wait, 1978).

The statistical aspects of simulation are the focus of Ripley (1987) and Morgan (1984) and the books by Kleijnen (1974a, b), which also contain valuable surveys of applications of variance reduction in operations analysis. Algorithmic and mathematical issues of pseudo-random number generation, and the testing of pseudo-random number generators are discussed in Kennedy and Gentle (1980, Ch. 6) and Knuth (1981, Ch. 3). Another earlier and valuable general reference for simulation is Hammersley and Handscomb (1964). Applications of simulation given in that book are, in particular, to problems that arise in physics. Valuable and simple descriptions of variance reduction techniques are also given. A later reference to the use of simulation in physics is Binder (1984); see, particularly, the review chapter by Binder and Stauffer (1984). A very detailed description of techniques for generating nonuniform random variables is given in DeVroye (1986). Generation of multivariate random variables is considered in Johnson (1987).

A rather mathematical extension of many ideas of variance reduction and variate generation, in particular to multivariate situations, is given by Rubinstein (1981). There are also many recent survey articles and complete issues of technical journals devoted to applications of simulation in various fields. Thus, the journal *Operations Research* has a special issue on simulation, and in particular, on systems simulation (*Operations Research*, Vol. 31, No. 6). Also, the Association for Computing Machinery (ACM) published in the *Communications of the ACM* (Vol. 28, No. 45, April 1985), a special section entitled "Computing at the Frontiers of Theoretical Physics." Further afield, *Byte* (the Small Systems Journal) devoted a large portion of an issue (Vol. 10, No. 10, October 1985) to the topic of "simulating society." In particular, the article in that issue "Why Models Go Wrong," by T. R. Houston, is well worth reading.

The present book differs from those books and articles referenced above in several ways:

1. There are very few books on applications of simulations *in* statistics, even though it is a more and more frequently used tool of analysis. We stress in this book the common methodology of simulation in statistics, science, and operational analysis, and give applications in all of these fields.

2. This book is organized so that the first four chapters can be used as an introduction to simulation at a quite elementary level. More sophisticated techniques such as variance reduction are presented at a more advanced technical level in subsequent chapters. Systems simulation and the generation and analysis of stochastic sequences are presented separately in Volume II.

3. We have emphasized modern statistical techniques for data analysis and presentation.

4. We have emphasized graphical methods for analyzing simulation experiments. Not only is this eminently practical, because simulation outputs are already in a computer, it is essential if the full benefit of an analysis is to be achieved.

5. FORTRAN programs for microcomputers are available as a supplement to implement the analyses given in the book. The supplement, entitled *Enhanced Simulation and Statistics Package*, consists of an extensively documented and illustrated manual, as well as a floppy disk with compiled programs and illustrative drivers.

The initial release of this software supplement (Lewis, Orav, and Uribe, 1984) ran under Microsoft FORTRAN Version 3.13; the second release (Lewis, Orav, and

Uribe, 1986) was available to run under Digital Research FORTRAN 77 or IBM's Professional FORTRAN (Version 1.14 or later); the latest release (Lewis, Orav, and Uribe, 1988) is available to run under IBM's Professional FORTRAN (Version 1.14 or later) or equivalently, Ryan-McFarland FORTRAN (Version 2.0 or later); FORTRAN implementations for other computers are available from the authors. The main program in these supplements is SUPER SIMTBED, a program for the graphical output analysis of multifactor simulations. The use of this program is detailed in Chapters 8, 9, and 10.

Use of This Book in Teaching

This book is designed for use for several different audiences: elementary level, engineers and operations analysts, statisticians, and computer scientists.

• *Elementary Level*   At the most basic level, the first four chapters on crude simulation make up a complete course of one quarter's duration. There are many exercises at the end of Chapter 4; all but the simplest of these should be skipped. If possible, the student should have an elemental idea of how pseudo-random number generators work and what is available; Chapter 5 addresses this area. Although the chapter is not theoretical, sections on detailed properties of pseudo-random number generators can be skipped.

• *Engineers and Operations Analysts*   Chapters 1 through 4 can be covered, in a senior or first-year graduate level course, in 3 weeks. Most of the exercises should be attempted. Chapter 5, on pseudo-random number generation, can be covered in about a week.

The core of the remainder of the course is Chapters 8, 9, and 10. Some of the ideas on descriptions of quantifications of univariate samples in Chapters 6 and 7 should be touched on, especially the graphical methods in Chapter 7. However, an attempt to cover all the material would constitute a short statistics course and would detract from the main topic, namely simulation. The best plan is to discuss the two continuing examples in these two chapters.

If time permits, Chapter 11 on variance reduction can be covered in about 4 weeks. Sections 11.1 and 11.2, on antithetic variates and control variables, require less mathematical sophistication than the latter part of the chapter and would suffice for an introduction to the topic of variance reduction.

The topic of systems simulation, which is covered in Volume II, can be studied in a second quarter.

• *Statisticians*   The course for statisticians would be similar to that for engineers and operations analysts, except that the students should already be familiar with the material in Chapters 6 and 7 on descriptions and quantifications of univariate samples. These chapters could be assigned as a review after the completion of Chapters 1 through 4 in the first 3 weeks. Chapter 5 on uniform pseudo-random variable generation should be covered in total, and it should be possible and profitable to present all of Chapter 11 on variance reduction. Again, after the introduction to crude simulation, Chapters 8, 9, and 10 are the core of the course.

• *Computer Scientists*   The course for computer scientists would be similar to the first quarter for engineers and operations analysts. However, the statistical material in Chapters 6 and 7 would be of less direct interest and might be skipped. The course might also go more deeply into the theory of uniform pseudo-random variable generation than is possible in Chapter 5; the second volume of Knuth (1981) would be appropriate for this.

References

Binder, K. (1984). *Applications of the Monte Carlo Method in Statistical Physics.* Springer: Berlin.

Binder, K., and Stauffer, D. (1984). A Simple Introduction to Monte Carlo Simulation and Some Specialized Topics. In *Applications of the Monte Carlo Method in Statistical Physics*, K. Binder, Ed. Springer: Berlin.

Bratley, P., Fox, B. L., and Schrage, L. E. (1983). *A Guide to Simulation*, 1st ed. Springer: Berlin.

Bratley, P., Fox, B. L., and Schrage, L. E. (1987). *A Guide to Simulation*, 2nd ed. Springer: Berlin.

Crane, M. A., and Lemoine, A. J. (1977). *An Introduction to the Regenerative Method for Simulation Analysis.* Springer: Berlin.

DeVroye, L. (1986). *Non-uniform Random Variate Generation.* Springer: Berlin.

Fishman, G. S. (1973). *Concepts and Methods in Discrete Event Digital Simulation.* Wiley: New York.

Fishman, G. S. (1978). *Principles of Discrete Event Simulation.* Wiley: New York.

Hammersley, J. M., and Handscomb, D. C. (1964). *Monte Carlo Methods.* Methuen: London.

Iglehart, D., and Shedler, G. S. (1980). *Regenerative Simulation of Response Times in Networks of Queues.* Springer: New York.

Johnson, M. (1987). *Multivariate Statistical Simulation: A Guide to Selecting and Generating Continuous Multivariate Distributions.* Wiley: New York.

Kennedy, W. J., Jr., and Gentle, J. E. (1980). *Statistical Computing.* Dekker: New York.

Kleijnen, J. P. C. (1974a). *Statistical Techniques in Simulation: Part I.* Dekker: New York.

Kleijnen, J. P. C. (1974b). *Statistical Techniques in Simulation: Part II.* Dekker: New York.

Knuth, D. E. (1981). *The Art of Computer Programming*, Vol. 2: *Seminumerical Algorithms*, 2nd ed. Addison-Wesley: Reading, MA.

Korn, G. A., and Wait, J. V. (1978). *Digital Continuous-System Simulation.* Prentice-Hall: Englewood Cliffs, NJ.

Law, A. M., and Kelton, W. D. (1982). *Simulation Analysis and Modeling.* McGraw-Hill: New York.

Lewis, P. A. W., Orav, E. J., and Uribe, L. (1984). *Introductory Simulation and Statistics Package.* Wadsworth & Brooks/Cole: Pacific Grove, CA.

Lewis, P. A. W., Orav, E. J., and Uribe, L. (1986). *Advanced Simulation and Statistics Package.* Wadsworth & Brooks/Cole: Pacific Grove, CA.

Lewis, P. A. W., Orav, E. J., and Uribe, L. (1988). *Enhanced Simulation and Statistics Package.* Wadsworth & Brooks Cole: Pacific Grove, CA.

Morgan, B. J. T. (1984). *The Elements of Simulation.* Chapman & Hall: London.

Ripley, B. D. (1987). *Stochastic Simulation.* Wiley: New York.

Rubinstein, R. Y. (1981). *Simulation and the Monte Carlo Method.* Wiley: New York.

Modeling and
Crude Simulation

1

Definition of Simulation

We discuss here briefly the definition and scope of simulation and its several aspects. Because terminology varies in the literature, the definitions that follow are common but not universal. To avoid confusion and a lengthy lesson in semantics, alternative interpretations and nuances are not presented; however, these can be found in detail in Kleijnen (1974, Ch. 1).

An idea that will come up repeatedly is that of a system. A *system is a collection of interacting parts*, and much of the effort in simulation is directed toward determining the result of such interactions. A computer is an example of a system, as is the whole complex of the computer users in their interaction with the computer. Some systems are called *deterministic* because, given the same *input* and starting situation, the *output* will be uniquely determined. But there are also systems where random, unpredictable events enter the interaction. These are called *stochastic* or *random* systems, and they are the type of systems considered in this book.

Next we discuss the very important idea of a model.

A *model* is an abstract representation of a (real-world) physical, social, or other system in terms of mathematical equations, flow diagrams, computer programs, or algorithms.

The models considered in this book are distinguished by the fact that their abstract representations contain some (random) stochastic element. For example, in considering the reliability of an electronics system—perhaps the "time to failure" of the system—it is well known that the system fails when a certain combination of its components fails and that the times to failure of the components are random (i.e., nondeterministic). Consequently, the time to failure of the system is random.

We are now in a position to define what we mean by a simulation.

DEFINITION OF SIMULATION Simulation is essentially a controlled statistical sampling technique (experiment) that is used, in conjunction with a model, to obtain approximate answers for questions about complex, multifactor probabilistic problems. It is most useful when analytical and numerical techniques cannot supply answers.

Thus, one wants to obtain an estimate of a quantification, under given conditions, of the system—for example, expected time to failure or the probability that the system will survive (not fail in performing) a 10 hour task—by producing repeated realizations of the system, generally on a digital computer. It is also essential to note that the estimate must have enough *precision* to answer relevant questions about the system (or to allow comparisons between two or more systems or a single system under different conditions). Here, precision is usually defined to be the ratio of the standard deviation of the estimate to the absolute value of the estimate.

This procedure is sometimes called *empirical sampling*, or Monte Carlo in the wide sense (Kleijnen, 1974, p. 6), and it is important to note that it is different from a "physical (or analogue) simulation" such as that performed in the flight simulators used to train civil aviation pilots or that performed on airplane models in wind tunnels. It is also different from the "simulation" that is the numerical solution of partial differential or integral equations that model deterministic physical systems. Simulation is, in fact, a *controlled statistical experiment, performed on a digital computer.* This is in contrast to real experiments performed in agricultural fields or on patients taking different drugs.

EXAMPLE   1.1 A Simple Analogue Simulation and Its Digitization

Suppose that a very odd and rough-shaped figure is inscribed in a square. The figure could be thought to represent an ore deposit and the square an aerial photograph covering 1 square mile. To determine the extent of the ore deposit, the map could be put on a wall and darts could be thrown at it haphazardly, so that the chance of a dart landing on any point on the map is "just as likely" as its falling on any other point. If we found that the proportion of darts thrown that fell into the odd-shaped figure in the map was p, then, intuitively, the area of the odd-shaped figure would be estimated by p times the area of the aerial photograph, which is $(1 \times p)$ square miles.

Clearly, the more darts thrown (always removing the previous dart) the "better" or "more precise" the estimate of the area will be. However, this procedure is impractical and tedious, and the modern predilection is to "put it on a digital computer" so that the experiment can be done more frequently. One would then have to model the real situation by digitizing the map—that is, breaking it up into little squares, each of which would be labeled "1" if most of the area consisted of the odd-shaped figure, and "0" otherwise.

The problem that arises immediately lies in being able to "repeatedly pick little squares at random" in a digital computer. This is the biggest initial problem in a digital simulation (Monte Carlo) experiment; the generation of the pseudo-random numbers used to do this "random" selection is considered in Chapter 5.

EXAMPLE   1.2 Buffon's Needle: Physical Versus Monte Carlo Simulation

A classical problem proposed in 1733 by Georges-Louis Leclerc, Comte de Buffon, allows us to illustrate the difference between an analytical solution, a physical simulation, and a Monte Carlo simulation on a digital computer. Our goal will be to calculate the transcendental number π, which is the ratio of the circumference of a circle to its diameter.

Although there is no closed-form numerical value for π, we can compute an approximate *analytical* solution by using the identity

$$\pi = 4[\arctan(\tfrac{1}{2}) + \arctan(\tfrac{1}{3})]$$

and applying the Taylor series expansion

$$\arctan(\chi) = \sum_{i=0}^{\infty} (-1)^i \left(\frac{\chi^{2i+1}}{2i + 1} \right)$$

to approximate $\arctan(\tfrac{1}{2})$ and $\arctan(\tfrac{1}{3})$. Notice that neither the problem nor the solution involves any random components. A similar identity was used by Shanks and Wrench (1962) to calculate π to 100,000 decimal places.

An alternative way to view the problem as a physical simulation was described by Leclerc as follows: assume that a needle of length l is dropped "randomly" on a surface marked by parallel lines separated by d units, where it is assumed for simplicity that d is equal to or greater than l. Then it can be shown that the probability that the needle intersects one of the parallel lines is p, where

$$p = \frac{2l}{\pi d}$$

(Exercise 1.1 indicates how to prove this assertion).

This fact suggests the following *physical* simulation for the estimation of π:

1. Construct a large surface and draw horizontal parallel lines distance d apart.
2. Drop a needle of length l at random on the surface.
3. Repeat step 2 n times.
4. Estimate p by counting the number of times the needle crosses a line and dividing this total by n, call this estimate $\hat{p}$.
5. Estimate π by $\hat{\pi} = 2l/\hat{p}d$.

With enough patience, anyone can estimate π through this algorithm, and a description of an early effort by a hospitalized Civil War captain is given in Hall (1873).

Turning to our final approach to the problem, digital simulation, our definition of simulation requires that when no analytical solution exists (as is the case here, since no closed numerical form for π can be calculated), then the random counterpart to the problem—the random drop of a needle—should be implemented on a computer. In this case, the random drop of a needle can be mathematically modeled by specifying the pair of random variables Y and θ, where

$Y = $ the distance from the center of the needle to the closest parallel line

and

$\theta = $ the angle the needle makes with respect to the closest parallel line

and noting that Y should be uniformly distributed between 0 and $d/2$, and θ should, independently, be uniformly distributed between 0 and $\pi/2$, or equivalently, between 0 and π. Then, as indicated in Exercise 1.1, the needle crosses a parallel line if and only if $Y \leqslant (l/2)\sin \theta$, where l, as before, is the length of the needle.

This leads to the following computer simulation algorithm:

1. Generate a random variable Y_i uniform on $[0, d/2]$ and an independent random variable θ_i uniform on $[0, \pi/2]$ (the specifics of uniform random variable generation are given in Chapter 5).
2. Repeat step 1 *n independent* times, $i = 1, 2, \ldots, n$. Each time, keep track of whether $Y_i \leqslant (l/2)\sin \theta_i$.
3. Estimate p by counting the number of times that $Y_i \leqslant (l/2)\sin \theta_i$ and dividing this sum by n. Denote this estimate by $\tilde{p}$.
4. Estimate π by $\tilde{\pi} = 2l/\tilde{p}d$.

The advantages of such a computer simulation over a physical simulation include the speed of the computer and the "true" randomness with which the "needle" can be dropped.

Further extensions of Buffon's needle, which measure and increase the accuracy with which π can be estimated, are given in Exercises 1.2 and 1.3.

Continuing with the development of ideas in this chapter, the *purpose of simulation is generally to facilitate choice* among several different and competing schemes for either:

1. A practical operational situation, such as which of several queue disciplines to use or which of several inventory ordering policies to use, or
2. A statistical procedure, such as least-squares regression versus robust regression.

A complication is that each scheme may perform differently for different values of inherent factors. Thus, one scheme may be better than another only in a particular range of values of the factors. In fact, simulation is often used to determine whether a scheme or statistical procedure is sensitive to a change in factor values.

The technique of simulation is distinguished by the fact that it, like applied mathematics and statistics, is used in a broad spectrum of fields, including civil engineering, hydrology, biostatistics, physics, and operational analysis. In addition, simulation uses techniques from many applied and pure fields. It is worth setting out these aspects of simulation, most of which will be considered later on, and the disciplines involved.

Aspects of Simulation	**Disciplines Involved**
Probabilistic modeling of systems	Probability concepts
	Stochastic processes
	Physical and engineering concepts
	Human factors
	Operational concepts
	Systems analysis
Generation of uniform random numbers	Number theory
	Probability theory
	Computer science
Generation of nonuniform random numbers (e.g., normals)	Applied probability
	Numerical analysis
	Computer science

Organization of models and data in
computers (e.g., systems simulation
packages such as GPSS, and
SIMSCRIPT)

Computer science

Repeated sampling (crude simulation)

Basic statistics
Graphics

Design of simulations and analysis of
output data.

Statistical estimation theory
Experimental design
Stochastic processes (ergodic theory)
Graphics

These aspects of simulation will become more meaningful in subsequent chapters, which cover modeling and the basic ideas of simulation.

Exercises   **1.1.** **a.** Using the notation and description of Buffon's needle from Example 1.2, show that if Y is the distance from the center of the needle to the closest parallel line and θ is the angle between the needle and the closest line, then the needle crosses that line if and only if

$$Y \leqslant \left(\frac{l}{2}\right) \sin \theta$$

Note that the range of Y is from zero to $d/2$.

   b. Argue that if the needle is dropped at random, then Y must be uniformly distributed (see Chapter 4, if necessary, for definitions), between 0 and $d/2$, and θ must be uniformly distributed between 0 and π.

   c. From facts (a) and (b), prove that the probability p that the needle touches or crosses a line is given by $p = 2l/\pi d$.

   d. Is there any fact that would make it impossible to implement this scheme for estimating the value of π?

1.2. The *relative error* or *precision* of an estimate like $\hat{p}$ is defined to be the ratio of the standard deviation of the estimate $\hat{p}$ to the expected value of $\hat{p}$. We assume that this expectation is not zero, because the concept of relative error then would be meaningless. Remembering the restriction $d \geqslant l$, prove that the simulated estimates $\hat{p}$ and $\tilde{p}$ of p have minimum precision if we take $d = l$.

1.3. Morgan (1984) indicates that if $d = l$, then $\mathrm{Var}(\hat{p}) = p(1 - p)/n = [(2/\pi)(1 - 2/\pi)/n] = .2313/n$, where n is the number of random needles "dropped." However, if two needles are attached as a cross and dropped, and the number of parallel line intersections counted, then the resulting estimate of p has a variance of only $\approx .0995/n$, if n such random crosses are "dropped." Read and discuss the section in Morgan (1984) and the article by Pearlman and Wichura (1975) for ways to improve Buffon's needle to provide more precise estimates of π.

1.4. Contrast the methods above for computing π to the approximate analytical solution derived through the Taylor series expansions. You could program and run alternatives and compare precision or sample variances.

References

Hall, A. (1873). On an Experimental Determination of π. *Messengers of Mathematics, 2,* 113–114.

Kleijnen, J. P. C. (1974). *Statistical Techniques in Simulation: Part I.* Dekker: New York.

Morgan, B. J. T. (1984). *Elementary Simulation.* Chapman & Hall: London.

Pearlman, M. D., and Wichura, M. J. (1975). Sharpening Buffon's Needle. *American Statistician, 29,* 157–163.

Shanks, D., and Wrench, J. W. (1962). Calculation of π to 100,000 Decimals. *Mathematics of Computation, 16,* 76–99.

2

Golden Rules and Principles of Simulation

Although simulations may utilize a large part of the resources of any computing system, the field of simulation is in many cases held in low repute as an applied discipline. This is generally because of unwise or unnecessary use of simulation techniques. To remedy this, some basic rules are worth stating.

1. *Never simulate unless you have to*: that is, calculate what you can mathematically (analytically or numerically), making use of special structures of the problem and approximations to the problem.

2. *Careful modeling is very important*; as always, it is useless to find the right answer to the wrong problem, or to an inadequate approximation to the problem. Similarly, the right quantification of the answer is important. Thus in designing a dam, and deciding on its ability to withstand the force of a flood, one is not interested in the average height of a river, but in its *maximum* height in a fixed period, or the amount of time the river spends above a given level in a fixed period.

This need to *model* before proceeding represents one large difference between a simulation experiment and a physical experiment (e.g., in medicine). A medical experiment takes place in a fixed, *real* setting, where no models of patients or of drug reactions are needed. Only afterward is there a need for a statistical model to analyze the results of the experiment.

As in simulation, modeling is used as a fundamental tool of the analytic approach in, say, physics or operations research. Sir Maurice Kendall summed up the role of modeling in applied science when he said "Models are for thinking with." The role of modeling in simulation is broader, encompassing also decision making, as remarked earlier in connection with the purposes of simulation.

3. *Validation* of the model and *verification* of the computer program that implements the simulation is important. Validation of the model can sometimes be achieved by consulting people who are familiar with the process being simulated. At other times the simulation itself will provide clues in that outputs from the simulation can be compared to real-life outputs. Verification and debugging of the computer program is a vexing problem in simulations, especially in very complex processes. One solution is to run special cases of the simulation for which the answers are known.

4. Unlike a mathematical solution, the answer one obtains from a simulation is an *estimate* based on one or more realizations of a random variable. It is *absolutely* necessary to obtain some idea of the precision of the estimate. This is because one usually is comparing systems with different values of inputs or concomitant variables, and valid comparisons can be made only if one knows the precision of the estimates.

5. Finally, it is worthwhile noting again the breadth of the topic. Simulation is *used* to answer questions in statistics, probability, engineering, physics, business, wargaming, and so on. Each particular problem requires special knowledge in order to make the model and assumptions realistic. Such requirements cannot be anticipated or handled by any single text, *but we emphasize that you should think about each situation and be willing to go to the sources you need.*

On the other side, simulation *uses* methodology that is primarily statistical but also includes mathematical and probabilistic analysis, numerical analysis, computer science, and computing science. Techniques or "tricks" from these fields have been adapted to make simulations run more efficiently and produce estimates that are more precise. Such techniques, at various degrees of sophistication, are covered throughout the book. With respect to the point of efficiency, it is the advent of *large-scale digital computers* that makes simulation useful, and it is the continuing evolution in terms of speed and capacity of digital computers that will make simulation a more and more important methodology.

6. Even more finally is the ultimate golden rule. BE SURE, BEFORE YOU START, THAT YOU KNOW WHAT QUESTIONS YOU WANT TO ANSWER. This aspect of simulation will become clearer in the examples we discuss in Chapter 3.

Modeling: Illustrative Examples and Problems

This chapter and the next chapter, on aspects of crude simulation, do not have to be read sequentially; the modeling is described first to emphasize the points about adequate modeling made previously. One reading strategy might be to read one or two of the examples and then go to Chapter 4. However, the examples in the present chapter should eventually be read because they also serve to introduce notation and basic ideas for the rest of the book. Thus, several distribution function models are presented, as well as the distribution for minima and the idea of order statistics. Details on some of the distribution function models introduced here are given in Chapter 6 (Tables 6.1.2 and 6.1.3).

3.1 The Modeling Aspect of Simulation

One of the key aspects of a simulation—usually the first step—is the derivation of a model for the situation to be considered. This process is called modeling. A *model*, as defined in Chapter 1, is an abstract representation of a (real-world) physical, social, or other system in terms of mathematical equations, flow diagrams, computer programs, or algorithms.

An *algorithm* (see Knuth, 1974) is a finite set of rules that gives a sequence of operations for solving a specific type of problem. In addition it must terminate after a finite number of steps, and each step of the algorithm must be precisely defined.

Models and the process of modeling are best understood by example. Thus we now give five examples of stochastic systems and statistical situations one might want to model and simulate. These examples illustrate some of the points made in Chapters 1 and 2 and also serve to introduce terminology. *Note that the examples are primarily pedagogical and not completely realistic.* In actuality there are always complications. The examples do show, however, that even in simple situations there are aspects of models that are difficult, if not impossible, to represent adequately on a digital computer.

3.2   Single-Server, Single-Input, First-In/First-Out (FIFO) Queue

The FIFO queue is a congestion or queueing system in which a single server (e.g., a sole-practitioner lawyer, doctor, or plumber; a toll booth collector; a computer) provides service to a stream of customers. In this idealized system customers arrive one after another in a single stream at times $T_n, n = 0, 1, 2, \ldots,$ where $0 = T_0 < T_1 < T_2 < \cdots < T_n < \cdots,$ and the differences in the times of arrival are $X_n = T_n - T_{n-1}$ $(n = 1, \ldots)$. Note that tied arrival times are, by assumption, not allowed here. Allowing ties would create a queue with bulk arrivals, which might be of interest, but is not considered here. Successive service times are $S_0, S_1, \ldots,$ so that the customer arriving at $T_0 = 0$ has service time S_0, and so on. The service protocol or discipline is *first come, first served*: that is, the customers queue in order of arrival and the server always provides service to the person at the head of the queue. Thus customers always leave before customers who arrive after them. Hence the name, first-in/first-out (FIFO).

For illustrative purposes we are interested in looking at a single quantity W_n, the *waiting time* (sometimes called queueing time) of the nth customer to arrive, *defined here as the time from the customer's arrival until the time to the beginning of the customer's service*. Let $W_0 = 0$, so that we start with the queue empty, and the customer who arrives at time zero does not have to wait. The *physical specification* of the queue given in Figure 3.2.1 allows one to write down a compact, recursive, and easily computable expression for W_{n+1} as follows:

$$W_{n+1} = \begin{cases} \text{time when the } n\text{th customer leaves the system minus the arrival} \\ \text{time of the } (n + 1)\text{st customer } \textit{if} \text{ the first quantity is bigger than} \\ \text{the second;} \\ \\ 0 \quad \text{otherwise (} n\text{th customer leaves the system before} \\ \qquad (n + 1)\text{st arrives)} \end{cases}$$

$$= [T_n + W_n + S_n - T_{n+1}]^+ = [W_n + S_n - X_{n+1}]^+ \tag{3.2.1}$$

$$= [W_n + Z_n]^+ \qquad (n \geqslant 0) \tag{3.2.2}$$

where $[x]^+$ means the maximum of x and zero, and $Z_n = S_n - X_{n+1}$. (Draw a picture if the derivation is not clear.)

Note that the random variables W_n are "attributes" of the system for which we want to find some quantification—for example, $E(W_n)$ or $\text{prob}(W_n = 0)$. There are also other attributes of possible interest, such as N_n, the number of customers in the queue at time $T_n + W_n + S_n$, the time at which the nth customer leaves the queue. These attributes of the system will be discussed later; they are, however, not all as easily computed as W_n.

Now the model has been specified *physically* only and is not complete until we have "modeled" or specified the joint probability structure of the stochastic sequences $\{X_n\}$ and $\{S_n\}$: that is, we must provide a distribution function model for these sequences.

The MI|GI|1 (see below) FIFO single-server queue models these probabilistic aspects of the system by assuming the following.

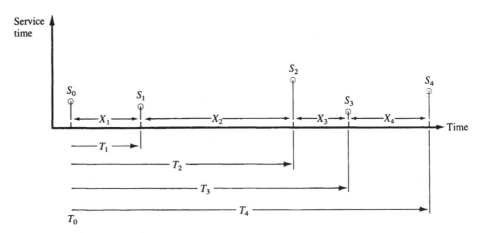

Figure 3.2.1   Arrival process for a FIFO queue and related interarrival and service times.

1. That the two sequences are stochastically independent.
2. That the $\{S_n\}$ sequence consists of independent, identically distributed (iid) nonnegative random variables with arbitrary distribution function:

$$G(x) = P(S_n \leqslant x) \qquad (x \geqslant 0, n = 0, 1, \ldots)$$

3. That the input is a homogeneous Poisson process. One definition of this process is that the times between arrivals, X_n, are independent and identically distributed (iid) with *exponential* distribution function, for $n = 1, 2, 3, \ldots$

$$
\begin{aligned}
F_{X_n}(x) = F_X(x) = P(X \leqslant x) &= 1 - e^{-\lambda x} \qquad & x \geqslant 0, \lambda > 0 \\
&= 0 & x < 0
\end{aligned}
\tag{3.2.3}
$$

The notation "$F_{X_n}(x) = F_X(x)$" means that the service times X_n have the same distribution for all n. For equivalent definitions of a homogeneous Poisson process in terms of the counting process, see Ross (1980), Cinlar (1976), and Lewis and Shedler (1979).

Thus in the abbreviation M|G|1 the M refers to Poisson arrivals, the G refers to general service time distribution $G(x)$, the 1 is the number of servers, and the independence assumptions are implicit. The preferred notation MI|GI|1 is used to denote that successive service and interarrival times are independent. Note that a decision still has to be made on a parametric form for $G(x)$; this could be made on physical grounds or from data considerations.

The items above are very special assumptions for a particular queueing system that *might* be reasonable and for which complete or partial analytical results [e.g., $E(W_n)$] are known in several cases (Feller, 1971, p. 194; Cohen, 1969). Analytical results become almost impossible when dependence is introduced—for example, between successive service times—and in this kind of situation, *simulation is necessary. The analytical results for the simple queue are not then totally useless; they give some idea of results that might pertain with the more complicated assumptions and can be used for variance reduction and programming validation in the simulation.*

Some analytical results are almost always necessary for a successful simulation. Thus for the simple queue we use the fact that W_n goes to W, the stationary or steady-state waiting time, as n approaches infinity (in some sense) if the expected service time $E(S)$ is less than the expected interarrival time $E(X)$. That is, if the *traffic intensity* $t = E(S)/E(X) < 1$, then one is simulating in these cases a *stable queue* for which there are special techniques, to be discussed in Volume II. Does a similar convergence result hold in the more complicated queue in which dependence is introduced, say, in the service times? Knowledge of the convergence or a lack of convergence of W_n to a stable situation makes a lot of difference in how the simulation is performed, the questions that must be asked, and the dimensionality of the answer. This is because we can then look at the single equilibrium distribution rather than all the waiting time distributions for each customer.

The case in which $E(S) \geqslant E(X)$ can also be considered; in this case it is known that for the FIFO queue, the average waiting time increases as n increases. Hence the simulation may be concerned with W_n for more than one value of n, and in particular how $E(W_n)$ changes as a function of n. In this case, analytical approximations for $E(W_n)$ can be obtained quite easily; see Cox and Smith (1961, pp. 65–68) for example.

In this single-server, single-input FIFO queueing system there are other quantities that might be of interest and could be studied by simulation. One might want to investigate the following quantities.

1. The number N_n of customers in the queue at the time at which the nth customer leaves the queue (it is important to say whether it is just before or after the customer leaves the queue).

2. The number N_t of customers in the queue at a time t after the start of the process. In fact N_n is just a (random) sampling of N_t at departure times.

3. The alternating process $\{B_n, I_n\}$ of successive busy and idle times. A formal mathematical definition of this bivariate sequence is given in Chapter 11 and in Volume II. However it is clear that if there is, by definition, a customer arriving at time zero, the server is initially busy and will continue to be busy until some random time, B_0, later when he completes a service and finds no customers waiting for service. He then sits idle for a random time I_0, which is the time until the next arrival of a customer. Then the next busy period, B_1, commences, and so forth. The busy–idle sequence is really of interest only in the stable case $(t < 1)$, since otherwise the server eventually becomes busy almost all the time.

The MI|GI|1 queue, as noted above, is an idealization for which many analytical results are known. In practice many departures and differences from this simple model occur. The arrival process may be nonstationary—for instance, a nonhomogeneous (time-dependent) Poisson process (Lewis and Shedler, 1979), in which the rate of arrivals peaks at "rush hour." Also, certain types of preemptive behavior may occur in the queueing discipline, as when emergency cases break in on the schedule of a physician. Again, service may occur in separate periods over time, as for a job submitted to a computer. The service is said to be time-sliced. In this latter type of situation the interest is not so much in waiting time as in the time to leave the queue, sometimes called the response time, R_n, for the nth job.

3.3 Multiple-Server, Single-Input Queue

When we generalize the foregoing example to allow for k servers, we no longer have a "natural" queueing discipline as we did in the MI|GI|1 case. Some interesting facets of modeling realistic systems become apparent when we try to specify the queueing discipline for this system.

1. In one commonly used queueing discipline, customers queue in front of the server of their choice. How do we model this choice?
   a. One possibility is to let the customer join the shortest queue if a unique shortest queue exists.
   b. Otherwise let the customer join, with equal probability, one of the other server queues with the smallest number of customers waiting. This includes the case where more than one server is idle.

   This will be called the multiple-line discipline. A complication here is that, in reality, customers become impatient and switch from one server to another. *It is extremely difficult to derive a realistic model for this behavior.* (Try it.)
2. Another queueing discipline, slowly coming into use, is to let each server, upon becoming free, take the customer at the head of a single queueing station. This will be called the single-line discipline. We are sure *you* prefer this, but why?

The answer can be explored by simulation. For many cases of the input parameters, the average waiting time $E(W_n)$ (in steady state) is not much different for queueing disciplines (1) and (2), but the probability of waiting a long time is much larger in the first case. Thus in comparing these two situations one is interested in waiting times, but the correct quantification requires the *distribution* of the *waiting time*, not its *average value*.

This example will be continued later, in particular in the examples of Chapter 6. A fairly complete description of a simulation of this queue will be given, as a function of several factors, in Chapter 8.

3.4 An Example from Statistics: The Trimmed t Statistic

Assume that we have n random variables $X_1, \ldots, X_i, \ldots, X_n$, and consider the t statistic, or function of these n random variables, defined by

$$T_n = \frac{\bar{X} - \mu_0}{S/n^{1/2}} \tag{3.4.1}$$

where μ_0 is a known constant,

$$\bar{X} = \text{sample mean} = \frac{1}{n} \sum_{i=1}^{n} X_i \tag{3.4.2}$$

and

$$S = \text{sample standard deviation} = \left[\frac{1}{n-1} \sum_{i=1}^{n} (X_i - \bar{X})^2 \right]^{1/2} \tag{3.4.3}$$

When the X_i's are iid Normal random variables with mean μ_0 and standard deviation σ [i.e., when they are $N(\mu_0, \sigma)$], so that

$$F_X(x) = P(X_i \leqslant x) = \frac{1}{\sqrt{2\pi}\sigma} \int_{-\infty}^{x} \exp\left[-\frac{1}{2}\left(\frac{z - \mu_0}{\sigma} \right)^2 \right] dz \tag{3.4.4}$$

$$= \int_{-\infty}^{(x-\mu_0)/\sigma} \frac{e^{-v^2/2}}{\sqrt{2\pi}} \, dv = \Phi\left(\frac{x - \mu_0}{\sigma} \right) \tag{3.4.5}$$

then the distribution of the t statistic T_n is known and tabulated for all n (see any statistics text and in particular Conover, 1980).

The statistic is useful for testing the location hypothesis H_0: mean $= \mu_0$, against the alternative H_1: mean $\neq \mu_0$ when σ is unknown. That is, σ is a nuisance parameter. It is also useful for giving a confidence interval estimate for the mean of a normal population with unknown standard deviation (see Chapter 6, Section 6.1.4).

EXERCISE   Review the use of the t statistic in these contexts (see, e.g., Conover, 1980, Ch. 5, and Chapter 9 of this book).   ∎

The *trimmed t statistic* $T_{n,1}$ is used instead of T_n to guard against errors of measurement, and so on, which might produce very extreme values (outliers) that could seriously affect the t test. Thus, let the random variables X_i when ordered be denoted by $X_{(1)} < X_{(2)} < \cdots < X_{(n)}$. These random variables are called the *order statistics* for the random sample $X_1, X_2, \ldots, X_n$, and we are assuming that the random variables are continuous so that there are no "tied" values (i.e., X_i's with the same value). In particular, we have

$$X_{(i)} = i\text{th smallest of the } X_i\text{'s} \tag{3.4.6}$$

and the special cases

$$X_{(1)} = \min(X_1, \ldots, X_n) \tag{3.4.7}$$

and

$$X_{(n)} = \max(X_1, \ldots, X_n) \tag{3.4.8}$$

One version of the *trimmed t statistic* is defined by Tukey and McLaughlin

(1963) to be a trimming of $X_{(1)}$ and $X_{(n)}$:

$$T_{n,1} = \left\{ \frac{\frac{1}{n-2} \sum_{i=2}^{n-1} X_{(i)} - \mu_0}{\left[\frac{1}{n-3} \sum_{i=2}^{n-1} (X_{(i)} - \bar{X}_{n,1})^2 \right]^{1/2}} \right\} n^{1/2} \tag{3.4.9}$$

where $\bar{X}_{n,1}$ is the mean of the sample after trimming as in the numerator of the expression for $T_{n,1}$. This statistic should be compared to the t statistic; the factors $\{1/(n-2)\}$ and $\{1/(n-3)\}$ in the numerator and denominator are largely arbitrary, as is the multiplier $n^{1/2}$. One could multiply by $(n-2)^{1/2}$; for large n the difference is negligible.

Questions to Be Asked and Answered by Simulation in This Problem

1. What is the distribution of $T_{n,1}$? (It should be "similar" to that of T_n, and this might be useful in variance reduction, as will be shown in Chapter 11). Note that n is a parameter in the simulation and one might be required to tabulate the distribution (or several quantiles) for various values of n. A *quantile* is a value t_α, such that for given probability α, $0 < \alpha < 1$,

$$P(T_{n,1} \leqslant t_\alpha) = \alpha \tag{3.4.10}$$

2. Another question might be, At what point n is $T_{n,1}$ approximately Normally distributed? And, what are the mean and the standard deviation of $T_{n,1}$? One would surmise, because of the minor difference between T_n and $T_{n,1}$, that $T_{n,1}$ is, like T_n, asymptotically $N(0, 1)$.

3. One might also be required to give expressions for, say, the 0.95 quantile of $T_{n,1}$ as a *function* of n. This brings in regression analysis and other techniques of statistics for output analysis, as discussed in Chapters 8 and 9.

4. If the X_i are non-Normal, then $T_{n,1}$ might be less sensitive to departures from Normality than T_n (i.e., more "robust"). A simulation experiment can reject or validate this contention by examining the distribution and power of T_n and $T_{n,1}$ for several distributions for the X_i and several alternative hypotheses.

5. Note that $T_{n,1}$ is a very nonlinear function of the X_i's. More drastic trimming might also be of interest: drop off the two highest and two lowest observations to get $T_{n,2}$ or, perhaps, drop off a certain percentage δ of observations from each end.

A very large simulation study of robust techniques in statistics has been carried out by Andrews et al. (1972). For details on the trimmed t statistic, see Tukey and McLaughlin (1963).

3.5   An Example from Engineering: Reliability of Series Systems

In this example we have n components in a structure (e.g., links in a chain or transistors in a computer), and it is assumed that the system fails if any one of the components fails. This is sometimes called a *series* system. Another example of a series system would be a jury with no alternate jurors.

A further physical assumption is that $X_1, X_2, \ldots, X_i, \ldots, X_n$, the times to failure, are independent random variables. (An alternative is some kind of coupling, i.e., such that two can fail simultaneously, but this is difficult to specify exactly, as discussed in Chapter 10.) One might then also assume, either from data or from physical considerations, that the X_i, $i = 1, \ldots, n$, all have identical two-parameter Gamma distributions

$$
\begin{aligned}
F_X(x) &= \frac{1}{\Gamma(k)} \int_0^{kx/\mu} v^{k-1} e^{-v} \, dv \qquad (x \geq 0; k > 0, \mu > 0) \\
&= 0 \qquad\qquad\qquad\qquad\quad (x < 0)
\end{aligned}
\tag{3.5.1}
$$

with probability density function (pdf)

$$
\begin{aligned}
f_X(x) &= \frac{dF_X(x)}{dx} = \left(\frac{k}{\mu}\right)^k \frac{x^{k-1} e^{-kx/\mu}}{\Gamma(k)} \qquad x \geq 0 \\
&= 0 \qquad\qquad\qquad\qquad\qquad\quad x < 0
\end{aligned}
\tag{3.5.2}
$$

where the integral in Equation 3.5.1 is the incomplete Gamma function and $\Gamma(k)$, the integral from 0 to ∞, is the *complete Gamma function*. For $k = 1$ this reduces to the *exponential* distribution given at Equation 3.2.3. This distribution will recur throughout the book. (Note that this parametrization of the Gamma distribution in terms of the parameter μ, which is the mean of X, and the shape parameter k, is different from that given in Table 6.1.2. The rate parameter β, used there, equals k/μ).

Now let

$$
\begin{aligned}
M_n &= X_{(1)} = \min(X_1, \ldots, X_n) \\
&= \text{time to failure of the system}
\end{aligned}
\tag{3.5.3}
$$

We may be required to find the average time to failure $E(M_n)$, or the complete distribution of M_n, or just a quantile x_α of the distribution. *A quantile x_α is*, for given α, where $0 < \alpha < 1$, the smallest value such that

$$
P(M_n \leq x_\alpha) = \alpha
\tag{3.5.4}
$$

so that if M_n has an absolutely continuous density function $F_{M_n}(x)$ with inverse $F_{M_n}^{-1}(\alpha)$, then

$$
x_\alpha = F_{M_n}^{-1}(\alpha)
\tag{3.5.5}
$$

In this problem there is a great possibility for analytical solution, since the distribution of the minimum is, in principle, known. Thus, using the assumed independence of the X_i's,

$$F_{M_n}(x) = 1 - P(M_n > x) = 1 - P(\text{all } X_i\text{'s} > x)$$
$$= 1 - \{1 - F_X(x)\}^n \qquad (3.5.6)$$

Extreme value theory gives the forms of the distribution of M_n for large n; as n approaches infinity, $F_{M_n}(x)$, suitably normalized, approaches an asymptotic distribution that does not depend on n (Mood, Graybill, and Boes, 1974, Ch. 6).

Simulation is likely to be useful for moderately large n when the problem of numerically integrating to find, for example, $E(M_n)$ can be severe. It is also not known how well the asymptotics apply; that is, it is not certain for what values of n the asymptotic distribution is approximately correct. If there is dependence between the X_i's, then Equation 3.5.6 no longer holds and simulation is essential.

Ordered samples, and maxima and minima, occur frequently in simulation and will be discussed throughout the text. The problem of efficient ordering of data in computers has received much attention (see Knuth, 1973) and is well understood. The problem of efficiently generating random variables with the distribution of a maximum (or minimum) from a population with given distribution is another problem in simulation methodology with a viable solution; it is discussed in Chapter 4. Efficient generation of any order statistic $X_{(i)}$ or any group of order statistics has also been considered in the literature.

3.6 A Military Problem: Proportional Navigation

This example is more detailed and realistic than the five that precede it; it may be more meaningful after Section 3.7 and Chapter 4 have been read.

When designing an automatically guided rocket or missile, it is necessary to include an on-board navigation system that will detect and respond to movements of an airborne target according to some strategy. Typical situations include missiles that are launched to intercept incoming aircraft and rockets that are sent to dock with (or destroy) orbiting satellites. In the first scenario, we expect the aircraft to take evasive maneuvers. In the second, it may not be possible to predict exactly where a given satellite will be at some future time because of random perturbations in its predicted path, or because of random errors in the measurement of its path. Two possible navigational strategies to guide the missile to intercept its target will be considered.

The easiest and most direct strategy for navigation is to periodically aim the missile directly at the target. Then, at the end of every fixed time interval of length Δt, the guidance system aboard the missile again aims the missile directly at the target. This strategy often requires very extreme lateral acceleration of the missile and constant changes in flight path, even if the target does not maneuver at all. Figure 3.6.1 shows a typical chase with no target maneuvers. If the speed of the missile is sufficiently greater than that of the target, the missile will, with high probability, complete its mission. However, errors in the guidance system (i.e., in locating the target accurately), combined with the constant need to maneuver, produce a "jumpy" missile that may fail, lose the target, or run out of fuel before intercepting the target.

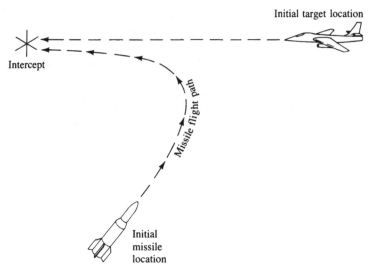

Figure 3.6.1 Illustration of updating method of guiding a missile to a target.

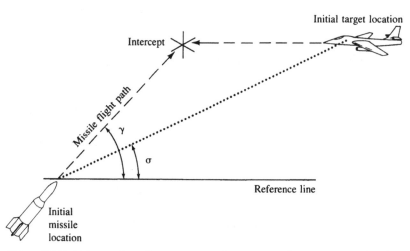

Figure 3.6.2 Interception of nonmaneuvering target using a constant bearing course.

Our second approach is represented by the navigation strategy usually used by both the U.S and Soviet military, called proportional navigation (Murtaugh and Criel, 1966). This strategy is based on the idea of aiming ahead of the incoming target. If the target does not maneuver and the velocities of the target and the missile are known, it is possible to calculate an intercept point that allows the missile to maintain constant bearing: that is, given the velocities, the angle γ of the missile's flight path with respect to an arbitrary, fixed-angle reference line is constant during its flight. Figure 3.6.2 shows the interception of a nonmaneuvering target using a constant bearing course. If the reference line through the head of the missile and parallel to the ground remains at a fixed angle to the body of the missile, we require

γ to be chosen so that with σ and γ remaining fixed during the flight of the missile, there is interception. If the target maneuvers, the angle σ changes continuously in time and, according to proportional navigation, the angle of the missile's flight path should change proportionally, according to the law

$$\frac{d\gamma}{dt} = N\frac{d\sigma}{dt} \tag{3.6.1}$$

where N is a fixed constant called the navigation ratio. For practical purposes, the continuous time changes in γ and σ are reduced to the observed differentials at each time separation Δt:

$$\frac{(\gamma_{n\Delta t} - \gamma_{n\Delta t-\Delta t})}{\Delta t} = N\frac{(\sigma_{n\Delta t} - \sigma_{n\Delta t-\Delta t})}{\Delta t} \tag{3.6.2}$$

The randomness that needs to be modeled in this simulation scenario enters in two ways:

- *Errors in the guidance system.* These are modeled as follows:

$$\sigma_{n\Delta t} = \sigma_{n\Delta t}^{A} + \sigma_{n\Delta t}^{E} \tag{3.6.3}$$

  where, at time $n\Delta t$, $\sigma_{n\Delta t}^{A}$ is the actual angle of the target to the reference line, $\sigma_{n\Delta t}$ is the angle recorded by the guidance system of the missile, and $\sigma_{n\Delta t}^{E}$ is error introduced by the guidance system. With reference to Figure 3.6.2, we have $\sigma = \sigma_{n\Delta t}^{A}$.
- *Maneuvering by the target.* In some cases, though, if enemy maneuvers are well known, we may want to concentrate on a few fixed flight paths rather than random flight paths.

Some Questions That Simulation Can Help Answer
1. Is proportional navigation better than direct pursuit? (The definition of "better" is subjective and can include the concepts of *hit probability, time to intercept, distance to intercept,* or *distance of miss.*)
2. What values of N should we use in proportional navigation to get the most effective system?
3. What maneuvers by the target are most effective against a missile using proportional navigation?

Notice that in this example we require a great deal of input from outside physical sources before any type of simulation should be attempted. We need to know, for instance:

1. What are reasonable specifications (i.e., velocity, maneuverability, range, kill radius) of target and missile?

2. What factors determine which system is better?

3. How do we translate $d\gamma/dt$ into lateral acceleration requirements for the missile? In particular, how long does the maneuver take?

4. What are reasonable maneuvers for the target?

5. How should we model the distribution of the $\sigma_{n\Delta t}^E$? Is there correlation between the successive errors? Does the distribution depend on the actual angle of the target to the reference line, $\sigma_{n\Delta t}^A$? For instance, is the guidance system poor at extreme angles?

6. Are there any theoretical results that will help us? (Using differential equations based on the lateral acceleration capability of the missile and typical plane and missile velocities, it can be shown that N should be around 3 if the missile is chasing the tail of the target and around 6 if the missile and target are approaching head-on.)

The actual modeling in this case can be given roughly as the following sequence of steps.

1. Determine all appropriate specifications, assumptions, and the initial locations (X_0^M, Y_0^M) and (X_0^T, Y_0^T) of the missile and target, respectively.

2. Decide on a rule that guides the movement of the target: at each time $n\Delta t$, we need to determine the current location $(X_{n\Delta t}^T, Y_{n\Delta t}^T)$, from the previous flight history, $\{(X_{n\Delta t - k\Delta t}^T, Y_{n\Delta t - k\Delta t}^T)\}$, $k = 1, 2, \ldots, n$. The rule may be based on either random or fixed target maneuvers.

3. Decide on a rule that guides the movement of the missile: at each time $n\Delta t$, we need to determine the current location $(X_{n\Delta t}^M, Y_{n\Delta t}^M)$ from previous flight history, $\{(X_{n\Delta t - k\Delta t}^M, Y_{n\Delta t - k\Delta t}^M)\}$, $k = 1, 2, \ldots, n$, and from the movements of the target. The rule will depend on the maneuvering of the target, on guidance error, and on the lateral acceleration capabilities of the missile. If we are testing proportional navigation, the rule will include the condition $d\gamma/dt = (d\sigma/dt)N$ for some prespecified N.

4. After both target and missile have moved, compute the distance between the two:

$$D_n = [(X_{n\Delta t}^T - X_{n\Delta t}^M)^2 + (Y_{n\Delta t}^T - Y_{n\Delta t}^M)^2]^{1/2} \tag{3.6.4}$$

5. If the distance is less than the kill radius of the missile, terminate the simulation and record whatever output values you have decided are pertinent. If the missile has missed or become disabled, terminate the simulation and record output values. Otherwise, let both missile and target continue their flight paths. Alternatively, if lethal radius is unknown, continue the engagement until closest point of approach.

3.7　Comments on the Examples

3.7.1　The Role of a "Time" Evolution Parameter

In all the examples the "sample size" n is present. In some instances it may be that one is required to study output random variables (e.g., W_n, $T_{n,1}$, M_n, D_n), which are statistics or functions of n other random variables, for a given n. This could be particularly true in statistical problems. In general, however, one wants to study the output random variable *as a function of n*. Since the successive values—for example, the waiting times (W_n, W_{n+1}) in Section 3.2, or the distances (D_n, D_{n+1}) in Section 3.6—are generally correlated because of the way the simulation is run (and also functionally related), we are dealing with a *discrete-time-parameter stochastic process* or *time series*. Thus all the theory of regression analysis and *time series* is relevant. The interest in the time evolution is particularly strong in operations research problems such as the queueing problems of Sections 3.2 and 3.3.

Note that n is sometimes time and sometimes a sample size parameter.

3.7.2 Stability and Convergence Considerations

Note too that it makes a difference in difficulty of simulation if the sequence "settles down" or converges to a stationary distribution after an initial transient. If the sequence does "settle down," then we sometimes need consider only the one distribution that exists after it has settled, not a separate distribution for each value of n. This convergence is always true for the t statistic T_n, in Section 3.4, no matter what μ_0 and σ_0 are, if the X_i's have finite mean and standard deviation, but we need to study both the transient and the stable distributions. Convergence holds for W_n only under special conditions on the "traffic intensity," which was defined for the MI|GI|1 queue in Section 3.2. In most cases we study *stable* queues, but transient queues are also of interest in operations analysis work. For example, the build-up and decay of a queue of customers for a departing airplane is often of interest.

One might also want to study how fast the system converges (e.g., as $n^{1/2}$ or $n^{1/4}$) and, for example, how big n can become before the 0.01 quantile, $x_{0.01}$, of M_n in Section 3.5 is "too small." Put another way, we would ask how big n can be before, say, the probability that the time to failure is less than or equal to 10 hours, is greater than 0.01.

We discuss these rather difficult evolutionary problems, which require sophisticated statistical methodology, in more detail later and in particular in Volume II; *for the first part of the book we assume that we are studying the variables* W_n, $T_{n,1}$, M_n, D_n, *and so on, at a fixed value n*, or at most at a few values of n.

Notice that in some types of simulation, such as proportional navigation, the physical setting alone precludes questions of convergence: although we may be interested to see how D_n behaves in time, this behavior always terminates when D_n becomes smaller than the kill radius or when n becomes so large that the missile's fuel runs out.

3.7.3 Assumptions and the Use of Simulation in Experimentation

Generally, in statistics problems, assumptions are given and the problems are simple; in engineering and physics problems it becomes more difficult to delineate assumptions deductively or experimentally. Simulation may then be used as an experimental tool, as in the usual scientific method, to validate hypotheses: loosely speaking we put our hypotheses into a computer and see if the output mirrors reality. This approach has, for example, been used to invalidate several theories of galactic evolution.

In operations research, however, not only are the problems much more complex and therefore much more costly to simulate, but the role of modeling is much more important. It is in particular very difficult to specify interactions, dependencies, and distributions in the system. In most cases extensive data analysis is necessary *before* model building takes place.

An example we have already seen occurs in a bank or other facility with several servers: one can have customers either (1) join whichever line they wish and jump from line to line or (2) join one line, with the customer at the head of the line going to the first server to become free.

It is not hard, as we have seen, to describe the second situation, but for the first case, the probabilistic specification of the switching might be quite complex, if it is at all possible.

A simulation may be used to determine which system is "better," but one would generally also want to validate the assumptions by comparing the simulation ouptut from each system to real data.

3.7.4   Simulation Detail: Macro or Micro?

Another problem to be faced in complex simulations is the *detail* of the simulation model. In general, the more detail there is in a simulation model, the more difficult it is to model the probabilistic aspects of these details and validate them with data. This is the situation in Section 3.3, where we modeled the vacillations of customers as they switch from server to server. Do they switch because the queues in front of the other servers are shorter, or because of the length of service of the customer at the head of their server's queue?

Deciding at what level to model a system is an art; the one thing that can be said is that microscopic detail in a macroscopic model is seldom useful.

3.7.5   Verifying and Validating a Simulation

Another important problem is that of verifying or debugging a complex simulation algorithm and getting it implemented on the computer. A useful technique is to write the simulation program so that in special cases there are known analytical solutions. Your computer output should be compared to the analytical results in those special cases. As an example, if you are simulating an MI|GI|1 queue (Poisson arrival, general service time distribution, single-server queue: see Section 3.2), you should check your program by running an MI|MI|1 queue (Poisson arrival, exponential service time, single server), since for this case many analytical results exist. Methods for determining how good the agreement is between the simulation output and the known result are given in Chapter 6.

References

Andrews, D. F., Bickel, P. J., Hample, F. R., Huber, P. J., Rogers, W. H., and Tukey, J. W. (1972). *Robust Estimates of Location: Survey and Advances.* Princeton University Press: Princeton, NJ.

Cinlar, E. (1976). *Introduction to Stochastic Processes*. Prentice-Hall: Englewood Cliffs, NJ.

Cohen, J. W. (1969). *The Single Server Queue*. North-Holland: Amsterdam.

Conover, W. J. (1980). *Practical Nonparametric Statistics*, 2nd ed. Wiley: New York.

Cox, D. R., and Smith, W. L. (1961). *Queues*. Methuen: London.

Feller, W. (1971). *An Introduction to Probability Theory and Its Applications*, Vol. II, 2nd ed. Wiley: New York.

Knuth, D. E. (1973). *The Art of Computer Programming*, Vol. 3: *Sorting and Searching*. Addison-Wesley: Reading, MA.

Knuth, D. E. (1974). *The Art of Computer Programming*, Vol. 1: *Fundamental Algorithms*, 2nd ed. Addison-Wesley: Reading, Ma.

Lewis, P. A. W., and Shedler, G. S. (1979). Simulation of Nonhomogeneous Poisson Processes by Thinning. *Naval Reserve Logistical Quarterly*, *26*, 403–413.

Mood, A. M., Graybill, F. A., and Boes, D. C. (1974). *Introduction to the Theory of Statistics*. McGraw-Hill: New York.

Murtaugh, S. A., and Criel, H. E. (1966). Fundamentals of Proportional Navigation. *IEEE Spectrum*, pp. 75–85.

Ross, S. M. (1980). *Introduction to Probability Models*, 2nd ed. Academic Press: New York.

Tukey, J. W., and McLaughlin, D. H. (1963). Less Vulnerable Confidence and Significance Procedures for Location Based on a Single Sample: Trimming/Winsorization 1. *Sankhyā* A, *25*, 331–352.

4

Crude (or Straightforward) Simulation and Monte Carlo

4.0 Introduction: Pseudo-Random Numbers

We describe in this chapter the basic elements necessary to perform a simulation experiment on a digital computer. We call this *crude* or *straightforward simulation* because everything that goes beyond it is a refinement that uses special aspects of the particular simulation to gain speed and/or more efficient use of the computer.

Refinements in simulations are desirable in order to obtain more precision in the answers from the simulation in a fixed amount of computing time. Special aspects of the simulation can include, for example, stability of the process being simulated and the "discrete event" aspect of certain stochastic systems that arise in operations research. As another example of refinements, in Volume II we will look at much faster and simpler methods for generating nonuniform random variables than the inverse probability integral transform given in this chapter.

To start a simulation, we always need a computer-generated source of *pseudo-random numbers* $\{U_i\}, i = 1, 2, \ldots$: that is, random numbers that are assumed to be independent *and* to have uniform marginal distributions. (This topic is addressed more fully in Chapter 5.) Formally, this means that for any finite collection of k U_i's, namely $\{U_{i_1}, \ldots, U_{i_k}\}$, we have

$$
\begin{aligned}
P(U_{i_1} \leqslant u_1; U_{i_2} \leqslant u_2; \ldots; U_{i_k} \leqslant u_k) &= P(U_{i_1} \leqslant u_1) \times \cdots \times P(U_{i_k} \leqslant u_k) \\
&= u_1 \times u_2 \times \cdots \times u_k \\
&= \prod_{i=1}^{k} u_i
\end{aligned}
\tag{4.0.1}
$$

where the following conditions pertain.

1. The last line holds when all the u's lie between 0 and 1.
2. The last line is replaced by 0 if any of the u's is less than 0.
3. If any of the u's in the last line is greater than 1, it is replaced by 1.

For example, if $k = 1$, for all i we have

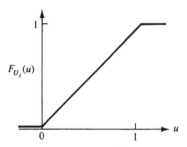

Figure 4.0.1 Distribution function of a uniform$(0, 1)$ random variable.

$$F_{U_i}(u) = P(U_i \leqslant u) = \begin{cases} 0 & u < 0 \\ u & 0 \leqslant u \leqslant 1 \\ 1 & u > 1 \end{cases} \tag{4.0.2}$$

This distribution function of a random variable that is uniformly distributed between 0 and 1 is shown in Figure 4.0.1. Again if $k = 2$ then *any pair* of the U_i's, namely U_{i_1} and U_{i_2}, must be independent:

$$P(U_{i_1} \leqslant u_1; U_{i_2} \leqslant u_2) = \begin{cases} 0 & \text{if } u_1 \text{ or } u_2 \leqslant 0 \\ u_1 u_2 & \text{if } 0 \leqslant u_1 \leqslant 1; 0 \leqslant u_2 \leqslant 1 \\ u_1 & \text{if } 0 \leqslant u_1 \leqslant 1; u_2 > 1 \\ u_2 & \text{if } u_1 > 1; 0 \leqslant u_2 \leqslant 1 \\ 1 & \text{if } u_1 > 1; u_2 > 1 \end{cases} \tag{4.0.3}$$

Note that the indices i_1 and i_2 mean that this relationship must hold for *any pair*: that is, contiguous pairs U_i, U_{i+1}, pairs U_i, U_{i+2}, which are one apart, and so forth. This is a very strenuous requirement. It also says that if we look at successive pairs such as $(U_1, U_2), (U_3, U_4), \ldots$ and do a scatter plot (see Chapter 10) in the unit square, these points will be "uniformly" distributed in the square. In particular this means that if we divide the unit square into N smaller nonoverlapping squares of equal area, each square has the same probability $1/N$ of having any given point (pair) fall into it.

Figures 4.0.2, 4.0.3, and 4.0.4 show scatter plots for three pseudo-random number generators. The first is RANDU, a generator provided in the IBM Scientific Subroutine Package and still provided in the DEC VAX minicomputer; the second is LLRANDOMII (Lewis and Uribe, 1981), a generator used in APL and the Naval Postgraduate School Random Number Generator Package; the third is a generator proposed by Wichmann and Hill (1982) for computers with 16-bit arithmetic. All three figures, which are generated from 10,000 adjacent pairs (U_i, U_{i+1}), appear to be uniformly distributed in the plane.

Going beyond Equation 4.0.3, the definition of Equation 4.0.1 implies that if we look at triples U_i, U_{i+1}, U_{i+2}, then the distribution U_i and U_{i+2}, given that U_{i+1} lies between, say, 0.5 and 0.51, should again be the same as the distribution in Equation 4.0.3. In other words, independence implies that the value of U_{i+1} does not affect U_i and U_{i+2}. In Figures 4.0.5, 4.0.6, and 4.0.7 we give, for the same three

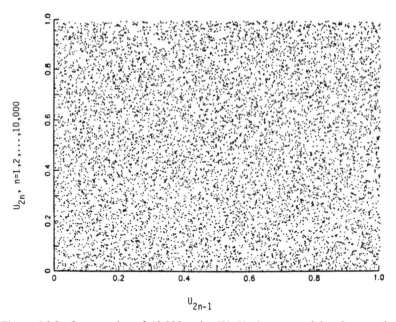

Figure 4.0.2   Scatter plot of 10,000 pairs (U_i, U_{i+1}) generated by the pseudo-random number generator RANDU. The seed used was 45813.

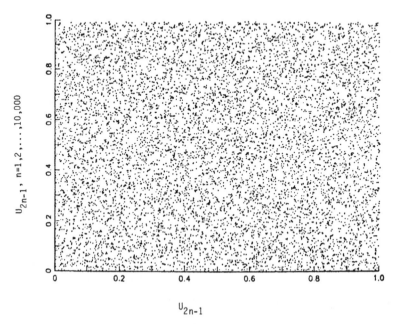

Figure 4.0.3   Scatter plot of 10,000 pairs (U_i, U_{i+1}) generated by the pseudo-random number generator LLRANDOM II. The seed used was 31854.

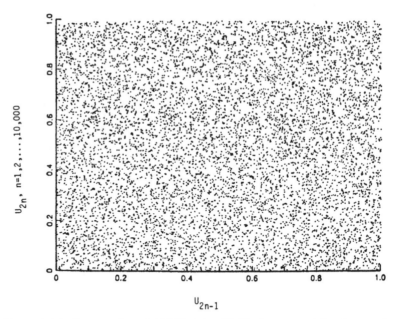

Figure 4.0.4 Scatter plot of 10,000 pairs (U_i, U_{i+1}) generated by the pseudo-random number generator proposed by Wichmann and Hill. The three seeds used were 10000, 20000, and 30000.

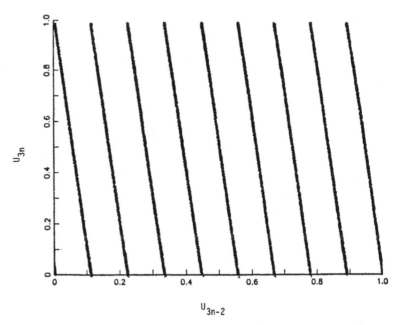

Figure 4.0.5 Scatter plots of U_i and U_{i+2} for 20,000 triples (U_i, U_{i+1}, U_{i+2}) in which $0.5 < U_{i+1} \leq 0.51$. These pairs are generated by the pseudo-random number generator RANDU. The seed used was 45813.

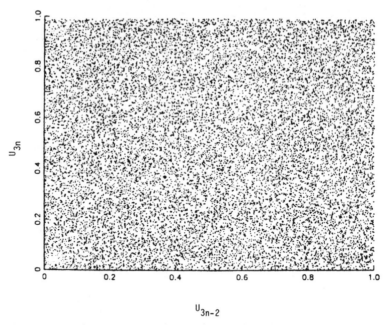

Figure 4.0.6 Scatter plots of U_i and U_{i+2} for 20,000 triples (U_i, U_{i+1}, U_{i+2}) in which $0.5 < U_{i+1} \leqslant 0.51$. These pairs are generated by the pseudo-random number generator LLRANDOMII. The seed used was 31854.

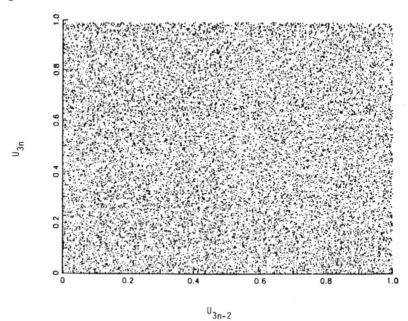

Figure 4.0.7 Scatter plots of U_i and U_{i+2} for 20,000 triples (U_i, U_{i+1}, U_{i+2}) in which $0.5 < U_{i+1} \leqslant 0.51$. These pairs are generated by the pseudo-random number generator of Wichmann and Hill. The three seeds used were 10000, 20000, and 30000.

generators, the scatter plots of 20,000 pairs $\{U_i, U_{i+2}\}$, given that $0.5 < U_{i+1} \leqslant 0.51$. It is clear that RANDU is not conditionally uniform! This point is pursued in Chapter 5.

Generation of *pseudo-random variables* (i.e., numbers that *simulate* a sequence of uniformly distributed iid random variables) is a very large topic touched on in Chapter 5. We note here only that while we always need pseudo-random variables, it is *almost never safe to assume* that a pseudo-random number generator on a computer is adequate. Thus one should have access to the testing and validation performed on the pseudo-random number generator; the checking should be done *before* any simulations are run. Two well-tested generators are the LLRANDOM II package from the Naval Postgraduate School (Lewis and Uribe, 1981) and the generators in Chapter G of the older IMSL (International Mathematics and Statistics Library) package, and Chapter 18 of the newer IMSL STAT/LIBRARY. The random number generators in IMSL are portable and will work on the large variety of computers (IBM, DEC, CDC, Burroughs, etc.) supported by IMSL.

4.1 Crude Simulation

The following is a prescription for crude or raw simulation; the steps in the prescription are very similar to the steps performed by many computer programs and simulation packages once the details of a model and the quantities to be simulated have been decided. For concreteness, the prescription is couched in terms of the simple FIFO queue of the example constituting Section 3.2.

1. Apply the *inverse probability method* (Section 4.2.5) to the pseudo-random numbers $U_1, U_2, \ldots$, to get the (nonuniform) random variables needed by the simulation model. Thus for the simple FIFO queue, we get n interarrival times $X_1, X_2, \ldots, X_n$ and n service times $S_0, S_1, S_2, \ldots, S_{n-1}$, by applying the inverse probability method to $U_1, U_2, \ldots, U_n$ and $U_{n+1}, \ldots, U_{2n}$, respectively.

2. Compute the desired function of the n (or more) random variables (e.g., W_n, the waiting time of the nth customer).

3. Repeat Steps 1 and 2 m times, using successive, independent streams of uniform random numbers, to get m *realizations* of the desired function. For example, we generate m observations of the waiting time of the nth customer,

$$W_n(1), W_n(2), \ldots, W_n(m)$$

4. *Estimate* the desired quantity; for example, if you want $E(W_n)$, use

$$\overline{W}_n = \frac{1}{m} \sum_{i=1}^{m} W_n(i) \tag{4.1.1}$$

5. When the estimate is a sample average, obtain an estimate of the precision of the estimate (e.g., $\overline{W}_n$), in the form of an *estimated standard deviation* of $\overline{W}_n$. Since

$\sigma_{\overline{W}_n} = \sigma_{W_n}/m^{1/2}$, the estimate is obtained by substituting the sample standard deviation of the $W_n(i)$ sample for σ_{W_n}:

$$S_{\overline{W}_n} = \frac{1}{m^{1/2}} \left[\sum_{i=1}^{m} \frac{(W_n(i) - \overline{W}_n)^2}{m-1} \right]^{1/2} \tag{4.1.2}$$

4.2   Details of Crude Simulation

4.2.1   The Parameter n: Time or Sample Size

Because the quantity being simulated is usually composed of n random variables or evolves with n time steps, we assume that some multiple of n uniform variates is used for its generation. This assumption is not a necessity, and it is used mainly for simplicity of exposition. In general, the more complex the system, the more *pseudo-random* numbers are required for generation of the statistic. In fact, as the methods for generation of the statistic become more and more sophisticated, the number of deviates required will itself probably become random. This is the case when acceptance–rejection schemes, as described in Volume II, are used.

The index n is only a generic indicator of the "size" of the statistic; later it will be important when we study how the statistic changes with this "size" or "time" index. *This dimension of "size" or "time" is present as a factor in most simulations.*

4.2.2   Reason for the Use of the Sample Mean

The sample average $\overline{W}_n$ may not be the best (point) estimate of $E(W_n)$ available. By best we mean smallest variance, since it is clearly an unbiased estimate of $E(W_n)$. It is used because, by the Gauss–Markov theorem, it is the least-squares estimator of $E(W_n)$ and therefore has minimum variance among all estimators that are linear in the observations. Moreover, to find a better estimator of $E(W_n)$, one must know the parametric probability distribution of W_n, and this will seldom be the case in a simulation.

4.2.3   Precision of the Estimate of the Sample Mean

Note from Equation 4.1.2 that the absolute precision of the simulation, as measured by the standard deviation, goes down as $1/m^{1/2}$. The derivation of this result is as follows. Recall that the $W_n(i)$ are, by construction, independent. Thus,

$$\mathrm{var}(\overline{W}_n) = \mathrm{var}\left[\frac{W_n(1) + \cdots + W_n(m)}{m} \right] = m \left[\frac{\mathrm{var}(W_n)}{(m^2)} \right]$$

Thus

$$\mathrm{var}(\overline{W}_n) = \sigma_{\overline{W}_n}^2 = \frac{\mathrm{var}(W_n)}{m} \tag{4.2.1}$$

and

$$\text{s.d.}(\overline{W}_n) = \sigma_{\overline{W}_n} = \frac{\sigma_{W_n}}{m^{1/2}} \tag{4.2.2}$$

Of course σ_{W_n} is usually not known and must be estimated if specific results are needed. However Equation 4.2.2 gives the quantitative result that to reduce the precision by a factor of 2, we have to take four times as many replications!! The same result holds even when other characteristics of W_n are being estimated. For example, quantiles of W_n use order statistics derived from the sample of m W_n's. Again the variance of this estimate is $O(1/m)$—that is, the variance decreases as $1/m$.

4.2.4   Estimation of the Variance of the Estimate of the Mean

You may want to compute other quantifications of the random variable W_n, such as the variance, skewness, or complete distribution. We consider this in Chapter 6. It is safe to say, however, that if you estimate the mean $E(W_n)$ using $\overline{W}_n$, you will always want to estimate the sample variance $S_{W_n}^2$ because, from Equation 4.2.1, you can then *estimate* the variance of the estimate $\overline{W}_n$ as

$$S_{\overline{W}_n}^2 = \frac{S_{W_n}^2}{m} = \frac{1}{m} \times \frac{1}{m-1} \sum_{i=1}^{m} (W_n(i) - \overline{W}_n)^2 \tag{4.2.3}$$

(The properties of $S_{W_n}^2$ that make it desirable as an estimator of $\sigma_{W_n}^2$ are again left to Chapter 6.)

This result, that one can use the actual data to estimate the sample variance $\text{var}(\overline{W}_n)$ of a sample average $\overline{W}_n$ is of fundamental importance in statistics. It is also of fundamental importance to the output analysis of simulations. The estimate of the standard deviation of $\overline{W}_n$ is then $S_{W_n}/m^{1/2}$.

4.2.5   The Inverse Probability Method for Generating Random Variables

The inverse probability method for generating random variables is different in detail for continuous and discrete random variables; hence these are treated separately.

i. The Continuous Case

Theorem   Let X have a *continuous* distribution $F_X(x)$, so that $F_X^{-1}(\alpha)$ exists for $0 < \alpha < 1$ (and is hopefully computable). Then the random variable $F_X^{-1}(U)$ has distribution $F_X(x)$, if U is uniformly distributed on $(0, 1)$, as at Equation 4.1.2.

Proof $P(F_X^{-1}(U) \leqslant x) = P(F_X(F_X^{-1}(U)) \leqslant F_X(x))$

because $F_X(x)$ is monotone. Thus

$$P(F_X^{-1}(U) \leqslant x) = P(U \leqslant F_X(x))$$
$$= F_X(x)$$

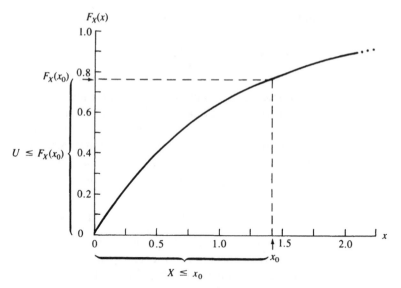

Figure 4.2.1   Illustration of the inverse probability method for generating random variables.

The last step follows because U is uniformly distributed on $(0, 1)$. Diagrammatically, we have that $\{X \leqslant x\}$ if and only if $\{U \leqslant F_X(x)\}$, an event of probability $F_X(x)$; see Figure 4.2.1.   □

As long as we can invert the distribution function $F_X(x)$ to get the inverse distribution function $F_X^{-1}(\alpha)$, the theorem assures us we can start with a (pseudo-) random uniform variable U and turn it into a random variable, $F_X^{-1}(U)$, which has the required distribution, $F_X(x)$.

a. Continuous Example: The Exponential Distribution   Consider the exponential distribution defined previously at Equation 3.2.3 as

$$\alpha = F_X(x) = 1 - e^{-\lambda x} \qquad \lambda > 0, x \geqslant 0$$
$$= 0 \qquad\qquad x < 0 \qquad\qquad\qquad (4.2.4)$$

Then for the inverse distribution function we have

$$x = -\frac{1}{\lambda}\ln(1 - \alpha) = F^{-1}(\alpha)$$

Thus if U is uniformly distributed on 0 to 1, then $X = -(1/\lambda)\ln(1 - U)$ has the distribution of an exponential random variable with parameter λ. We say, for convenience, that X is exponential(λ). Note that if U is uniform$(0, 1)$, then so is $(1 - U)$, and the pair U and $(1 - U)$ are interchangeable in terms of distribution. Hence, $X' = -(1/\lambda)\ln(U)$ is exponential. However, the two variables X and X' are correlated and are known as an *antithetic* pair.

EXERCISE   Find the correlation ρ between X and X', the antithetic pair of exponential(λ) random variables, where

$$\rho = E\left\{\frac{X' - E(X')}{\sigma(X')} \times \frac{X - E(X)}{\sigma(X)}\right\}$$ ∎

b. Continuous Example: The Normal and Gamma Distributions   For both these cases there is no simple functional form for the inverse distribution function $F_X^{-1}(\alpha)$, but because of the importance of the Normal and Gamma distribution models, a great deal of effort has been expended in deriving good approximations.

The Normal distribution is defined through its density,

$$f_X(x) = \frac{1}{\sqrt{2\pi}\sigma}\exp\left[\frac{-(x - u)^2}{2\sigma^2}\right]$$ (4.2.5)

so that

$$F_X(x) = \int_{-\infty}^{x}\frac{1}{\sqrt{2\pi}\sigma}\exp\left[\frac{-(v - u)^2}{2\sigma^2}\right]dv$$ (4.2.6a)

The Normal distribution function $F_X(x)$ is also often denoted $\Phi(x)$ when the parameters u and σ are set to 0 to 1, respectively. The distribution has no closed-form inverse, $F^{-1}(\alpha)$, but the inverse is needed so often that $\Phi^{-1}(\alpha)$, like logarithms or exponentials, is a system function on many computers.

The inverse of the Gamma distribution function, which was seen previously (Equation 3.5.1),

$$F_X(x) = \frac{1}{\Gamma(k)}\int_0^{kx/u} v^{k-1}e^{-v}dv \qquad x \geqslant 0, k > 0, u > 0$$ (4.2.6b)

is more difficult to compute because its shape changes radically with the value of k. It is sometimes available on computers, but it is not always a numerically reliable function.

It is of course in principle possible to compute the inverse probability distribution function by numerical methods (e.g., Newton's method), especially since the derivative of $F^{-1}(\alpha)$ is $1/f_X(\alpha)$, which is generally known. This method is generally very slow. More sophisticated methods for generating Normal and Gamma variates that do not require the inverse distribution function and are relatively simple are described in Volume II. In fact, many of these methods were developed because of the difficulties of computing inverse probability distribution functions (Abramowitz and Stegun, 1964; Kennedy and Gentle, 1980). A simple method for generating random variables with Normal distributions is given in Section 4.2.6.

c. Continuous Example: The Logistic Distribution   A commonly used symmetric distribution, which has a shape very much like that of the Normal distribution, is the (standardized) logistic distribution:

$$F_X(x) = \frac{e^x}{1 + e^x} = \frac{1}{1 + e^{-x}} \qquad -\infty < x < \infty \tag{4.2.7}$$

with probability density function

$$f_X(x) = \frac{e^{-x}}{(1 + e^{-x})^2} \qquad -\infty < x < \infty \tag{4.2.8}$$

Note that $F_X(-\infty) = e^{-\infty}/(1 + e^{-\infty}) = 0$ and $F_X(\infty) = 1$ by using the second form for $F_X(x)$.

The inverse is obtained by setting

$$\alpha = \frac{e^x}{1 + e^x} \tag{4.2.9}$$

Then $\alpha + \alpha e^x = e^x$, or $\alpha = e^x(1 - \alpha)$.

Therefore,

$$x = F_X^{-1}(\alpha) = \ln \alpha - \ln(1 - \alpha)$$

and the random variable is generated, using the inverse probability integral method, as follows:

$$X = \ln U - \ln(1 - U) \tag{4.2.10}$$

ii. The Discrete Case

Let X have a *discrete* distribution $F_X(x)$—that is, $F_X(x)$ jumps at points x_k, $k = 0$, 1, 2, ... Usually we have the case that $x_k = k$, so that X is *integer* valued.

Let the *probability function* be denoted by

$$p_k = P(X = x_k) \qquad k = 0, 1, \dots \tag{4.2.11}$$

The probability distribution function is then

$$F_X(x_k) = P(X \leqslant x_k) = \sum_{j \leqslant k} p_j \qquad k = 0, 1, \dots \tag{4.2.12}$$

and the *reliability* or *survivor* function is

$$R_X(x_k) = 1 - F_X(x_k) = P(X > x_k) \qquad k = 0, 1, \dots \tag{4.2.13}$$

The survivor function is sometimes easier to work with than the distribution function, and in fields such as reliability, it is habitually used.

The inverse probability integral transform method of generating discrete random variables is based on the following theorem.

Theorem Let U be uniformly distributed in the interval $(0, 1)$. Set $X = x_k$ whenever $F_X(x_{k-1}) < U \leqslant F(x_k)$, for $k = 0, 1, 2, \ldots$, with $F_X(x_{-1}) = 0$. Then X has probability function p_k.

Proof By definition of the procedure,

$$X = x_k \qquad \text{if and only if} \quad F_X(x_{k-1}) < U \leqslant F_X(x_k)$$

Therefore

$$
\begin{aligned}
P(X = x_k) &= P(F_X(x_{k-1}) < U \leqslant F_X(x_k)) \\
&= F_X(x_k) - F(x_{k-1}) \\
&= p_k
\end{aligned}
$$

by the definition of the distribution function of a uniform $(0, 1)$ random variable.

$\square$

Thus the inverse probability integral transform algorithm for generating X is to find x_k such that $U \leqslant F_X(x_k)$ and $U > F_X(x_{k-1})$ and then set $X = x_k$. An example of the implementation of this algorithm is given in Figure 4.2.2.

In the discrete case, there is never any problem of numerically computing the inverse distribution function, but the search to find the values $F_X(x_k)$ and $F_X(x_{k-1})$ between which U lies can be time-consuming; generally, sophisticated search procedures are required. In implementing this procedure, we try to minimize the

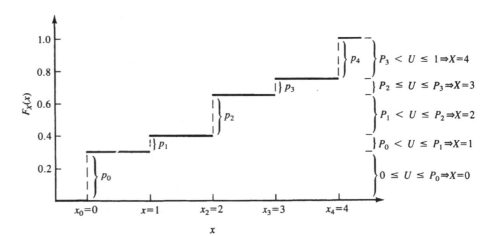

Figure 4.2.2 Illustration of the use of the inverse probability integral transform to generate a discrete random variable that takes on the values 0, 1, 2, 3, and 4 with probabilities p_0, p_1, p_2, p_3, and p_4. In the figure label, $P_0 = p_0, P_1 = p_0 + p_1$, $P_2 = p_0 + p_1 + p_2$, and so forth.

number of times one compares U to $F_X(x_k)$. If we want to generate many values of X, and $F_X(x_k)$ is not easily computable, we may also want to store $F_X(x_k)$ for all k rather than recompute it. Then we have to worry about minimizing the total memory to store values of $F_X(x_k)$.

a. Discrete Example: The Binary Random Variable   To generate a binary-valued random variable X that is 1 with probability p and 0 with probability $1 - p$, the algorithm is:

1. If $U \leqslant p$, set $X = 1$.
2. Else set $X = 0$.

b. Discrete Example: The Discrete Uniform Random Variable   Let X take on integer values between and including the integers a and b, where $a \leqslant b$, with equal probabilities. Since there are $(b - a + 1)$ distinct values for X, the probability of getting any one of these values is, by definition, $1/(b - a + 1)$. If we start with a continuous uniform $(0, 1)$ random number U, then the discrete inverse probability integral transform shows that

$$X = \text{integer part of } [(b - a + 1)U + a]$$

The proof is left as an exercise. Note that the continuous random variable $[(b - a + 1)U + a]$ is uniformly distributed in the open interval $(a, b + 1)$.

c. Discrete Example: The Geometric Distribution   Let X take values on zero and the positive integers with a geometric distribution. Thus

$$P(X = k) = p_k = (1 - \rho)\rho^k \qquad k = 0, 1, \ldots; \quad 0 < \rho < 1 \tag{4.2.14}$$

and

$$P(X \leqslant k) = F_X(k) = 1 - \rho^{k+1} \qquad k = 0, 1, \ldots; \quad 0 < \rho < 1 \tag{4.2.15}$$

To generate geometrically distributed random variables then, you can proceed successively according to the following algorithm:

1. Compute $F_X(0) = 1 - \rho$. Generate U.
2. If $U \leqslant F_X(0)$ set $X = 0$ and exit.
3. Otherwise compute $F_X(1) = 1 - \rho^2$.
4. If $U \leqslant F_X(1)$ set $X = 1$ and exit.
5. Otherwise compute $F_X(2)$, and so on.

Note that if $U \leqslant F_X(1)$ in Step 4, then because of the previous steps we have that $F_X(0) < U$.

Again it is easier here to use the survivor function $R_X(x_k)$ defined at Equation 4.2.13 because, for this geometric distribution $R_X(k) = \rho^{k+1} = \rho R_X(k - 1)$, with $R_X(-1)$ set to one. Thus $R_X(k)$ can be computed recursively. Also note that $F_X(x_{k-1}) < U \leqslant F_X(x_k)$ is equivalent to $R_X(x_{k-1}) > 1 - U \geqslant R_X(x_k)$. Thus since

$1 - U$ is uniform $(0, 1)$ it is better to use, in the algorithm, the condition that $X = x_k$ for which $R_X(x_{k-1}) > U \geqslant R_X(x_k)$. Note that $R_X(-1)$ is, by definition, 1.

Searching from the bottom—that is, from $F_X(0)$—is not essential, but it is faster on the average in this particular case. Notice that the theorem says we must make comparisons until we find the interval $[F_X(x_{k-1}), F_X(x_k)]$ into which U falls; it does not dictate the order in which we make the search. From an expository and programming standpoint it is easiest to start at the bottom. However, in terms of efficiency, we will see in Volume II that alternative strategies may be better.

In this case of a geometric distribution we can get a much faster algorithm by "inverting" $F_X(k)$. Thus note that for $k > 0$ we have the equivalences

$$
\begin{aligned}
F_X(k-1) &< U \leqslant F_X(k) \\
1 - \rho^k &< U \leqslant 1 - \rho^{k+1} \\
-\rho^k &< U - 1 \leqslant -\rho^{k+1} \qquad \text{(Subtract 1)} \\
\rho^k &> 1 - U \geqslant \rho^{k+1} \qquad \text{(Change sign)}
\end{aligned}
$$

and, since all the quantities are positive, the monotonic log transform applied throughout gives

$$
k \ln \rho > \ln(1 - U) \geqslant (k + 1)\ln \rho
$$

or

$$
k < \frac{\ln(1 - U)}{\ln \rho} \leqslant k + 1 \qquad k = 0, 1, \ldots \tag{4.2.16}
$$

the change in equalities occurring because $\ln \rho$ is negative.

Thus we assert that finding k such that the bounds are satisfied is the same as setting k equal to the smallest integer greater than or equal to $\ln(1 - U)/(\ln \rho)$, minus one.

As an example, if $\ln(1 - U)/\ln \rho$ is 4.15, then the smallest integer greater than or equal to 4.15 is 5, and $X = 4$.

Since rounding up is easily done on a computer, this *may* be faster than a search, especially if ρ is small. *More importantly it can be done without storing* the $F_X(k)$'s.

Note too that $-\ln(1 - U)$ is a unit exponential random variable so that $\ln(1 - U)/\ln \rho$ is exponential with mean $-1/\ln \rho$; we are actually using the result that discretizing an exponential random variable with mean μ gives a geometric random variable with $\rho = e^{-1/\mu}$.

4.2.6   An Alternative Method for Generating Normal Random Variables

The inverse probability integral transform is a general method for generating nonuniform random variables, but it is not always computationally feasible. Other methods based on ideas and special results of applied probability theory are described in Volume II. It is, however, worth mentioning one important case here.

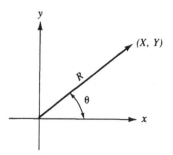

Figure 4.2.3   Relationship between the bivariate random variables X and Y and the equivalent polar pair (R, θ).

This is the case of Normally distributed random variables, for which the inverse distribution function $\Phi^{-1}(\alpha)$ is not readily computable.

A simple alternative scheme for generating Normal random variables is the Box-Muller technique. It is not the most efficient method for generating Normal random variables but is widely used because of its simplicity. It derives from the following well-known result (see, e.g., Fishman, 1978, p. 411).

Let X and Y be independent standardized Normal variates (mean 0, variance 1) and let R and θ be the polar coordinates of the (X, Y) point in the plane, as shown in Figure 4.2.3. Then

$$R^2 = X^2 + Y^2 \tag{4.2.17}$$

and

$$\theta = \tan^{-1}\left(\frac{Y}{X}\right) \tag{4.2.18}$$

are independent random variables, with R^2 having an exponential distribution with mean 2 and θ being uniformly distributed on the interval $(0, 2\pi)$.

The inverse relationship is

$$X = R\cos(\theta)$$
$$Y = R\sin(\theta) \tag{4.2.19}$$

But we have already seen how to generate exponentially distributed random variables R^2 and uniformly distributed random variables θ. Thus the following procedure generates two independent standardized Normally distributed variables X and Y.

1. Generate U_1, U_2 as independent random variables uniformly distributed on $(0, 1)$.
2. Set

$$X = \{-2(\ln U_1)\}^{1/2} \cos(2\pi U_2) \tag{4.2.20}$$
$$Y = \{-2(\ln U_1)\}^{1/2} \sin(2\pi U_2) \tag{4.2.21}$$

This is a commonly used method of generating Normally distributed random variables. It is simple to program and requires no calculation of inverse functions or computer memory, but it can be time-consuming because it is necessary to take a logarithm, a square root, a cosine, and a sine for each two random numbers.

It is in fact possible to generate Normal variates almost as fast as it is possible to generate pseudo-random numbers. This is discussed in Volume II, where we discuss sophisticated methods for generating nonuniform random variables.

At this point you have everything you need to simulate any quantity or aspect of a stochastic system that you can model. We proceed now to an example in which these ideas are applied. We then discuss a slightly different tool used in crude simulation, namely generation of random permutations, and give an example of its use.

4.3   A Worked Example: Passage of Ships Through a Mined Channel

The modeling examples considered in Chapter 3 were used to illustrate difficulties of modeling and the need for careful thought about assumptions and objectives before proceeding with a simulation. We turn now to a much more specific situation, which illustrates the techniques of crude simulation. *This example, while given in the context of a military problem, is typical of a large class of problems in statistical geometry ranging from the classical problem of random division and random covering of the interval to the problem of random packing of spheres in three-dimensional space* (see, e.g., Kendall and Moran, 1963).

We will consider the situation of a fleet of ships attempting to pass through a channel of water that is protected by mines. Under a very restricted set of assumptions, it is possible to work out theoretically the probability that a given number of ships will pass undamaged through the mine field. Crude simulation, however, gives us much greater freedom over the assumptions and allows us to consider scenarios (at the cost of additional programming and computing time) for which analytical results are not available.

The main issues or assumptions that need to be considered in this mine field problem include the following:

1. What is the size of the channel?
2. How many mines should be placed in the channel?
3. Where should the mines be placed in the channel?
4. How do we determine when a mine is going to be activated by a ship?
5. Once a mine has been activated, how much damage, if any, will it do?
6. How many ships are in the fleet?
7. When and where does each ship in the fleet cross the channel?
8. What measure of performance should we look at to determine whether a mine field is effectively stopping the ships, or, from the other point of view, whether the ships are successfully passing through the channel?

In the analysis that follows, we limit the simulation model to a very simple situation:

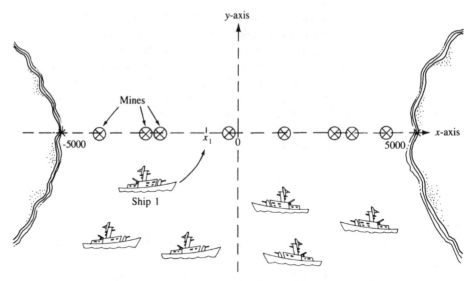

Figure 4.3.1   Passage of ships through a mine field.

1′. The channel will be one-dimensional (i.e., a line) of width 10,000 with the zero point located at the center and, hence, end points labeled − 5000 and 5000. In this way, mines are placed on the line and sense only along that line when a ship touches that line. Complications (and realism) due to radial sensing and time effects are thus removed (see Figure 4.3.1).

2′. The number of mines for each replication will be a Poisson(100) random variable.

3′. The mines will be placed in either of two ways: (a) equally spaced in the interval (− 5000, 5000), or (b) at points chosen independently according to a uniform (− 5000, 5000) distribution.

4′. If we let x denote the distance between a given ship and a given mine, then the probability that the mine activates is given by the function

$$A(x) = p\exp\left(-\frac{\pi x^2}{a^2}\right) \qquad 0 \leqslant p \leqslant 1; \quad a \geqslant 0 \tag{4.3.1}$$

We will consider two sets of constants, $(p, a) = (1.0, 400)$ and $(p, a) = (0.8, 200)$. The second set gives consistently lower probabilities of activation and corresponds to "less sensitive" mines, which might not detonate even if they were hit by the ship $(x = 0)$. Activation will be independent from mine to mine so that a ship may set off more than one mine.

5′. Again, we let x be the distance between ship and mine and define the probability of damage given activation according to the function

$$D(x) = q\exp\left(-\frac{\pi x^2}{d^2}\right) \qquad 0 \leqslant q \leqslant 1; \quad d \geqslant 0 \tag{4.3.2}$$

Ships are assumed to be in one of two states, either undamaged or damaged (unusable). (In this simple model they are never assumed to be partially damaged.) We also assume that damage is independent from mine to mine, that mines do not damage each other, and that a *mine can damage only one ship*. In the results that follow, we consider only the parameters $(q, d) = (1.0, 200)$.

6'. The fleet is composed of six ships.

7'. We will assume that the ships cross the channel successively and that the points at which the ships cross, denoted $X_1, X_2, \ldots, X_6$, are independent random variables chosen according to the truncated Normal density

$$f(x) = \frac{K_\sigma}{\sqrt{2\pi}\sigma} \exp\left(\frac{-x^2}{2\sigma^2}\right) I(|x|; 5000) \qquad (4.3.3)$$

Here, K_σ is chosen so that the density integrates to one, and $I(|x|; 5000)$ is the indicator function, which is one when $|x|$ is less than or equal to 5000, and zero otherwise.

Notice that the X_i can be generated on a computer by using simulated Normal $(0, \sigma)$ random variables but discarding any values less than -5000 or greater than 5000; more efficient methods can be obtained by using numerical evaluation of the inverse probability integral transform method.

The density, $f(x)$, clusters the ships toward the center of the channel, with their deviations from the center being influenced by the parameter σ. The usual intuition is that "channelization," or having all ships pass through the same point, is the optimal strategy. Hence, we would expect small values of σ to be favorable to safe passage of the ships. This intuition will be tested by comparing the ships' "strategies" $\sigma = 100$ versus $\sigma = 500$.

8'. To measure performance, we will have as the program output, after each replication, the value

$$Y(j) = \text{number of } undamaged \text{ ships} \qquad j = 1, 2, \ldots, m \qquad (4.3.4)$$

where m is the number of replications used. We then use $\bar{Y}$ and the empirical, discrete probability function of Y to decide between different cases of starting parameters. In addition, the effects of mine sensitivity are best understood by looking at performance on a ship-by-ship basis: very sensitive mines will be very effective against the first ship to attempt passage, but their numbers and therefore damage capability will be depleted by the sixth. Hence, to understand mine sensitivity better, we will also look at

$$P(i\text{th ship is damaged}) \qquad \text{for} \quad i = 1, 2, \ldots, 6$$

for each set of starting parameters.

The eight combinations of the three sets of input parameters and the corresponding results for $\bar{Y}$ and $\hat{P}(Y = 5 \text{ or } 6)$ are shown in the summary Table 4.3.1. All estimates are based on $m = 1500$ replications.

Table 4.3.1 Results of a simulation study of effects of three factors on the passage of ships through a mined channel

Case	Starting parameters	$\bar{Y}$	$\hat{P}(Y = 5$ or $6)$
1	Mine placement: Equally spaced Sensing: $(p, a) = (1, 400)$; entry: $\sigma = 100$	4.271	0.403
2	Mine placement: Equally spaced Sensing: $(p, a) = (1, 400)$; entry: $\sigma = 500$	2.026	0.006
3	Mine placement: Equally spaced Sensing: $(p, a) = (0.8, 200)$; entry $\sigma = 100$	3.545	0.137
4	Mine placement: Equally spaced Sensing: $(p, a) = (0.8, 200)$; entry: $\sigma = 500$	1.815	0.004
5	Mine placement: Uniform on $[-5000, 5000]$ Sensing: $(p, a) = (1, 400)$; entry: $\sigma = 100$	4.517	0.530
6	Mine placement: Uniform on $[-5000, 5000]$ Sensing: $(p, a) = (1, 400)$; entry: $\sigma = 500$	2.594	0.044
7	Mine placement: Uniform on $[-5000, 5000]$ Sensing: $(p, a) = (0.8, 200)$; entry: $\sigma = 100$	3.909	0.297
8	Mine placement: Uniform on $[-5000, 5000]$ Sensing: $(p, a) = (0.8, 200)$; entry: $\sigma = 500$	2.509	0.052

In every case the estimated standard deviation of $\bar{Y}$ was found to be less than 0.032 and the estimated standard deviation of $\hat{P}(Y = 5$ or $6)$, using Equation 4.2.3, was less than 0.013. Therefore, we can, with fairly strong confidence, make the following observations.

1. By comparing Cases 1, 2, 3, and 4 to Cases 5, 6, 7, and 8, respectively, we see that with all other parameters equal, placing the mines evenly in the channel will allow fewer ships to pass through the channel (more to be damaged). Equal spacing is therefore the preferred strategy for the person who lays the mines.

2. By comparing Cases 1, 3, 5, and 7 to Cases 2, 4, 6, and 8, respectively, we see that with all other parameters equal, having ships enter at as close to the same point as possible—that is, channelization—is the correct strategy from the ships' point of view. Hence the simulation quantifies and confirms the intuition cited in paragraph 7'.

3. By comparing Cases 1, 2, 5, and 6 to Cases 3, 4, 7, and 8, respectively, we see that with all other parameters equal, the lower sensitivity setting, $(p, a) = (0.8, 200)$, damages more ships than high sensitivity. The differences, because of sampling variation, are not always statistically significant, particularly in Case 6 versus Case 8, where the 95% confidence interval for the difference of the mean number of undamaged ships is approximately

$$[-0.006, 0.176]$$

(Confidence interval methodology is discussed in later chapters, and Chapter 8 covers graphical and formal statistical methods for analyzing multifactor statistical simulations.)

Probability of Damage Plotted by Ship Number

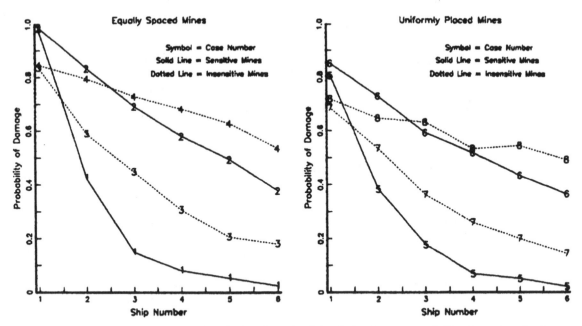

Figure 4.3.2   Display of results of simulating the passage of ships through a mined channel.

To understand what is happening, we can go to Figure 4.3.2, which shows $\hat{P}$(*i*th ship is damaged) plotted versus *i* for all eight cases. When the sensitivity is high, early ships are damaged with high probability, whereas later ships almost always get through. With low sensitivity, the probability of damage is much more constant with ship number. However, when the ships enter with wide dispersion (i.e., in Cases 2, 4, 6, and 8 when $\sigma = 500$), there are more mines within striking distance of a ship and "wasting" a few does not harm their effectiveness so much. It would be expected that either increasing the number of mines or decreasing the number of ships would lead us to look more favorably on highly sensitive mines. (Think about the extreme case of one ship attempting passage.)

To make this simulation more realistic, we could:

- Allow different forms for the functions $A(x)$ and $D(x)$.
- Allow a different distribution for the ships' entry points.
- Allow interaction between mines—for instance, the possibility of destroying each other (fratricide).
- Allow different mine-sensing schemes altogether—for instance, mines that do not explode until the *k*th sensing of a ship.
- Allow different degrees of damage to the ships.
- Allow ships of different types—for instance, a minesweeper, which would make the order of passage important.
- Allow mine sensing radially, in two dimensions, rather than just on the line. In this way multiple mines would not be exploding and damaging the same ship.

Unfortunately, this realism also brings in a time element, since each ship would have to be tracked at each time point while it was within the detection range of any of the mines.

Note too that the complexity of this example lies in the modeling of the situation on a digital computer. The statistical methodology used is basically that of straightforward simulation. This would change rapidly as more of the complications alluded to above were brought in, and the significance of multiple comparisons would need to be assessed. In fact, as more and more variations are brought in use, some kind of systematic statistical experimental design would be imperative.

4.4   Generation of Random Permutations

All the examples in the preceding sections have involved computing a statistic or random variable Y_n that is a function of other random variables X_1, X_2, ... In the simplest situations, the X_i are iid and are, themselves, generated as functions of iid uniform random variables; for instance, the X_i may be Normal $(0, 1)$ variates generated according to Section 4.2.6 and used to calculate a t statistic, $T_n = Y_n$. In other cases, however, the variables $\{X_1, \ldots, X_n\}$ required are *dependent* random variables. Sometimes the dependence is included implicitly through the algorithms —for instance, the fates of individual ships passing through a mine field in Section 4.3 are dependent by construction. Other simulations require exact dependence structures that can be built into the random variables as they are generated, as discussed in Chapter 10. Another broad class of problems requires a very particular type of dependence, in that $\{X_1, \ldots, X_n\}$ are a *random permutation* of the numbers $1, 2, \ldots, n$. It is the generation and use of random permutations that we consider in this section.

A *permutation* of $1, 2, \ldots, n$ is an arrangement of the numbers $1, 2, \ldots, n$ into the first, second, $\ldots$, nth positions. As an example, the six possible permutations of $1, 2, 3$ are:

$$(1, 2, 3), (1, 3, 2), (2, 1, 3), (2, 3, 1), (3, 1, 2), (3, 2, 1)$$

In general, there are $n! = n(n - 1)(n - 2)n \cdots 1$ permutations of the numbers $1, 2, \ldots, n$. This follows from the fact that there are n integers to fill the first position, leaving $n - 1$ integers to fill the second position, and so on, giving $n!$ distinct permutations.

A *random permutation of* $1, \ldots, n$ *is an equal probability selection of any one of the n! permutations*; that is, *any permutation can appear with probability* $1/(n!)$. Note then that while any X_i in $\{X_1, \ldots, X_n\}$ is marginally a discrete uniform variate on $1, \ldots, n$, the variables are dependent among themselves. This is easily seen by assuming that $X_1, X_2, \ldots, X_{n-1}$ are given; then X_n is totally determined as the only unassigned integer.

An equivalent definition of a random permutation is that X_1 is uniform on $(1, 2, \ldots, n)$; given X_1, then X_2 is uniform on the remaining integers, and so on down to X_n, which is completely determined by the values of $X_{n-1}, \ldots, X_2, X_1$. This observation gives a simple, direct algorithm for generating a random permutation of $(1, 2, \ldots, n)$. This is not the most efficient way to generate a random permutation, but it is certainly more efficient than computing all $n!$ permutations of $1, \ldots, n$ and picking out one of those! Note that $10! = 3,628,800$ and $20! \simeq 2.4329 \times 10^{18}$.

The foregoing idea for generating a random permutation is awkward and inefficient to put on a computer. The following method (Page, 1967; see also Knuth, 1981, p. 139, who ascribes it to Moses and Oakford, 1963) is very efficient and requires only $n + 1$ storage locations. It is described generally in terms of generating permutations of n objects $A_1, A_2, \ldots, A_n$, which could be the $n = 52$ cards in a deck or integers $1, 2, \ldots, n$.

Suppose the objects $A_1, \ldots, A_n$ are initially arranged in memory cells $1, 2, \ldots, n$ and let $U_1, U_2, \ldots$ as usual be iid uniform$(0, 1)$—that is, pseudo-random—numbers.

A random integer I_1 in the range $[1, n]$ is selected by $I_1 = [nU_1] + 1$, where $[x]$ is the integer part of x; then the contents of the memory cell indexed by I_1 and the contents of memory cell n are interchanged and, for the random permutation being generated, object A_{I_1} remains in memory cell n. Now I_2 is generated in $[1, n - 1]$ as $I_2 = [(n - 1)U_2] + 1$ and the contents of the memory cell indexed by I_2 are interchanged with the contents of memory cell $n - 1$. The rest of the procedure follows analogously; in all, $(n - 1)$ random integers are generated with a range that decreases from $[1, n]$ to $[1, 2]$. For a proof of the correctness of this procedure, see Page (1967).

We have chosen to concentrate on the generation of random permutations because of the many applications in statistics and operations research that require random permutations. Some experimental designs use a random permutation of the treatments. Investigations of scheduling strategies involving the order of job performance require repeated generation of random permutations. Card games, when simulated on a digital computer, use random permutations: the input to the card game is a shuffled deck of cards and the mathematical idealization of a shuffled deck requires that all 52! permutations of the 52 cards be equally likely. That is, the card on the top of the deck must have probability 1/52 of being any one of the 52 cards, and so on.

Another important application of random permutations is in nonparametric statistics. This area of statistical research involves the construction of a test statistic or estimator that, under a given null hypothesis, takes different values with equal probability, each value being based on a different permutation of the data. Ideally, we would compute the statistic at each permutation of the data and see how far out in the tail the statistic based on the observed data fell relative to this permutation distribution. In practice, however, calculation of the permutation distribution can be prohibitive and, classically, asymptotic approximations to the distribution have been used instead, as a reference. With the availability of fast computation, however, it has become practical in some cases to compute the full permutation distribution. In other larger problems, we randomly select a number of permutations from which we compute an approximate permutation distribution.

Another problem of interest is the generation of random samples of size r from the population of size n where $r < n$. (Notice that when $r = n$, we are back to the generation of random permutations.) This problem arises, for example, in simulation of sample survey schemes. Methods for generating these random subsamples can be found in Fan, Muller, and Rezucha, (1962); see, also, Knuth (1981, p. 137) for an algorithm.

EXAMPLE   **4.4.1 Kendall's Tau**

To illustrate the general ideas concerning permutation tests, we concentrate on a nonparametric test for correlation between the random variables (Y, Z) in an independent, identically distributed sample of n pairs. We denote the n pairs as $(Y_{(1)}, Z_1)$, $(Y_{(2)}, Z_2)$, ..., $(Y_{(n)}, Z_n)$, since we assume them to be ordered on the magnitude of the Y variables. Thus $Y_{(1)}, Y_{(2)}, ..., Y_{(n)}$ are the order statistics of the Y sample. The *rank* of Z_i is defined to be its index in an ordering of the Z sample. Thus if Z_i is the smallest of the Z_i's, its rank is 1. If it is the largest of the Z_i's, its rank is n. A nonparametric test for dependence between Y and Z is based on a *null permutation hypothesis. This says that if Y and Z are independent, then any of the n! orderings of the ranks of the Z variables is equally likely.* To quantify the dependence between Y and Z, we need a statistic that will assume extreme values if the Y and Z variables are associated (dependent). Observe that if Y and Z are dependent and very positively correlated, the ranks of the Z_i's will be ascending, and if they are negatively correlated the ranks will be descending.

Kendall's tau statistic (Conover, 1980, p. 256) quantifies dependence by scoring each pair (Z_i, Z_j), $i < j$, by 1 if $Z_i < Z_j$, and by -1 if $Z_i > Z_j$, and adding these scores. The added (total) score, S_n, will range from $n(n-1)/2$ (if the Z_i's are monotonically increasing) to $-n(n-1)/2$ (Z_i's are monotonically decreasing). For example, at $n = 3$ the total scores for each of the six possible permutations are as given in Table 4.4.1. Thus the total score statistic has value 3 with probability 1/6, value 1 with probability 1/3, value -1 with probability 1/3, and value -3 with probability 1/6. This is an exact distribution on which a test can be based—for example, reject independence if the statistic is 3 or -3, in which case the probability of a type I error (i.e., rejecting independence when it is true) is 2/6.

When $n = 4$, there are $4! = 24$ possible orderings of the Z ranks as shown in Table 4.4.2.

The total score statistic now takes on values from $k = -6 = -3!$ to $k = +6 = 3!$ with probabilities computed from Table 4.4.2 according to Table 4.4.3.

Generating this statistic, or statistics of this type, exactly for higher values of n is impractical even on the biggest computers. *Therefore, for, say n = 20, one might generate m = 10,000 random permutations from which the test statistic is computed*

Table 4.4.1   All possible permutations and associated total scores for $n = 3$

Permutation (ranks of Z_1, Z_2, Z_3)	Total score = S_n
1 2 3	3
1 3 2	1
2 1 3	1
2 3 1	−1
3 1 2	−1
3 2 1	−3

Table 4.4.2   All possible permutations and associated total scores for $n = 4$

Permutation	Total score	Permutation	Total score	Permutation	Total score	Permutation	Total score
4123	0	3124	+2	2143	+2	1423	+2
4132	−2	3142	0	2134	+4	1432	0
4231	−4	3241	−2	2431	−2	1234	+6
4213	−2	3214	0	2413	0	1243	+4
4312	−4	3412	−2	2314	+2	1342	+2
4321	−6	3421	−4	2341	0	1324	+4

Table 4.4.3   Distribution of total score statistic for $n = 4$

	k						
	−6	−4	−2	0	+2	+4	+6
$P\{S_4 = k\}$	1/24	3/24	5/24	6/24	5/24	3/24	2/24
	(0.0417)	(0.1250)	(0.2083)	(0.2500)	(0.2083)	(0.1250)	(0.0417)
$P\{S_4 \leqslant k\}$	0.0417	0.1667	0.3750	0.6250	0.8333	0.9583	1.0000

Table 4.4.4   Partial tabulation of results of 10,000 replications of total score statistic S_n at $n = 20$ with independent samples. (The maximum and the minimum values of S_{20} observed are 110 and −106, respectively)

k	$\hat{p}_k$	k	$\hat{p}_k$	k	$\hat{p}_k$	k	$\hat{p}_k$
−84	0.0020	−42	0.0105	0	0.0056	42	0.0127
−82	0.0010	−40	0.0134	2	0.0254	44	0.0093
−80	0.0009	−38	0.0118	4	0.0252	46	0.0070
−78	0.0007	−36	0.0131	6	0.0248	48	0.0094
−76	0.0008	−39	0.0141	8	0.0264	50	0.0071
−74	0.0011	−32	0.0159	10	0.0243	52	0.0076
−72	0.0020	−30	0.0159	12	0.0227	54	0.0063
−70	0.0023	−28	0.0163	13	0.0207	56	0.0057
−68	0.0026	−26	0.0218	16	0.0223	58	0.0046
−66	0.0028	−24	0.0268	18	0.0236	60	0.0036
−64	0.0025	−22	0.0200	20	0.0187	62	0.0035
−62	0.0038	−20	0.0218	22	0.0194	64	0.0029
−60	0.0047	−18	0.0230	24	0.0185	66	0.0027
−58	0.0051	−16	0.0223	26	0.0164	68	0.0020
−56	0.0053	−14	0.0256	28	0.0169	70	0.0014
−59	0.0060	−12	0.0242	30	0.0168	72	0.0018
−52	0.0068	−10	0.0221	32	0.0147	74	0.0020
−50	0.0067	−8	0.0230	34	0.0135	76	0.0022
−48	0.0085	−6	0.0278	36	0.0140	78	0.0028
−46	0.0066	−4	0.0249	38	0.0122	80	0.0003
−44	0.0089	−2	0.0241	40	0.0111	82	0.0008

and its distribution estimated. The interest here is not only the distribution at $n = 20$, but whether it can be approximated by a Normal distribution.

The results of generating 10,000 random permutations at $n = 20$ are presented in Table 4.4.4. Given in the even columns are the estimated probabilities $\hat{p}_k$, which

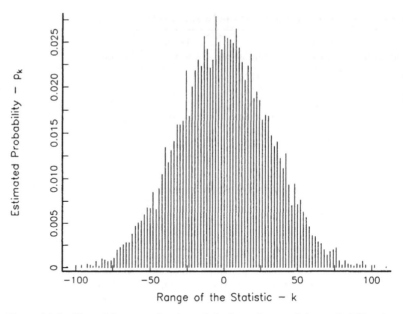

Figure 4.4.1   Plot of $\hat{p}_k$ versus k, where $\hat{p}_k$ is the estimate of the probability that $S_{20} = k$. The simulation experiment consisted of 10,000 replications of S_{20}, which is proportional to Kendall's statistic computed from two independent samples. The maximum observed value of S_{20} is 110, compared to the maximum possible value of $(n)(n-1)/2 = 190$. However, $S_{20} = 190$ occurs only with probability $1/(20!)$.

are the number of times $S_n = k$, divided by 10,000. An estimate of the standard deviation of $\hat{p}_k$, namely $\{\hat{p}_k(1 - \hat{p}_k)\}^{1/2}/100$ can then be calculated.

It is also true that S_n is a symmetric random variable with mean zero, and its variance $\sigma^2(S_n)$ equals $\{n(n-1)(2n+5)\}/18$ (Kendall, 1955). Since the mean and variance of S_n are known, a Normal approximation to S_n can be used, if it is established by simulation that for the given n, the approximation is close. (The asymptotic Normality has, in fact, already been established analytically.) Kendall's tau statistic is $S_n/\sigma(S_n)$. It is also common to divide S_n by its maximum $n(n-1)/2$ and use it as a type of correlation coefficient.

The estimated probabilities, $\hat{p}_k$, from Table 4.4.4 are plotted against k in Figure 4.4.1. Despite the raggedness, there is an indication of approximate Normality of the probability function. Testing for the Normality of outputs of simulations will be pursued further in Chapter 6.

Exercises   **4.1.** The FIFO single-server, single-input queue of the example of Section 3.2 is an idealization of many practical congestion systems. For example, the server may be a lawyer in solo practice or a plumber working alone.

    a. Analytical results in the MI|GI|1 case show that the expected waiting time in the steady state $E(W)$ is related to the traffic intensity $t = E(S)/E(X)$ (which must be less than 1 for stability) by

$$E(W) = \frac{E(S)}{1 - t}$$

Clearly the single server wants to be as busy as possible, which means that t must be very close to one; very large waiting times, on the other hand, would deter (anger) customers.

Generally lawyers or plumbers handle this by time slicing their services, doing a bit of a job at a time. How would you model this on a computer? That is, what physical and probabilistic assumptions would you make? Be sure to consider how the definition of waiting time may change.

b. Give a scenario you have experienced that is approximately a FIFO single-server queue and indicate any departures from the model, especially to handle the "almost infinite waiting time" problem.

4.2. Battleship is played by two people, A and B, as follows. On a 10 × 10 grid of squares, each player puts in 1 battleship 5 squares long, 2 aircraft carriers 4 squares long, 2 cruisers 3 squares long, 3 destroyers 2 squares long, and 2 submarines 1 square long. The ships can be horizontal or vertical, but not diagonal. Players fire in turn, and the firer is told when a hit has been scored and, when all squares of a ship have been hit, that a certain type of ship has been sunk. The game lasts until one player sinks all the opponent's ships.

It is clear that the time to completion is "random," since it would vary from game to game, and also with players. We wish to simulate the game to find the average length of a game.

a. Describe the output random variable. Is it discrete? Is it bounded? And if so, what are the bounds?

b. There are random elements in the game both in the sense of where the players put their ships and how they decide where to place successive shots. In addition, whether or not a shot scores a hit is a random variable. Describe a complete computer algorithm that will realistically simulate the game.

c. How would the randomness in the game change if we wanted to use it to measure the performance of a particular strategy for placement of ships?

4.3. Let U_1, U_2, and U_3 be uniformly distributed random variables (pseudo-random numbers).

a. Show that the distributions of

$$X_1 = \max(U_1, U_2)$$

and

$$X_2 = U_3^{1/2}$$

are the same, say $F_X(x)$. Based on this result, give two algorithms for generating a sequence of random variables with distribution $F_X(x)$.

b. If it takes $130\,\mu\text{sec}$ on a computer to take the square root of a number, $23\,\mu\text{sec}$ to compare two numbers, and $16\,\mu\text{sec}$ to generate a uniform random number, which is the fastest algorithm? Which takes less memory? Which would you prefer?

c. What is the inverse probability integral transform method for generating a random variable with the distribution function $F_X(x)$, which you found in part a?

4.4. A discrete random variable N has the *Poisson distribution* (with parameter $\lambda > 0$) if it takes on values $n = 0, 1, 2, \ldots$ with probability

$$p_n = P(N = n) = \frac{e^{-\lambda}\lambda^n}{n!} \qquad n = 0, 1, 2, \ldots$$

Give an algorithm for generating a Poisson(λ) random variable N using the *inverse probability integral transform* method (see subsection ii of Section 4.2.5). Assume that a sequence of uniform$(0, 1)$ (pseudo-random) random variables is available (i.e., $U_1, U_2, \ldots$). Evaluate your algorithm in terms of memory requirements and speed: that is, on the average, how many multiplications, additions, and comparisons do you have to make, to generate one Poisson(λ) variable N?

Note that the distribution function jumps at the points

$$x_0 = 0, x_1 = 1, \ldots, x_n = n, \ldots$$

and

$$F(n) = F(x_n) = P(N \leqslant n) = \sum_{i=0}^{n} \frac{e^{-\lambda}\lambda^i}{i!}$$

In designing your algorithm, can you make use of the fact that

$$p_0 = e^{-\lambda}$$

and that, for $n = 1, 2, \ldots$, the following holds:

$$p_n = p_{n-1} \times \frac{\lambda}{n}$$

4.5. You are told that the function $6x(1 - x)$ is a reasonable probability density function for a random variable X defined on the interval $[0, 1]$. That is,

$$f_X(x) = 6x(1 - x) \qquad 0 \leqslant x \leqslant 1$$
$$= 0 \qquad\qquad \text{elsewhere}$$

a. Verify that this is a probability density function (pdf). Also draw it!

b. Give two *simple* methods for generating a random variable with this pdf. Note that the pdf is symmetric, so that it might be easier to generate the density in two parts.

T4.6. Let the survivor function of a random variable X be given by

$$R_X(x) = P(X > x) = \begin{cases} 1 & x < 0 \\ 1 - x & 0 \leqslant x < 0.5 \\ 9/8 - 3x/2 + x^2/2 & 0.5 \leqslant x \leqslant 1.5 \\ 0 & x \geqslant 1.5 \end{cases}$$

Give a simple algorithm for generating X.

4.7. A continuous random variable X has survivor function (one minus the distribution function)

$$R_X(x) = P(X > x) = \frac{1}{1 + x} \qquad x \geqslant 0$$

Give an algorithm for generating X from a uniform random variable (i.e., a pseudo-random number).

4.8. A generalization of the survivor function given in Exercise 4.7 is to let

$$R_X(x) = P(X > x) = \left(\frac{1}{1 + x}\right)^\alpha \qquad x \geqslant 0, \quad \alpha > 0$$

a. Verify that this is a survivor function.
b. Find the pdf $f_X(x) = -dR_X(x)/dx$.
c. Give an algorithm for generating X from a uniform random variable.

The random variable X with this survivor function is said to have a *continuous Pareto distribution*. It occurs in many sociological, economic, and physical contexts.

d. What is the relationship of a Pareto-distributed random variable to an exponentially distributed random variable?

4.9. A Cauchy random variable X with scale parameter b and shift parameter a has the distribution function

$$F_X(x) = \frac{1}{2} + \frac{1}{\pi} \tan^{-1}\left(\frac{x - a}{b}\right) \qquad -\infty < x < \infty; b > 0; -\infty < a < \infty$$

a. Since you are given that this is a true distribution function, what can you say about the arctan function $\tan^{-1}(\cdot)$, as its argument ranges between $-\infty$ and $+\infty$?
b. How would you generate one random variable having this distribution function?
c. How many uniform random variables are required?

4.10. You require a random variable M, which has the distribution of the maximum of n random variables with distribution function $F_X(x)$. The inverse distribution function $F_X^{-1}(\alpha)$ can be computed. Give the inverse probability integral transform method for generating M.

4.11. Repeat Exercise 4.10 for a random variable that is the minimum of n random variables.

T4.12. The distribution function for the next-to-maximum $X_{(n-1)}$ of n iid random variables has distribution function:

$$F_{X_{(n-1)}}(x) = nF_X^{n-1}(x)[1 - F_X(x)] + F_X^n(x)$$

Can you find $F_{X_{(n-1)}}^{-1}(\alpha)$ in terms of $F_{X_{(n)}}^{-1}(\alpha)$? If not, what is an efficient way to generate $X_{(n-1)}$ by sorting or comparisons?

T4.13. You are required to find the probability p_0 that M, the maximum of 1000 Gamma(μ, k) variates, is less than a given value x_0 for given values of μ and k. You have a source of Gamma variates, perhaps LLRANDOMII (Lewis and Uribe, 1981).

a. Give a method for generating M. It should be the most "efficient" you can think of. Justify your answer. *Hint*: In general it takes n memory cells to sort n numbers. Can you do better if you want only the maximum?

b. You are required to give an answer to better than 1% precision, where "precision" means that

$$\frac{\hat{\sigma}_{\bar{M}}}{\bar{M}} \leqslant 0.01$$

You might be penalized if your answer is too accurate—say, better than 0.5%. Describe a way of accomplishing this (i.e., deciding on the number of replications).

T4.14. In shooting at a bull's-eye type of target, a score of 1 is given for a bull's-eye, and the scores 2, 3, 4, and 5 are given for successive rings outward. From experience, it is known that if X represents the score for one shot, then X has the following distribution:

$$P(X = 1) = 0.3; \quad P(X = 2) = 0.5; \quad P(X = 3) = 0.1; \quad P(X = 4) = 0.08;$$
$$P(X = 5) = 0.02$$

a. Assume you are given access to a random number generator for continuous random numbers between 0 and 1. Write a brief algorithm for simulating one value of X.

b. Assume you are given access to a generator of random integers between 1 and 100. Write a brief algorithm for simulating one value of X.

c. You are now interested in the actual distance from the bullet hole to the center of the target. If you miss to the left of the target, the distance is assigned a negative sign; if you miss to the right, the distance is recorded with a positive sign. Let Y represent that (signed) distance and notice that Y is a continuous random variable that can take values anywhere on the real line. Assume that Y has the following cumulative distribution:

$$F_Y(y) = \exp\left[-\exp\left\{ \frac{-(y - A)}{B} \right\} \right]$$

where A and B are given constants (this is called the Gumbel distribution; it arises in extreme value theory). Given access to a continuous uniform random number generator, how would you simulate one value of Y?

d. If you had access to an exponential random number generator only, how would you generate Y as in part c? Recall that if Z is exponential, mean 1, then:

$$F_Z(z) = 1 - \exp(-z) \quad z \geqslant 0$$
$$= 0 \quad z < 0$$

4.15. A binomial(n, p) random variable X has the probability function

$$p(k) = P(X = k) = \binom{n}{k} p^k (1 - p)^{n-k}$$

a. Give an algorithm based on the inverse probability integral transform method for generating X. Use any recursion relationships that exist between, say, $p(k)$ and $p(k + 1)$ to make the algorithm efficient.

b. Can you think of any probabilistic derivation of the binomial(n, p) random variable that would give a direct generation of X? Assume that you can use more than one uniform random variable to generate each X.

4.16. A *discrete* random variable N has a reliability function (survivor function)

$$R_N(k) = 1 - F_N(k) = P(N > k) = \frac{1}{1 + k - A} \quad k = A, A + 1, A + 2, \ldots$$

$$= 1 \quad k < A$$

where A is an integer. Give an algorithm for generating *one* random variable with this reliability function. Give only one algorithm. Assume that you can get uniformly distributed random numbers $U_1, U_2, \ldots$ and that your computer can add, multiply, divide, and compare two numbers.

ᵀ**4.17.** One can generalize the *discrete* random variable N in Exercise 4.16 by letting

$$R_N(k) = 1 - F_N(k) = \left(\frac{1}{1 + k - A}\right)^\alpha \quad k = A, A + 1, \ldots; \quad \alpha > 0$$

$$= 1 \quad k < A$$

a. Show that this is a survivor function. The random variable with this survivor function is called a discrete *Pareto-distributed* random variable.
b. Give a method for generating a random variable N by showing that an exact "inverse" analogous to that for the geometric distribution, can be found.

4.18. Assume you are given a uniform($0, 1$) random variable U. How would you simulate a value of X where X is a uniformly distributed random angle, in radians, between $-\pi$ and π? That is,

$$f_X(x) = \begin{cases} 1/2\pi & \text{if } -\pi < x \leqslant \pi \\ 0 & \text{if } x < -\pi \text{ or } x > \pi \end{cases}$$

4.19. Assume you are given two independent uniform($0, 1$) random variables U_1 and U_2.
a. Using U_1 and U_2, how would you generate X_1 and X_2 such that X_1 and X_2 are independent and have Normal($0, 1$) and Normal($1, 10$) distributions, respectively?
b. Using U_1 and U_2 and your knowledge of probability, how would you generate two independent chi-square random variables, each with 1 degree of freedom?

ᵀ**4.20.** Consider the following cumulative distribution function for a random variable X:

$$F_X(x) = \begin{cases} 0 & \text{if } x < 0 \\ \frac{1}{2} & \text{if } x = 0 \\ \frac{1}{2} + \frac{1}{2}[1 - \exp(-(\beta x)^k)] & \text{if } x > 0 \end{cases}$$

Equivalently, X is a positive random variable [i.e., $P(X < 0) = 0$] with a spike at 0 [i.e., $P(X = 0) = \frac{1}{2}$] and then has a Weibull tail [i.e., $F_{\text{Weibull}}(x) = 1 - \exp(-(\beta x)^k)$]. This distribution is useful if we are looking at the lifetime of an object—for instance, a lightbulb—that may be defective to begin with.

Assume that you are given only one value of U from a uniform($0, 1$) distribution. How would you simulate a value of X?

4.21. Given access to a uniform$(0, 1)$ random number generator:
   a. How would you generate an exponential(λ) random variable for any given value of λ?
   b. Using your knowledge of probability, how would you generate a Gamma$(2, \gamma)$ random variable?

4.22. Prove that if U is distributed uniform$(0, 1)$, then so is $(1 - U)$.

^{T}**4.23.** The distribution function of a double-exponential(1) random variable (also called a Laplace random variable) is given by:

$$F_X(x) = \begin{cases} \frac{1}{2}\exp(x) & \text{if } x \leq 0 \\ \frac{1}{2} + \frac{1}{2}[1 - \exp(-x)] & \text{if } x > 0 \end{cases}$$

   a. Show that the probability transform method gives the following rule to generate a double-exponential(1) random variable X from a uniform$(0, 1)$ random variable U:

$$X = \begin{cases} \ln(2U) & \text{if } U \leq \frac{1}{2} \\ -\ln[2(1 - U)] & \text{if } U > \frac{1}{2} \end{cases}$$

   Now assume that the following computer timings are true:

Comparisons take	$0.5\,\mu\text{sec}$
Additions or subtractions take	$1.0\,\mu\text{sec}$
Multiplies or divides take	$5.0\,\mu\text{sec}$
Logarithms take	$20.0\,\mu\text{sec}$
Uniform random variables take	$2.0\,\mu\text{sec}$

   Notice that the negation $-\ln[2(1 - U)]$ is the same as a subtraction from zero.
   b. What is the *expected* amount of time it would take to generate one value of X by the method above?

   Another way to generate a double-exponential(1) random variable is as follows. Take $E_1 \sim$ exponential(1) and $E_2 \sim$ exponential(1) with E_1 and E_2 independent. Let

$$X^* = E_1 - E_2$$

   c. Prove that X^* has a double-exponential(1) distribution. That is, prove that

$$P(E_1 - E_2 \leq x) = \begin{cases} \frac{1}{2}\exp(x) & \text{if } x \leq 0 \\ \frac{1}{2} + \frac{1}{2}[1 - \exp(-x)] & \text{if } x > 0 \end{cases}$$

   Hint: Use conditioning and evaluate the resulting integrals.
   d. If U_1 and U_2 are independent, uniform$(0, 1)$ random variables, we can generate independent exponentials,

$$E_1 = -\ln U_1 \quad \text{and} \quad E_2 = -\ln U_2$$

   so that

$$X^* = -\ln U_1 + \ln U_2$$
$$= \ln\left(\frac{U_2}{U_1}\right)$$

How long would it take to generate one value of X^*? Would you prefer the method that generates X or the method that generates X^*? Briefly, tell why.

[T]**4.24.** Show that the function

$$f(x; \alpha) = \begin{cases} \dfrac{1 + \alpha x}{2} & -1 \leqslant x \leqslant 1 \\ 0 & \text{otherwise} \end{cases}$$

is a probability density function for $-1 \leqslant \alpha \leqslant 1$.

a. Find the mean and variance of a random variable X with this pdf.

b. Show that for $\alpha \geqslant 0$ the random variable with this pdf is a mixture, with probability $(1 - \alpha)$, of a random variable that is uniformly distributed on $(-1, 1)$ and, with probability α, of a random variable on $(-1, 1)$ with pdf $(x + 1)/2$. Consequently show that if U_1 and U_2 are independent uniform$(0, 1)$ random variables, X can be generated by the following algorithm:

    1. If $U_1 \leqslant \alpha$, return $2U_1/\alpha - 1$.
    2. Else return $2\max[(U_1 - \alpha)/(1 - \alpha), U_2] - 1$.

c. Give a similar decomposition and algorithm for $\alpha < 0$.

d. Find $F(x) = \int_{-1}^{x} f(u; \alpha) \, du$ and $F^{-1}(p)$ and the generation algorithm for this random variable based on the inverse probability integral transform.

e. Compare the mixture algorithm and the algorithm of part d.

[T]**4.25.** In Example 4.3 it was mentioned that to generate a random variable X that has distribution function $F_X(x)$ and lies between a and b, with $a < b$, one could generate a sample of X's and throw out those that are not in the required interval.

a. What is the efficiency of this procedure? That is, what is the distribution of the number of rejects obtained in generating n of these truncated random variables?

b. Suggest a better method for generating the truncated X variables, using the inverse probability integral transform method.

[T]**4.26.** Compare the two algorithms for generating permutations in Section 4.4 in terms of speed, memory required, number of uniform variates needed, and ease of programming.

[T]**4.27.** Discuss how the generation of random permutations could be used to approximate the null permutation distribution of the Wilcoxon rank sum statistic (Conover, 1980, p. 215) for two samples of size r and $n - r$, respectively. If you could generate *directly* random subsamples of size r from a population of size n, how would you approximate this distribution? Contrast the two methods, in terms of computational efficiency.

References

Abramowitz, M., and Stegun, I., Eds. (1964). *Handbook of Mathematical Functions*, National Bureau of Standards Applied Mathematics Series 55. Government Printing Office: Washington, DC.

Conover, W. J. (1980). *Practical Nonparametric Statistics*, 2nd ed. Wiley: New York.

Fan, C. T., Muller, M. E., and Rezucha, I. (1962). Development of Sampling Plans by Using Sequential (Item by Item) Selection Techniques and Digital Computers. *Journal of the American Statistical Association*, 57, 387–402.

Fishman, G. S. (1978). *Principles of Discrete Event Simulation*. Wiley: New York.

Kendall M. G. (1955). *Rank Correlation Methods*, 2nd ed. Hafner: New York.

Kendall, M. G., and Moran, P. A. P. (1963). *Geometrical Probability*. Griffin: London.

Kennedy, W. J., Jr., and Gentle, J. E. (1980). *Statistical Computing*. Dekker: New York.

Knuth, D. E. (1981). *The Art of Computer Programming:* Vol. 2, *Seminumerical Algorithms*, 2nd ed. Addison-Wesley: Reading, MA.

Lewis, P. A. W., and Uribe, L. (1981). The New Naval Postgraduate School Random Number Package LLRANDOMII. Naval Postgraduate School Technical Report NPS55-81-005.

Moses, L. E., and Oakford, R. V. (1963). *Tables of Random Permutations*. Stanford University Press: Stanford, CA.

Page, E. S. (1967). A Note on Generating Random Permutations. *Applied Statistics, 16*(3), 273–274.

Wichmann, B. A., and Hill, I. D. (1982). Algorithm AS183: An Efficient and Portable Pseudo-random Number Generator. *Journal of the Royal Statistical Society, C*, 188–190.

Chapter

5

Uniform
Pseudo-Random Variable
Generation

5.0   Introduction: Properties of Pseudo-Random Variables

The topic of pseudo-random variable generation was presented in Chapter 4 as an element of crude simulation and, in fact, as the basis for most simulations. There we highlighted the two basic definitional properties: a uniform marginal distribution and independence, as defined at Equations 4.0.2 and 4.0.1, respectively. In addition, we saw how simple scatter plots could be used to partially examine these properties for given generators and to show that both good and bad generators are readily available.

Our goals in this chapter will be to look at, in more detail, how (and whether) particular types of pseudo-random variable generators work, and how, if necessary, we can implement a generator of our own choosing. Moreover, as we delve into aspects of sophisticated simulation in the chapters that follow, we will need to worry about issues such as the precision and speed of the generators. To accommodate all these goals, we have a list of requirements, more comprehensive than those in Chapter 4, for our uniform random variable generator:

1. A uniform marginal distribution
2. Independence of the uniform variates
3. Repeatability and portability
4. Computational speed

The third and fourth requirements are considered here for the first time. The first two properties are identical to those mentioned in Chapter 4, although certain probabilistic tricks considered here to make simulations run faster may also force us to be more careful in verifying these properties. Example 5.0.1 demonstrates such a situation.

EXAMPLE   **5.0.1 Recycling of Uniform Variates and "Bit-Stripping"**
In simulations, the process of making random decisions is often implemented by discretizing a uniform(0, 1) random variable through the inverse probability integral

transform for discrete random variables (see subsection ii of Section 4.2.5). As an example, assume that whether a machine is or is not functioning at a given time is a random event with associated probabilities 0.75 and 0.25, respectively. Then, to check in the simulation whether the machine is functioning, we would take U, a uniform$(0, 1)$ variate, and decide:

The machine is functioning if U is less than 0.75.

The machine is not functioning if U is greater than or equal to 0.75.

(This is the scheme given in Chapter 4 for generating binary random variables.)

Now, whenever a simulation calls for such discrete decisions, we have the option to *recycle* the uniform variate U through the use of the following theorem.

Theorem 5.0.1   Assume that U_0 is distributed as a uniform$(0, 1)$ random variable and that p is a number between 0 and 1.

a. Conditional on the event $U_0 < p$, we can define a new random variable, $U_1 = U_0/p$, such that U_1 has a uniform$(0, 1)$ distribution and is independent of the event $U_0 < p$.

b. Conditional on the event $U_0 > p$, we can define a new random variable, $U_2 = (1 - U_0)/(1 - p)$, such that U_2 has a uniform$(0, 1)$ distribution and is independent of the event $U_0 > p$.

The proof is left to the exercises (Exercises 5.1 and 5.2).   □

Note that U_1 and U_2 are not independent of the exact value of U_0, but rather of the events $U_0 < p$ and $U_0 > p$, respectively. In fact, in both cases, if the actual value of U_0 is given, then the values of U_1 and U_2 are fixed.

What we gain by such recycling is a new uniform variate, either U_1 or U_2, which we can use later in the simulation. For this scheme to be efficient, we presume that the recycling operation, either U_0/p or $(1 - U_0)/(1 - p)$, is faster than the generation of a new uniform variate from the uniform generator. This will depend on the computer under consideration, and, in particular, on the amount of time it takes to do a division and the way in which the pseudo-random number generator has been implemented. Exercise 5.4 indicates how recycling can be extended to uniformly distributed *discrete* variates that take on more than two values.

While recycling is entirely correct in theory for truly uniform$(0, 1)$ variates, we do take the chance of "bit-stripping" pseudo-random variates. A pseudo-random variate is only approximately uniformly distributed, with limited precision, so that repeated divisions by p (or $1 - p$) will eventually leave us working with low-order bits. For instance, if we take $p = 0.5$ and work on a binary 32-bit computer, then 32 divisions by p will bring us to the least significant binary bit of U. As we will show later, for certain generators these least significant bits may not be as random as they should be, and this may affect the results of our simulation. Notice that if the recycling step above is embedded in a conditional loop, we may strip the original variate many times without being aware of it. (See, e.g., the implementation by Brent, 1974, of a scheme by von Neumann to generate Normally distributed random

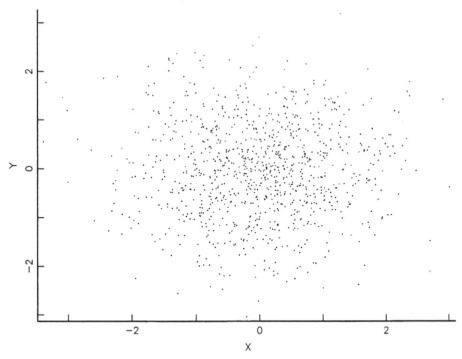

Figure 5.0.1 Scatter plot of 1000 pairs of "Normal" variates generated with the random number generator LLRANDOMII using the modified Box–Muller generator. *There is no stripping of the random numbers.*

variables; this scheme is in fact very sensitive to the properties of the uniform random number generator that is used.)

A direct, though artificial, example is now given to illustrate the effects of bit-stripping. Pairs of supposedly independent Normal random variables were generated using the Box–Muller technique (see Section 4.2.6). The actual algorithm used is a variant, to be described in Volume II, in which the cosine and sine of the uniform random variable in Equations 4.2.20 and 4.2.21 are generated by an acceptance–rejection scheme using two uniform random variables. (Acceptance–rejection is discussed in Volume II). This scheme depends critically on the independence and uniformity of the two uniform random variables.

In Figure 5.0.1 we show a scatter plot of 1000 pairs of "Normal" variates generated with the 31-bit random number generator LLRANDOMII (Lewis and Uribe, 1981). Properties of this random number generator were examined in Figure 4.0.3, and its structure is detailed later in this chapter. The scatter plot in Figure 5.0.1 shows no obvious departures from independence.

In Figure 5.0.2 the simulation experiment in Figure 5.0.1 is repeated, but each random number used in the generation is stripped of its highest 16 bits. This is achieved by recycling the random number 16 times after comparing it to $p = 0.5$. No obvious departure from independence is visible in the scatter plot of

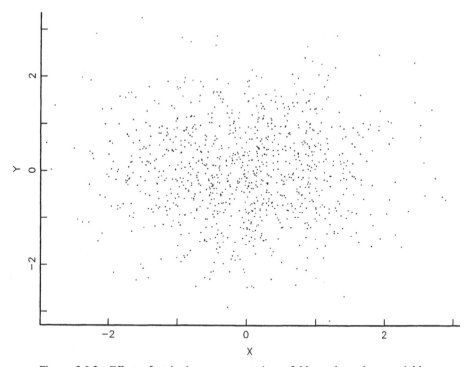

Figure 5.0.2 Effect of stripping on generation of Normal random variables. Scatter plot of 1000 pairs of "Normal" variates generated with the random number generator LLRANDOMII using the modified Box–Muller generator. The modification consists of generating the sine and cosine of a uniform random deviate with an acceptance–rejection scheme. *There is 16-bit stripping of the random numbers.*

Figure 5.0.2. Moreover, tests of marginal Normality (to be discussed in Chapter 6) all accept the null hypothesis that they are Normally distributed.

Figure 5.0.3 repeats Figure 5.0.1 but the uniform random number generator is RANDU. This generator was shown previously (Figure 4.0.5) to have problems in higher dimensions. There is a slight indication of pattern in the scatter plot in Figure 5.0.3, indicating that more formal tests of independence in bivariate Normal populations need to be applied. In Figure 5.0.4 the random numbers from RANDU have been stripped 10 times, resulting in the scatters, which are clearly not indicative of independent Normal random variables.

It is interesting that despite the patterns in the scatter plot in Figure 5.0.4, the variables are still marginally Normally distributed. This is illustrated in Figure 5.0.5. Details of these plots are discussed in Chapter 6. Suffice it to say here that the data points in the Normal probability plot (lower left) should form approximately a straight line. The departures from linearity are due to sampling fluctuations and are not significant. The same plot after stripping RANDU of 16 high-order bits shows gross departures from marginal Normality in the bivariate pairs.

Thus, we have demonstrated that the first two requirements discussed above, a uniform marginal distribution and independence of the uniform random variables,

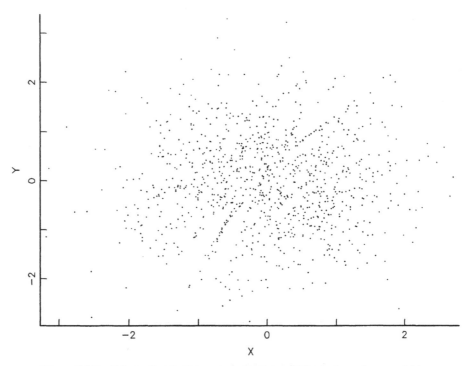

Figure 5.0.3 Effect of stripping on generation of Normal random variables. Scatter plot of 1000 pairs of "Normal" variates generated with the random number generator RANDU using the modified Box–Muller generator. The modification consists of generating the sine and cosine of a uniform random deviate with an acceptance–rejection scheme. *There is no stripping of the random numbers.*

can be destroyed by probabilistic tricks that make simulations run faster. This is due to the finite capacity, usually 32 bits, of computers.

The third criterion for our uniform generator, *repeatability*, is the ability to generate exactly the same stream of uniform random variates at any time in any simulation. This will be important when comparing the outputs of simulations run under two different sets of initial conditions, since the precision of the difference of the two estimates will be increased if the same input random variables are used. Such use of common variates is discussed in detail in Chapter 11, which is entitled "Variance Reduction." In addition, if our simulation produces unexpected output values, we may want to examine the input to decide whether the inputs were simply anomalous (but within statistical limits), whether the simulation is not running correctly, or whether our prior expectations of the results were simply incorrect. Being able to recreate our original random inputs makes this search procedure possible. Finally, the debugging of simulation programs is simplified if we can use the same random inputs, hence observe the exact same outputs on any run of the program.

The related aspect of *portability* means that exactly the same stream of random

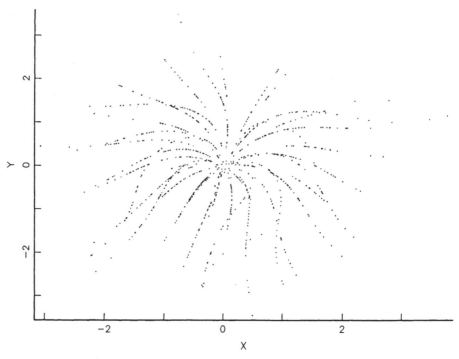

Figure 5.0.4 Effect of stripping on generation of Normal random variables. Scatter plot of 1000 pairs of "Normal" variates generated with the random number generator RANDU using the modified Box–Muller generator. The modification consists of generating the sine and cosine of a uniform random deviate with an acceptance–rejection scheme. *There is 10-bit stripping of the random numbers.*

inputs can be generated at any time *and on any computer.* This property is necessary if large simulation programs are to be transferred and used at many computer sites without being concerned about the accuracy of the results. The issues involved with portability, however, are usually computer science questions and are very machine dependent. For that reason, we will not consider portability in detail, though we will point out pitfalls and references to relevant material as necessary (see, e.g., Schrage, 1979; Gentle, 1981).

The final requirement, *computational speed*, is directly tied to the precision of our final simulation output. By increasing the speed of the uniform generator, we can generate more output in the same amount of time and thereby decrease the variance of our final output estimates. One way to achieve greater speed is to program random number generators in assembly language, rather than in higher level languages like FORTRAN. Unfortunately, an assembly language is usually specific to a given type of computer, so that the achievement of speed in this way is antithetic to the requirement for portability. It is also interesting to note (Volume II) that nonuniform random numbers such as Normals and Gammas can now be implemented in typically 1.1 to 1.5 times the time it takes to generate a uniformly distributed random variable. Thus the speed of generation of these nonuniform random numbers depends basically on the speed of the uniform generator.

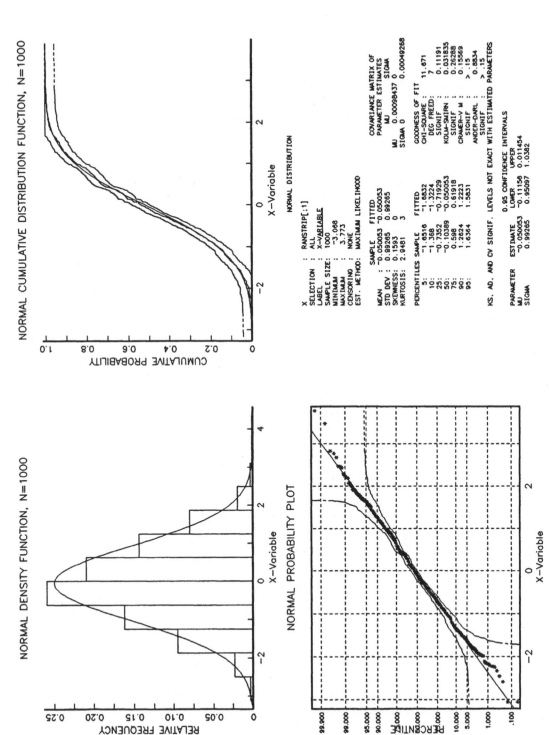

Figure 5.0.5 Investigation of the supposedly Normal marginal distribution corresponding to the scatter plot in Figure 5.0.4. The plots will be explained further in Chapter 6, but no significant departure from Normality is detected. Sample size is 1000.

These four requirements for pseudo-random number generators will be examined and sometimes balanced against each other throughout this chapter, as we look at the various methods for the computer generation of pseudo-random variates. By seeing the mechanics of the algorithms, you will be able to understand the proper use of generator "seeds" and the issues involving numerical precision and speed. By examining the historical growth and current proliferation of generators, you will be able to ask reasonable questions about the performance of given generators and to understand the inherent limits of classes of generators. We purposely avoid, however, many theoretical results and descriptions of statistical tests of generators. These are already well described in many other sources (e.g., Kennedy and Gentle, 1980; Knuth, 1981; and Bratley, Fox, and Schrage, 1983) and will be useful only if you have research interests in developing new methods for generating pseudo-random number generators. If you simply wish to use a random number generator, then we would urge you to find a generator for which there is documentation of thorough and satisfactory testing; our recommendations are given in Section 5.3.

5.1   Historical Perspectives

In simulation, uniform variates are used in a number of ways:

1. *Straightforward use of the uniform variates.* For example, a machine may be known to be equally likely to fail at any time within one year; a uniform variate will determine when during the year the machine fails in a simulation of that machine's performance.

2. *Generation of random variables from discrete distributions.* Such variables can be either qualitative, such as the decision process in Example 5.0.1, or quantitative, such as variates from a Poisson or geometric distribution. The inverse probability integral transform method for generating discrete random variables is given in subsection ii of Section 4.2.5.

3. *Generation of random variables from continuous distributions.* In subsection i of Section 4.2.5, the inverse transform method was used to convert uniform variates into variables from certain other continuous distributions. Likewise, more sophisticated methods in Volume II will also employ uniform variates in quite complex algorithms to generate continuous random variables.

4. *Generation of dependent sets or sequences of random variables.* Independent uniform variates can be transformed into dependent variables from other distributions using methods such as those shown in Chapter 10.

In fact, in general, any time a random event occurs in a simulation, a uniform generator will be involved. At times, that involvement will be direct as in the cases above; at other times the use will be transparent—you may call a Normal generator that has its own uniform variate generator embedded within it. *You should always be aware that the validity (and often speed) of your simulation rests heavily on the validity (and speed) of your uniform generator.* This is why so much attention is paid in this chapter to the development, testing, and efficient implementation of uniform generators. If we can start with a valid uniform generator—that is, one that simulates a sequence with the properties given by Equations 4.0.1 and 4.0.2—then all other random events in a simulation should follow the correct probability laws. There will be less need to check our Normal generator, our Poisson generator, and so on. If, on the other hand, our uniform generator is defective, its effects on the results of the simulation will be complex and unpredictable. It is always better and easier to use a good uniform generator (e.g., one of those recommended in Section 5.3) than to worry about the consequences of using a bad one.

Historically, people first looked to "random" sources to supply random numbers, as opposed to the deterministic algorithms that are used today for pseudo-random number generation. The first serious recorded effort was made by Tippett (1927), when he used census reports as a source of random digits. Unfortunately, such data are neither uniform nor independent and, instead, mechanical and electronic devices were employed later by Kendall and Babbington-Smith (1939a), the Rand Corporation some years later (Rand Corporation, 1955), and the British Premium Savings Bond Lottery in 1939 (Kendall and Babbington-Smith, 1939b). While such devices inherently produce random results—thereby satisfying our requirements 1 and 2—it is very hard to detect subtle malfunctions; moreover, the sequences that come out are not repeatable, and the generators are portable only in a very awkward physical sense. Today, such mechanical random number generators are still in use—for instance, in many U.S. state lotteries (numbered balls in a rotating drum determine winners), in situations where repeatability is not an issue, and in conditions that permit frequent checks for randomness. Such generators are, however, clearly not suitable for simulation work on high-speed digital computers.

A solution to the foregoing problems with mechanical generators was found by recording lists of the output from such machines in tables. This approach resulted in a table of 100,000 random digits from Kendall and Babbington-Smith (1939a) and the more familiar Rand table with 1 million random digits (Rand Corporation, 1955). For many applications in applied statistics—for instance, the randomization of patients in clinical trials, or for small simulations in which the computer has sufficient memory to store such a table—tabled uniform variates are an appropriate and efficient source of random variates. In larger simulations where at least 1 million random digits (or, equivalently, 100,000 10-digit uniform variates) are needed, such tabled variates are not appropriate and, for the types of simulation considered in this book, are not used.

Instead, attention has focused on algorithms to generate pseudo-random variables. Clearly, *an algorithm produces a fixed, deterministic sequence that can at best be called "pseudo-random"* if the output behaves, according to statistical tests, like a truly random sequence. The surprising fact that deterministic sequences generated by certain algorithms can *simulate* random sequences has stimulated much philosophical discussion of "randomness"; see, for example, the discussion in Knuth (1981, pp. 142–169).

An early attempt to mimic such random behavior was the mid-square method of von Neumann (1951).

EXAMPLE   5.1.1 The Mid-Square Method

To generate a random integer between 0 and 99,999, go through the following steps:

1. Choose any nonzero five-digit integer as a *seed* or starting value, say

   $$seed = 12345$$

2. Square the seed and place as many zeros in front of the result as necessary to make it a 10-digit integer. (Notice that the square of any five-digit number must contain 10 or fewer digits.) For our example,

   $$seed^2 = 16,604,025 = 0,016,604,025$$

3. Take the "middle" five digits as the desired random integer. Since there are in total an even number of digits, we define "middle" as the third through the seventh digits.

   $$random\ integer = 16604$$

4. For further random integers, repeat the steps above, using the last random integer as the seed in each loop.

   For instance, the second random integer would be found by squaring 16,604 to get 0,275,692,816, whose middle five digits give the result 75692.

   The mid-square method is not, and should not be, currently used because of defects in the output sequence with respect to randomness and uniformity of the variates. As an example, certain values of the seed will cause the generator to repeat the same value over and over again—that is, to have a cycle of length one, or in some cases, to degenerate to producing all zeros. For instance, the number 3792 is self-reproducing in the four-digit mid-square method, since $3792^2 = 14,379,264$. On the other hand, people have found (empirically) mid-square generated sequences that behave very randomly and do not degenerate or repeat until a large number of variates have been generated. Further details can be found in Knuth (1981, pp. 3–5) and Jansson (1966, Section 3A). The method does point out some of the general principles of algorithmic generators:

- The output can behave in a very "random" fashion.
- This apparent randomness must be established theoretically or verified empirically by statistical testing, not presumed.
- We must know something about the theoretical properties of the generator. For instance, how many variates can be generated before the sequence degenerates or repeats? In other words, we must know the cycle length (i.e., its period); making such a determination empirically can be very difficult.

   These rules can be seen as a plea to avoid "ad hoc" generators (see also Knuth, 1981), and they apply directly to the more fully examined classes of generators described in the next section.

5.2 Current Algorithms

The generation of pseudo-random variates through algorithmic methods is a mature field in the sense that a great deal is known theoretically about different classes of algorithms, and in the sense that particular algorithms in each of those classes have been shown, upon testing, to have good statistical properties. In this section, we describe the main classes of generators and then, in Section 5.3, we make specific recommendations about which generators should be implemented.

To avoid confusion later, it is worth noting that almost all algorithmic generators are inherently pseudo-random *integer* generators and only indirectly pseudo-random uniform(0, 1) generators. Thus, in an IBM System/370 computer, which has 32-bit registers and uses one bit for the sign of the number, the goal is to generate integers in the range 1 to $2^{31} - 1$. The conversion from integers to uniform variates is done simply by dividing by one plus the largest possible integer (e.g., 100,000 in Example 5.1.1 or 2^{31} in the IBM System/370 computer). Hence, in the algorithms that follow, we discuss either random uniform or random integer generation, as appropriate.

5.2.1 Congruential Generators

The most widely used and best understood class of pseudo-random number generators are those based on the linear congruential method introduced by Lehmer (1951). Such generators are based on the following formula:

$$U_i = (aU_{i-1} + c)\bmod m \tag{5.2.1}$$

where the U_i, $i = 1, 2, \ldots$ are the output random integers; U_0 is the chosen starting value for the recursion, called the *seed*; and a, c, and m are prechosen constants. Notice that to convert to uniform(0, 1) variates, we need only divide by the *modulus* m; that is, we use the sequence $\{U_i/m\}$.

The following properties of the algorithm are worth stating explicitly:

1. Because of the "mod m" operation (for background on modular operations, see, e.g., Knuth, 1981), the only possible values the algorithm can produce are the integers 0, 1, 2, ..., $m - 1$. This follows because, by definition, $x \bmod m$ is the remainder after x is divided by m.

2. Because the current random integer U_i depends only on the previous random integer U_{i-1}, once a previous value has been repeated, the entire sequence after it must be repeated. Such a repeating sequence is called a *cycle*, and its *period* is the cycle length. Clearly, the *maximum period* of the congruential generator of Equation 5.2.1 is m.

For given choices of a, c, and m, a generator may contain many short cycles (as shown in the examples below), and the cycle you enter will depend on the seed you start with. Notice that a generator with many short cycles is not a good one, since the output sequence will be one of a number of short series, each of which may not be uniformly distributed or randomly dispersed on the line or the plane.

Moreover, if the simulation is long enough to cause the random numbers to repeat because of the short cycle length, the outputs will not be independent.

3. If we are concerned with uniform$(0, 1)$ variates, the finest partition of the interval $(0, 1)$ that this generator can provide is $\{0, 1/m, 2/m, \ldots, (m - 1)/m\}$. This is, of course, not truly a uniform$(0, 1)$ distribution since, for any k in $(0, m - 1)$, we have $P\{k/m < U < (k + 1)/m\} = 0$, not $1/m$ as required by theory for continuous random variables.

However, the same type of shortcoming will be true of any computer algorithm because of the limited precision of the computer. For instance, on a machine with 32 bits of accuracy, the finest partition of $(0, 1)$ by any method is $\{0, 1/2^{32}, 2/2^{32}, \ldots, (2^{32} - 1)/2^{32}\}$.

4. Choices of a, c, and m (as well as the particular machine's arithmetic), will determine not only the fineness of the partition of $(0, 1)$ and the cycle length, and therefore, the uniformity of the marginal distribution, but also the independence properties of the output sequence.

Properly choosing a, c, and m is a science that incorporates both theoretical results and empirical tests. The first rule is to select the modulus m to be "as large as possible," so that there is some hope to address point 3 above and to generate uniform variates with an approximately uniform marginal distribution. However, simply having m large is not enough: one may still find that the generator has many short cycles, or that the sequence is not approximately independent.

As examples, consider either

$$U_i = 2U_{i-1} \bmod 2^{32}$$

where a seed of the form 2^k creates a loop containing only integers that are powers of 2, or

$$U_i = (U_{i-1} + 1) \bmod 2^{32}$$

which generates the nonrandom sequence of increasing integers. Therefore, the second equation gives a generator that has the maximum possible cycle length but is useless for simulating a random sequence.

Fortunately, once a value of m has been selected, theoretical results exist that give conditions for choosing values of the multiplier a and the additive constant c such that *all* the possible integers, 0 through $m - 1$, are generated before any are repeated (i.e., we have the maximal cycle length of m). Notice that this does not eliminate the second counterexample above, which already has maximal cycle length, but is a useless random number generator.

Theorem 5.2.1   A linear congruential generator will have maximal cycle length m, if and only if:

    a. c is nonzero and is relatively prime to m (i.e., c and m have no common prime factors).

    b. $(a \bmod q) = 1$ for each prime factor q of m.

    c. $(a \bmod 4) = 1$ if 4 is a factor of m.

Proof See Knuth (1981, p. 16).   □

As a mathematical note, c is called relatively prime to m if and only if c and m have no common divisor other than 1, which is equivalent to c and m having no common prime factor.

A related result concerns the case of c chosen to be 0. This case does not conform to condition a in Theorem 5.2.1; a value U_i of zero must be avoided because the generator will continue to produce zeros after the first occurrence of a zero. In particular, a seed of zero is not allowable. By Theorem 5.2.1, a generator with $c = 0$, which is called a *multiplicative congruential generator* in contrast to the general case (Equation 5.2.1) of a *mixed congruential generator*, cannot have maximal cycle length m. However, by Theorem 5.2.2, it can have cycle length $m - 1$.

Theorem 5.2.2 If $c = 0$ in a linear congruential generator, then $U_i = 0$ can never be included in a cycle, since the 0 will always repeat. However, the generator will cycle through all $m - 1$ integers in the set $\{1, 2, \ldots, (m - 1)\}$ if and only if:

a. m is a prime integer and
b. a is a primitive element modulo m

Proof See Knuth (1981, p. 19).   □

A formal definition for primitive elements modulo m, as well as theoretical results for finding them, are given in Knuth (1981). In effect, *when m is a prime, a* is a primitive element if the cycle is of length $m - 1$. The results of Theorem 5.2.2 are not intuitively useful, but for our purposes, it is enough to note that such primitive elements exist and have been computed by researchers (see, e.g., Table 24.8 in Abramowitz and Stegun, 1965). Hence, we now must select one of two possibilities:

• Choose a, c, and m according to Theorem 5.2.1 and work with a generator whose cycle length is known to be m.

• Choose $c = 0$, take a and m according to Theorem 5.2.2, use a number other than zero as the seed, and work with a generator whose cycle length is known to be $m - 1$. A generator satisfying these conditions is known as a *prime-modulus multiplicative congruential generator* and, because of the simpler computation, it usually has an advantage in terms of speed over the mixed congruential generator. Note that since this generator will not output a zero, functions of the uniform number generator output such as logarithms can be evaluated without special time-consuming checking.

Another method for speeding up a random number generator that has $c = 0$ is to choose the modulus m to be computationally convenient. For instance, consider $m = 2^k$. This is clearly not a prime number, but on a binary computer the modulus operation becomes a "bit-shift" operation in machine code. In such cases, Theorem 5.2.3, or a generalization of it (see Knuth, 1981, p. 20) gives a guide to the maximal cycle length.

Theorem 5.2.3   If $c = 0$ and $m = 2^k$ with $k > 2$, then the maximal possible cycle length is 2^{k-2}. This is achieved if and only if two conditions hold:

    a. a is a primitive element modulo m.
    b. The seed is odd.

Proof   See Knuth (1981, p. 19). □

Notice that we sacrifice some of the cycle length (although this may be insignificant on a 32-bit machine) and, as we will see later (Theorem 5.2.4), we also lose some randomness in the low-order bits of the random variates.

Having used any of Theorems 5.2.1, 5.2.2, or 5.2.3 to select triples (a, c, m) that lead to generators with sufficiently long cycles of known length, we can ask which triple gives the most random (i.e., approximately independent) sequence. Although some theoretical results exist for generators as a whole—for instance, computations for the correlation between U_i and U_{i+1} in terms of a, c, and m—these are generally too weak to eliminate any but the very worst generators. Marsaglia (1985) and Knuth (1981, Chap. 3.3.3) are good sources for material on such results.

There are, however, some useful general results for congruential generators. The first is an observation by Marsaglia (1976) on the effect of the parameter c on the "randomness" of the sequence. Assume that we have a generator with $c = 0$, giving us relatively fast computational speed, and that this generator has sufficient cycle length. Then changing the value of c serves to transform the output from

$$\text{seed, seed} + [(a - 1)\text{seed}], \text{seed} + [(a - 1)\text{seed}(a + 1)],$$
$$\text{seed} + [(a - 1)\text{seed}(a^2 + a + 1)], \ldots$$

to

$$\text{seed, seed} + [(a - 1)\text{seed} + c], \text{seed} + [(a - 1)\text{seed} + c][a + 1],$$
$$\text{seed} + [(a - 1)\text{seed} + c][a^2 + a + 1], \ldots$$

(all mod m). Hence this "linear" transformation (by "nonzero" c) in some sense does not increase the "randomness" properties of the sequence and would not justify the increase in computing time. A second result can be stated as a theorem, with generalizations given in Knuth (1981, pp. 12–14).

Theorem 5.2.4   If $U_i = aU_{i-1} \bmod 2^k$, and we define

$$Y_i = U_i \bmod 2^j \qquad 0 < j < k$$

then

$$Y_i = aY_{i-1} \bmod 2^j$$

In practical terms, this means that the sequence of j-low-order binary bits of the U_i sequence, namely Y_i, cycle with cycle length at most 2^j. In particular, the sequence of the *least* significant bit (i.e., $j = 1$) in $\{U_1, U_2, U_3, U_4, \ldots\}$ must behave as $\{0, 0, 0, 0, \ldots\}$, $\{1, 1, 1, 1, \ldots\}$, $\{0, 1, 0, 1, \ldots\}$, or $\{1, 0, 1, 0, \ldots\}$.

Proof   See Knuth (1981, pp. 12–14).   □

Such nonrandom behavior in the low-order bits of a congruential generator with non-prime-modulus m is an undesirable property, which may be aggravated by techniques such as the recycling of uniform variates discussed in Example 5.0.1.

It has been observed (Hutchinson, 1966) that prime-modulus multiplicative congruential generators with full cycle (i.e., when a is a positive primitive element) tend to have fairly randomly distributed low-order bits, although no theory exists to explain this.

The final result we give puts a limit on the validity of all congruential generators.

Theorem 5.2.5   If our congruential generator produces the sequence $\{U_1, U_2, \ldots\}$, and we look at the following sequence of points in n dimensions:

$$\{U_1, U_2, U_3, \ldots, U_n\}, \{U_2, U_3, U_4, \ldots, U_{n+1}\}, \{U_3, U_4, U_5, \ldots, U_{n+2}\}, \ldots$$

then the points will all lie in fewer than $(n!\, m)^{1/n}$ parallel hyperplanes.

Proof   See Marsaglia (1976).   □

As an example, if $m = 2^{32}$, then in $n = 10$ dimensions, the output of the generator lies on at most 41 hyperplanes.

The scatter plot in Figure 4.0.5 of the sequence $\{U_1, U_2\}, \{U_2, U_3\}, \ldots$ generated from RANDU shows that with poor generators the number of hyperplanes can be *much smaller* than the maximum, $(n!\, m)^{1/n} = (2 \cdot 2^{32})^{1/2}$, prescribed by this theoretical result, and this is a major problem with congruential pseudo-random number generators.

Although we must recognize that this planar structure is inherent in all congruential generators, we hope that with intelligent choices for a, c, and m, the structure is not striking until very high dimensions, where it will not affect the simulation results in systematic ways.

Given these known limitations of congruential generators, we are still left with the question of how to choose the "best" values for a, c, and m. To do this, researchers have followed a straightforward but time-consuming procedure:

1. Take values a, c, and m that give a sufficiently long, known cycle length and use the generator to produce sequences of uniform variates.
2. Subject the output sequences to batteries of statistical tests (see Section 5.5) for independence and a uniform marginal distribution. Document the results.
3. Subject the generator to theoretical tests. In particular, the spectral test of Coveyou and MacPherson (1967) (see also Knuth, 1981) is currently widely used and recognized as a very sensitive structural test for distinguishing between good and bad generators. Document the results.
4. As new, more sensitive tests appear, subject the generator to those tests. Several such tests are discussed in Marsaglia (1985).

In general, this testing procedure applies to generators of all types, not just congruential ones. Generators that have gone successfully through this process are considered good, valid generators and can be used for simulation work. Examples are given in Section 5.3.

5.2.2   Shift-Register Generators

An alternative class of pseudo-random generators are the shift-register or Tausworthe generators, which have their origins in the work of Golomb (1967). These algorithms operate on n-bit, pseudo-random binary vectors, just as congruential generators operate on pseudo-random integers. To return a uniform$(0, 1)$ variate, the binary vector must be converted to an integer and divided by one plus the largest possible number, 2^n.

Example 5.2.1 illustrates the algorithm for a classical type of shift-register generator, where the binary-vector seed is put through successive bit-shift and exclusive-or operations to transform it into the first random binary vector (which then seeds the next call to the generator).

EXAMPLE   5.2.1   A Shift-Register Generator

Assume that we wish to generate random 7-bit binary vectors, or equivalently, random integers between 1 and $2^7 - 1 = 127$. The algorithm we will use consists of the following steps:

a. Choose a seed (in binary form):   Seed $= (1, 0, 0, 1, 1, 0, 1)$
b. Shift the seed to the right 2 bits: $(0, 0, 1, 0, 0, 1, 1)$
c. Perform an exclusive-or addition of lines a and b: $(1, 0, 1, 1, 1, 1, 0)$
d. Shift the sum to the left 3 bits: $(1, 1, 1, 0, 0, 0, 0)$
e. Perform an exclusive-or addition of lines c and d: $(0, 1, 0, 1, 1, 1, 0)$

The result, $(0, 1, 0, 1, 1, 1, 0)$, is our first binary random vector, which we would use as the next seed if more than one random vector were desired.

In the algorithm in Example 5.2.1, one free parameter was the length of the binary vector, which we chose to be 7. The longer the vector, the finer the partition of the interval $(0, 1)$ when we finally convert to uniform$(0, 1)$ variates, and the greater the possible cycle length of the generator. The other two free parameters are the amount that we shift to the right (chosen to be 2) and the amount we shift to the left (chosen to be 3). These parameters perform similar functions to the multiplier a and the increment c in a congruential generator; choosing them wisely would assure a long cycle of variates, as well as pseudo-independence of the sequence. We return to these issues shortly.

The form of the algorithm in Example 5.2.1 is popular because the bit-shift and exclusive-or operations are particularly simple and efficient on a binary computer when programming in machine language. More generally, though, a shift-register generator can be characterized by the use of a binary matrix T, which is used to multiply the binary vector seed b, to produce the binary pseudo-random vector bT. Successive random vectors follow according to the sequence:

$$\{bT, bT^2, bT^3, bT^4, \dots\} \tag{5.2.2}$$

where all arithmetic is performed modulo 2 and the addition of binary vectors is the exclusive-or of the vectors. In this formulation, the right-shift 2 operation in Example 5.2.1 can be represented by the matrix:

$$R^2 = \begin{bmatrix} 0 & 1 & 0 & 0 & 0 & 0 & 0 \\ 0 & 0 & 1 & 0 & 0 & 0 & 0 \\ 0 & 0 & 0 & 1 & 0 & 0 & 0 \\ 0 & 0 & 0 & 0 & 1 & 0 & 0 \\ 0 & 0 & 0 & 0 & 0 & 1 & 0 \\ 0 & 0 & 0 & 0 & 0 & 0 & 1 \\ 0 & 0 & 0 & 0 & 0 & 0 & 0 \end{bmatrix}^2$$

the left-shift 3 operation can be represented by the matrix:

$$L^3 = \begin{bmatrix} 0 & 0 & 0 & 0 & 0 & 0 & 0 \\ 1 & 0 & 0 & 0 & 0 & 0 & 0 \\ 0 & 1 & 0 & 0 & 0 & 0 & 0 \\ 0 & 0 & 1 & 0 & 0 & 0 & 0 \\ 0 & 0 & 0 & 1 & 0 & 0 & 0 \\ 0 & 0 & 0 & 0 & 1 & 0 & 0 \\ 0 & 0 & 0 & 0 & 0 & 1 & 0 \end{bmatrix}^3$$

and then $T = (1 + R^2)(1 + L^3)$.

Exercise 5.8 looks at the computation of T for Example 5.2.1. From the general form of Equation 5.2.2, it is apparent that shift-register generators are simple linear transformations, albeit in a binary space and, therefore, are related to congruential generators. Notice that as with congruential generators, shift-register generators will cycle, since the current value depends only on the previous value. There are only 2^n different binary vectors of length n, so that the cycle can include at most 2^n elements; the zero vector will produce successive zeros so that the cycle can actually include at most $2^n - 1$ elements. Note however that n for Tausworthe generators can be greater than the word length of the computer being used.

In support of using shift-register generators, we have not only the efficiency and speed of the generators when programmed at the binary level, but also a well-developed theory to use in achieving the maximal cycle length of $2^n - 1$.

Theorem 5.2.6   Let T be an $n \times n$ nonsingular binary matrix and b a nonzero seed. Then for the shift-register generator to have maximal cycle length $2^n - 1$, it is necessary and sufficient that T have order $2^n - 1$ in the group of nonsingular $n \times n$ binary matrices.

Proof   See Marsaglia (1985), where the idea of the order of the binary matrix is defined.

$\square$

As examples, if we consider shift-register generators of the type in Example 5.2.1, where only left and right shifts are involved, then for 31-bit vectors, the combinations (right-shift, left-shift) = $(3, 28)$, $(6, 25)$, and $(13, 18)$ all produce maximal cycle length; for 32-bit vectors, no combinations produce maximal cycle length (Marsaglia, 1976).

Given that the matrix T has been chosen to provide a maximal, known cycle length, the randomness of the output is tested through the same steps laid out for congruential generators. It should be noted that the same structural problems that affect congruential generators in high dimensions (see Theorem 5.2.5) are also inherent in shift-register generators. Graphical illustration of this can be found in Marsaglia (1976).

5.2.3 Fibonacci Generators

The final major class of generators to be considered are the lagged Fibonacci generators, which take their name from the famous Fibonacci sequence, $U_i = U_{i-1} + U_{i-2}$. This recursion is reminiscent of the congruential generators, with the added feature that the current value depends on the *two* previous values.

The integer generator based directly on the Fibonacci formula

$$U_i = (U_{i-1} + U_{i-2}) \bmod m$$

has been investigated, but not found to be satisfactorily random (Knuth, 1981, p. 26). A more general formulation, however, can be given by the equation

$$U_i = U_{i-r} \cdot U_{i-s} \qquad r \geqslant 1, s \geqslant 1, r \neq s \tag{5.2.3}$$

where the symbol $\cdot$ represents an arbitrary mathematical operation. We can think of the U_i as either binary vectors, integers, or real numbers between 0 and 1, depending on the operation, $\cdot$, involved. As examples:

1. The U_i are real and $\cdot$ represents either mod 1 addition or subtraction.
2. The U_i are $(n-1)$-bit integers and $\cdot$ represents either mod 2^n addition, subtraction, or multiplication.
3. The U_i are binary vectors and $\cdot$ represents any of binary addition, binary subtraction, exclusive-or addition, or multiplication.

Early investigations (Green, Smith, and Klem, 1959; or T. G. Lewis and Payne, 1973) into certain combinations of the triples $(r, s, \cdot)$ produced little in the way of useful generators. Recently, though, attention has refocused on such generators because of work by Marsaglia (1985) showing that certain combinations produce output that is random according to stringent criteria. In addition, a good deal is known about cycle length for integer generators:

- Using mod $m = 2^n$ addition or subtraction, there are known conditions under which it is possible to achieve a cycle length of $(2^r - 1)2^{n-1}$.
- Using mod $m = 2^n$ multiplication and odd seeds, there are known conditions under which it is possible to achieve a cycle length of $(2^r - 1)2^{n-1}$.

The conditions are given in detail in Knuth (1981) and Marsaglia (1985). Note that with the scheme of Equation 5.2.3, the recurrence of a number in the generated sequence does not necessarily start the cycle all over again.

Clearly, the increased cycle length of a well-chosen Fibonacci generator over that of a congruential generator could translate into a great advantage in terms of uniformity and randomness. As drawbacks, however, the Fibonacci generators require specification of the parameters r, s, and m, a choice for the operation $\cdot$, and a number $(\max(r, s))$ of seeds. This last requirement is necessary to start the recursion shown in Equation 5.2.3. We do not yet feel that enough is known or published in the literature about how best to choose all these parameters; hence, we have not included a Fibonacci generator in Section 5.3. However, we do see this area as potentially fruitful, and refer readers to the remarks by Marsaglia (1985) for suggested generators.

Other generators that generalize even further on the Fibonacci idea by using a linear combination of previous random integers to generate the current random integer are discussed in Knuth (1981, Chap. 3.2.2). No such generator has emerged as superior to simpler existing congruential ones, and given the complexity of choosing the proper coefficients and seeds, we do not pursue the topic here.

5.2.4   Combinations of Generators (Shuffling)

Intuitively, it is tempting to believe that "combining" two sequences of pseudo-random variables will produce one sequence with better uniformity and randomness properties than either of the two originals. In fact, even though good congruential, Tausworthe, and Fibonacci generators exist, combination generators may be better for a number of reasons.

1. Theoretical results (Rosenblatt, 1975; Brown and Solomon, 1979; Marshall and Olkin, 1979; Marsaglia, 1985) support the claim that combinations of sequences have as good or better uniformity properties than the individual sequences. These results are diluted by the assumption that the individual sequences are independent, but, if we choose sequences from very different generators (e.g., one congruential and one Tausworthe), the assumption may be approximately true.

2. If the periods of the original sequences are expressible as rd and sd, with d an integer, r and s integers that are relatively prime, the period of the combined generator is rst, where t is a divisor of d (Marsaglia, 1985). Hence, individual generators with short cycle lengths can be combined into one with a very long cycle. This can be a very compelling advantage, especially on computers with limited mathematical precision.

These potential advantages have led to the development of a number of successful combination generators and research into many others. One such generator, a combination of three congruential generators that exploits property 2 above, was developed and tested by Wichmann and Hill (1982) and is presented in Section 5.3.

Another generator, Super-Duper, developed by G. Marsaglia, combines the binary form of the output from the multiplicative congruential generator with multiplier $a = 69{,}069$ and modulus $m = 2^{32}$ with the output of the 32-bit Tausworthe generator using a left-shift of 17 and a right-shift of 15. This generator performs well, though not perfectly (Marsaglia, 1985), and suffers from practical drawbacks that are mentioned later.

A third general variation, a *shuffled generator*, randomizes the order in which a generator's variates are output. Specifically, we consider one pseudo-random variate generator that produces the sequence $\{U_1, U_2, \ldots\}$ of uniform(0, 1) variates, and a second generator that outputs random integers, say between 1 and 16. The algorithm for the combined, shuffled generator is as follows:

1. Set up a "table" in memory of locations 1 through 16 and store the values U_1, $U_2, \ldots, U_{16}$ sequentially in the table.
2. Generate one value, V, between 1 and 16 from the second generator.
3. Return the U variate from location V in the table as the desired output pseudo-random variate.
4. Generate a new U variate and store it in the location V that was just accessed.
5. If more random variates are desired, return to Step 2.

Notice that the size of the table (16 in the example) can be any value, with larger tables, intuitively, creating more randomness but requiring more memory allocation. Also, for simplicity, the same generator can be used to produce both the U_i sequence and the V_i sequence (after converting to random integers in the correct range). Such shuffled generators can perform well, especially if the base generator for the U_i's is, itself, a good generator. One implementation using the LLRANDOMII generator of Section 5.3 to generate both the U_i's and the V_i's and creating a 128-element table is described in Lewis and Uribe (1981). In fact, the value V is the lowest 6 bits of the number U. This relies on the empirically derived result that for prime-modulus multiplicative generators with full period, the low-order bits tend to be quite random.

This method of shuffling by randomly accessing and filling a table is due to MacLaren and Marsaglia (1965). Another scheme, attributed to M. Gentleman in Andrews et al. (1972), is to permute the table of 128 random numbers before returning them for use. The permutation is accomplished as in Section 4.4 with the second generator. This scheme would be time-consuming, and we know of no case in which it has been systematically implemented.

The use of this type of combination of generators has also been described in the contexts of simulation problems in physics by Binder and Stauffer (1984).

The drawbacks to combinations of generators include two practical considerations. The combination of two generators will take twice as long to produce the same number of random variates (similarly for combinations of three or more), and this may force an investigator to compromise the precision of the simulation output. Second, the coding of the generators is more difficult and, therefore, potentially open to error. As an example, a high-level language implementation of Super-Duper requires that a Tausworthe generator be coded in integer or real arithmetic, rather than its natural binary form; the alternative, programming in assembly or machine code, will be prohibitive for many potential users. In addition, there is no quantitative measure of how much better (if at all) a combination generator is than a good single generator. All algorithmic generators are inherently flawed, and it is left to the individual investigator to decide whether the intuitive marginal gain in randomness compensates for the practical drawbacks.

Note that the shuffling scheme used in LLRANDOMII is faster than the other combination schemes, but there is no theory to predict its period. The period (cycle length), however, must be greater than the period of the original generator.

5.2.5 Generators in Packages

In some instances, it may be either inconvenient or impossible to choose a desired pseudo-random variate generator. If you are working within a software package such as the discrete-event systems simulation program SIMSCRIPT or the data analysis program SAS, you cannot directly access alternative generators. However, you should still be aware of which generator is in use in the package and, by applying the ideas from the preceding sections, its potential limitations. In other cases—for instance, while using the IBM/PC BASIC or perhaps working on a computer with IMSL or FORTRAN libraries available—you must choose whether to take advantage of the convenience of the prewritten generator or to implement one of the recommended generators from Section 5.3. To help in these situations, we offer information about the generators currently in some of the major packages. For more up-to-date information or descriptions of generators in packages not listed here, you should contact the software supplier before deciding whether the generator meets your simulation needs.

a. SIMSCRIPT II.5 (Kiviat, Villanueva, and Markowitz, 1983) uses a prime-modulus multiplicative congruential generator with multiplier $a = 630,360,016$ and modulus $m = 2^{32} - 1$. The performance documented in Fishman and Moore (1985) appears good, with some weaknesses in three and six dimensions.

b. GPSS is a popular discrete-event simulation package with several implementations. One of these for microcomputers, GPSS/PC, uses a prime-modulus multiplicative congruential generator with the multiplier 397,204,094 proposed by Learmonth and Lewis (1974), and modulus $m = 2^{31} - 1$.

c. SIMULA (Birtwistle, Dahl, and Myhrhaug, 1979) uses a prime-modulus multiplicative congruential generator with multiplier $a = 5^{2p+1}$ and modulus $m = 2^n$, where p and n (and, therefore, the generator's properties) depend on the computer.

d. SAS (1982) uses either a prime-modulus multiplicative congruential generator with multiplier $a = 397,204,094$ and modulus $m = 2^{31} - 1$, or a shuffled prime-modulus multiplicative congruential generator with multiplier $a = 16,807$ and modulus $m = 2^{31} - 1$. Both generators have been tested and found to have good performance, as discussed in Section 5.3.

e. SPSS (1983) offers a prime-modulus multiplicative congruential generator with multiplier $a = 16,807$ and modulus $m = 2^{31} - 1$ (Lewis, Goodman, and Miller, 1969). This generator is discussed in Section 5.3.

f. APL (see, e.g., Katzen, 1970) in its various implementations uses the prime-modulus multiplicative congruential generator with multiplier $a = 16,807$ and modulus $m = 2^{31} - 1$.

g. The IMSL Library (1984) provides a choice of two prime-modulus multiplicative congruential generators, both with modulus $m = 2^{31} - 1$, and, as multipliers, either $a = 16,807$ or $a = 397,204,094$. The latest version of this subroutine library (IMSL, 1987) also provides for use of the multiplier $a = 950,706,376$.

h. LLRANDOMII (Lewis and Uribe, 1981) offers two prime-modulus multiplicative congruential generators with modulus $m = 2^{31} - 1$ and either of the two multipliers $a = 16,807$ or $a = 397,204,094$. Shuffled versions of both are also available as options. The generator uses a division simulation algorithm for the mod operation (see Exercises 5.10 and 5.11) and assembly language coding to achieve great speed. The latest version of this program also allows the use of several of the Fishman and Moore (1982, 1985) multipliers.

i. The *Enhanced Simulation and Statistics Package* (Lewis, Orav, and Uribe, 1988), the latest in a series of software packages, now being offered in conjunction with this book, uses the prime-modulus multiplicative congruential generator with multiplier $a = 16,807$ and modulus $m = 2^{31} - 1$.

j. The NAG Library, which is produced by the National Algorithms Group in Great Britain and is available throughout the world, is similar in scope to the IMSL Library. It uses a prime-modulus multiplicative congruential generator with $m = 2^{59}$ and $a = 13^{13}$. The origins of this generator are obscure, although it is mentioned in Sylwestrowicz (1982) and Ripley (1983). It requires quadruple-precision integer arithmetic to implement on most computers, and thus would not seem to be suitable for most purposes. Furthermore, nothing is documented about its statistical properties.

k. Random number generators that come with microcomputers should, in general, be considered suspect. One can look to the January 1983 issue of *CALL-A.P.P.L.E.* magazine for a discussion of problems with the Apple generator (Sparks, 1983). A scatter plot of output pairs from the IBM/PC BASIC generator analogous to, for instance, Figure 4.0.2, is shown in Figure 5.2.1. Although Figure 5.2.1 was made on a very-high-resolution monitor, the striations in the plot are barely evident with only the graphics capability of the IBM/PC. However, the slice in the scatter cube of triples shown in Figure 5.2.2 makes it clear that the IBM/PC generator is completely defective. No details of this generator are published, but it has been established empirically that the generator has cycle length 2^{16}.

l. Suprisingly enough, the random number generator supplied with the VMS operating system for DEC VAX machines is RANDU, which, as we have seen previously, is highly defective. Since the VAX is probably the most widely used scientific computing machine in the world, this highlights the care that must be taken to ascertain the properties of random number generators in even the best of circumstances. The version of RANDU in the VAX machine uses $c = 0$, $a = 2^{16} + 3$, and $m = 2^{31}$. The fact that m is a power of 2 and the binary form of a contains only 3 one-bits makes this generator very fast. The speed is clearly antithetic to good statistical properties, as seen in Figures 4.0.5 and 5.0.4.

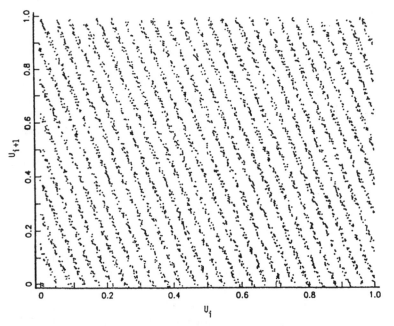

Figure 5.2.1   Scatter plot of 2001 successive pairs of $\{U_i, U_{i+1}\}$ produced by the IBM/PC BASIC random number generator.

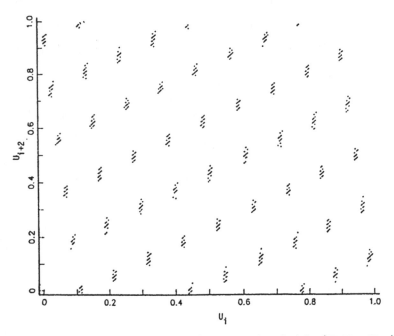

Figure 5.2.2   Scatter plot of a slice in the scatter plot of triples $\{U_i, U_{i+1}, U_{i+2}\}$ produced by the IBM/PC BASIC random number generator. The 2000 plotted pairs are $\{U_i, U_{i+2}\}$ in triples for which $0.50 < U_{i+1} \leqslant 0.51$.

5.3   Recommendations for Generators

The algorithmic nature of pseudo-random variate generators makes each of them imperfect in some way. Hence, we cannot recommend any generator as being "good" in an absolute sense. Indeed, as new algorithms and tests are developed, our standards will change and currently acceptable generators will become outdated. Therefore, the generators we present here are ones that have been tested extensively and found to have reasonably good properties in general situations. As the field of random number generation develops and improved generators become available, these recommendations should still provide adequate pseudo-random variates, though they will no longer be the "best" available.

Be sure to read Section 5.4 for information about numerical and speed considerations before implementing any generator. If the algorithm works inefficiently or incorrectly on your computer, you must switch to an alternative generator.

GENERATOR 1   This generator, often referred to as LLRANDOMII in this book, is actually one of the pseudo-random variate generators available in the LLRANDOMII simulation package (Lewis and Uribe, 1981) developed at the Naval Postgraduate School in Monterey, California. It is a prime-modulus, multiplicative congruential generator that is widely used as part of the APL and IMSL packages. The generator is attributed to Lewis, Goodman, and Miller (1969), and its testing is documented (Learmonth and Lewis, 1974; Learmonth, 1976). The formula is given by:

$$U_{i+1} = (16{,}807\,U_i)\bmod(2^{31} - 1)$$

Advantages of this generator include the following.

- The simplicity of the algorithm, particularly for implementation in a high-level language.
- Its widespread use. Although not necessarily proof that the generator should be used, the testing that has been done and long-term satisfaction with its performance argue that the generator can be trusted. Hence, if you use and refer to this algorithm, people will recognize it and generally believe its results.
- A long cycle length of 2,147,483,646 ($= 2^{31} - 2$), which includes all possible integer values that can be generated on a 31-bit machine except 0 and $2^{31} - 1$. This will be sufficient for most simulations.

Disadvantages of the LLRANDOMII generator include the following:

- This is not the "best" prime-modulus multiplicative generator with modulus $2^{31} - 1$ that is known today. More details are given in the discussion of Generator 2.
- The generator may have a problem with dependencies in high dimensions. If such structure is important in your simulation, you should consider Generator 2.

GENERATOR 2   Actually, we refer here to a group of generators, all of the same prime-modulus multiplicative, congruential class as Generator 1. The modulus is the same prime for each, $2^{31} - 1$, but the multipliers we present are the five "best" multipliers, all positive primitive roots of $2^{31} - 1$, found by Fishman and Moore (1985) after an exhaustive computer test. In all, more than 267 million possible multipliers that allow maximal cycle length for a prime-modulus multiplicative congruential generator with modulus $2^{31} - 1$ were tested using the most sensitive tests currently available. The "best" multipliers were chosen on the basis of these tests. The form of the generators is:

$$U_{i+1} = (aU_i)\mod(2^{31} - 1)$$

where the five "best" values of a are:

$$a = 950,706,376$$
$$a = 742,938,285$$
$$a = 1,226,874,159$$
$$a = 62,089,911$$
$$a = 1,343,714,438$$

Notice that the multiplier 16,807 from Generator 1 is not on this list, although it was a candidate. Some defects in the lattice structure of this generator, as measured by the lattice test discussed in Section 5.5, show up in the three-dimensional structure of the generator. What effect this suboptimality of 16,807 has on any particular simulation is difficult or impossible to assess. We also note that another multiplier that appeared often in Section 5.2.5, $a = 397,204,094$, is also not on the list. This multiplier was proposed (Learmonth and Lewis, 1974) as an alternative to 16,807 when the high-dimensional properties of 16,807 came under question. As with 16,807, the multiplier 397,204,094 has passed many tests previously and been found satisfactory, if not the best available for simulation work.

All these generators have the same long cycle length (2,147,483,646), since all the a's given above are positive primitive roots of $2^{31} - 1$. They also have the apparent advantage of producing uniform variates that are more random than the variates produced by Generator 1. One possibly fatal drawback, however, concerns implementation: it may not be possible to evaluate the product aU_i correctly, even in double-precision arithmetic, on some computers. This point is discussed in detail in Section 5.4.

GENERATOR 3 This combination congruential generator was devised by Wichmann and Hill (1982) and tested at the National Physical Laboratory; a Pascal implementation is given in Wichmann and Hill (1987). It is an excellent generator in its own right, and serves as an alternative to Generators 1 and 2 when computer word size is a problem or a very long cycle length is necessary. To produce a uniform$(0, 1)$ variate, U_{i+1}, in the particular implementation of the general scheme given by Wichmann and Hill, we require the output from 3 separate congruential generators:

$$X_{i+1} = (171 X_i)\mod 30269$$
$$Y_{i+1} = (172 Y_i)\mod 30307$$
$$Z_{i+1} = (170 Z_i)\mod 30323$$

and then define

$$U_{i+1} = \left\{ \left(\frac{X_{i+1}}{30269} \right) + \left(\frac{Y_{i+1}}{30307} \right) + \left(\frac{Z_{i+1}}{30323} \right) \right\} \bmod 1$$

Note that three "seeds" are needed to start the generator. The fact that U_{i+1} is uniformly distributed follows from the generalization of Theorem 5.3.1.

Theorem 5.3.1   Assume that V and W are independent, uniform$(0, 1)$ random variables. Then

$$U = (V + W) \bmod 1$$

is distributed as a uniform$(0, 1)$ variate.

Proof   The proof is straightforward and is left to the reader (Exercise 5.9).   □

The advantages of the Wichmann and Hill generator include:

• Since the moduli of the individual congruential generators are all prime, the cycle length can be as long as $(30,269) \times (30,307) \times (30,323) = 2.782 \times 10^{13}$. However, it was shown later that the cycle length is actually one-quarter of this value (Wichmann and Hill, 1984). If we take this to be approximately 2^{42}, we can see that the cycle length is more than 2000 times longer than the cycle length of $2^{31} - 2$ for Generators 1 and 2.

• Since each individual congruential generator uses a relatively small multiplier and modulus, the computations can be done with smaller computer word sizes. In particular, Wichmann and Hill (1982) show an implementation using 16-bit integer arithmetic.

The main potential disadvantage to this generator (as with all combination generators) is the element of speed. Since three congruential generators must be run, and each must be converted to a uniform$(0, 1)$ number, the generator will usually take about three times as long as either Generator 1 or 2. (On some microcomputers, this problem may be offset by the ability to work purely with integers.) Comparisons in terms of "randomness" of the output are not possible, since the tests used to choose the five multipliers for Generator 2 do not apply to combination generators.

Although we feel that any of these three recommendations are viable, if an ordering is desired, we would choose Generator 2 (any of the five generators) as our preferred algorithm because of the improved randomness of the output. This choice presumes that the computations required by Generator 2 are possible. If, as is often the case on small computers, the multiplications are not done correctly when the multiplier is large (e.g., 1,343,714,438 or even 62,089,911), we would choose Generator 1 because of its speed advantage over Generator 3. The special cases in which Generator 3 is preferred have already been explained above.

5.4 Computational Considerations

Now that a number of pseudo-random variate generators that produce suitably random output have been identified, we can turn to computational considerations that may further delineate our preferences.

Portability in its broad-case sense, as defined in Section 5.0, will usually not be an issue unless you are designing a package to be used on many, possibly unspecified, computers. In this case, the computing issues become complex, time-consuming (see, e.g., Schrage, 1979, and Gentle, 1981) and irrelevant for most researchers programming simulations in a high-level language on a single computer.

Note that the generators in the IMSL package are designed to be very portable and, in fact, produce the same stream of random numbers for any of the machines for which the IMSL package is implemented. One aspect of portability, however, is relevant to everyone: *The computer you are working on must be able to evaluate your algorithm correctly.* This may seem obvious, but there are many potential hidden problems, as demonstrated by the next two examples.

EXAMPLE **5.4.1 Word Size for Congruential Generators**
Consider the formula for Generator 1:

$$U_{i+1} = (16,807 U_i) \bmod (2^{31} - 1)$$

Since all the U_i's are integer valued, between 1 and $(2^{31} - 2)$, we may be tempted to code the algorithm using "long" 32-bit integers. This would save time by avoiding floating-point computations. Unfortunately, the algorithm would be incorrect, since the product, $16,807 U_i$, could not be held in a 32-bit register for large values of U_i. Consider, for instance, $U_i = 2^{31} - 2$; then the product is larger than 2^{32}. Unless the compiler made you aware of such overflows, the sequence of uniforms that you used would not be those that had been tested and found to be random. Note that a correct version of the algorithm could be based on 64-bit double-precision real arithmetic.

Similar problems would occur if you attempted to use the Wichmann and Hill generator directly with 16-bit operations.

Both these generators can be programmed using only integers (see Schrage, 1979, for Generator 1; Wichmann and Hill, 1982, for Generator 3), but the algorithms are more complex and, as shown later, may offer no more speed than if 64-bit, double-precision real operations were used. Even then, McLeod (1985) found *problems in implementing the Wichmann and Hill algorithm on Prime computers.* This is because the Prime-400 computer uses chopped arithmetic with only 23 bits for the representation of the fractional part of a real variable.

EXAMPLE **5.4.2 Precision for Congruential Generators**
Consider using the first of the formulas for Generator 2,

$$U_{i+1} = (950,706,376 U_i) \bmod (2^{31} - 1)$$

and coding the algorithm in FORTRAN on an IBM PC/AT with an 80287 floating-point coprocessor. Since $U_i < 2^{31}$ and $950{,}706{,}376 < 2^{31}$, it seems reasonable to expect that any possible product, $950{,}706{,}376\,U_i < 2^{62}$, should be correctly computed using 64-bit double-precision arithmetic. Unfortunately, a hand calculation shows that if we take

$$U_i = 2^{31} - 2 = 2{,}147{,}483{,}646$$

then

$$(950{,}706{,}376)(2{,}147{,}483{,}646) = 2{,}041{,}626{,}394{,}607{,}926{,}896$$

while the IBM PC/AT returns a value of

$$(905{,}706{,}376)(2{,}147{,}483{,}646) = 2{,}041{,}626{,}394{,}607{,}926{,}780$$

This inaccuracy in the low-order bits demonstrates the limited precision of the multiply operation, a condition that exists for many micro- and minicomputers.

A technical reference guide may specify the amount of precision your computer can retain, hence the largest multiplier you can hope to use. If, as is often the case, such a guide is unavailable, you should experiment for yourself: compare a hand calculation of $2^{31} - 2$ times your multiplier to the answer given by your computer. (Be aware that overflows are not noted by some compilers.)

We have concentrated throughout the examples on some problems specific to congruential generators. You must ask analogous questions when implementing alternative algorithms.

Although assuring the correctness of the algorithm is an issue you *must* deal with, increasing the speed of the computation is a factor that is also often desirable. If you are installing a random number generator on a mainframe or even a mini that will be accessed by many users for large simulations, it is worthwhile to put some initial time and effort into implementing the most efficient algorithm possible. Optimally, the generator should be written in assembly code. As an example, the LLRANDOMII package, which includes Generator 1, is written in assembly code for IBM series 360 and 370 computers. Generator 1 in LLRANDOMII takes approximately 3 seconds to generate 1 million uniform$(0, 1)$ variates on an IBM 370/3033 computer. In contrast, the portable FORTRAN implementation of Generator 1 in IMSL Subroutine GGSUB requires 9 seconds for the identical computation.

An assembly language version of the congruential generator in Super-Duper is given for microcomputers with 8087 math coprocessors by Lee (1985) and can be easily adapted for other congruential generators.

If the expertise or desire to program in assembly code is lacking, use a high-level language as efficiently as possible. The specifics will depend on the language and the computer, but, as examples of good programming habits, you can:

• Use assigned values for constants, not calculated values. For instance, for $2^{31} - 1$ use $2{,}147{,}483{,}647$, not $[(2**31) - 1]$.

• Use operations that are fast on your computer. For instance, to convert from a uniform integer in $\{1, 2, \ldots, 2^{31} - 2\}$ to a uniform$(0, 1)$ variate, check whether it is faster to divide by $2^{31} - 1$ or to multiply by (the assigned constant) $1/(2^{31} - 1)$. (On an IBM PC with a math coprocessor, it makes no difference.)

• Use any tricks that are available. For instance, Payne, Rabung, and Bogyo (1969) give a way to emulate the double-precision mod$(2^{31} - 1)$ operation. This division simulation algorithm is used in LLRANDOMII in the assembly language version of Generator 1 to achieve the speed of 3 μsec per uniform random number. See Exercise 5.10 for details.

Fortunately, you will often find that even a simple, straightforward implementation of a generator in a high-level language will provide enough speed for a particular simulation. As an example, the accompanying program listing shows a FORTRAN implementation of Generator 1 that was used to generate uniform$(0, 1)$ variates on an IBM PC/XT with the 8087 floating-point coprocessor.

```
      SUBROUTINE LRNDPC( DSEED, U,N )
C
      INTEGER           N, I
      REAL              U(N)
      DOUBLE PRECISION  D31M1, DSEED
C
      DATA    D31M1  /2147483647.D0/
C
      DO 5 I=1,N
      DSEED = DMOD( 16807.D0*DSEED, D31M1 )
5     U(I) = DSEED / D31M1
      RETURN
      END
```

This program took 1 minute and 36 seconds to generate 100,000 uniform$(0, 1)$ variates on the IBM PC/XT. The timing was comparable to either the double-precision or long-integer version of Schrage's portable version of Generator 1, and about three times faster than straightforward implementation of Wichmann and Hill's Generator 3. Any of these implementations is probably fast enough for most simulations. If, however, the same comparisons are made on an IBM PC/XT without the floating-point coprocessor, we found that 100,000 uniform$(0, 1)$ variates require:

1. About 24 minutes using the foregoing program
2. More than 100 minutes using Schrage's implementation (Schrage, 1979) in double precision
3. About 9 minutes using Schrage's implementation in long integers
4. About 30 minutes for the Wichmann and Hill generator using either long or short integers

In this case, the algorithm and the implementation make a great deal of difference, although the obvious recommendation would be to purchase a math coprocessor.

5.5   The Testing of Pseudo-Random Number Generators

The area of uniform generator testing, both theoretical and statistical, is an active field of research that is extremely important to everyone involved with simulation. The actual testing, however, is difficult and extremely time-consuming: it is not a practice we would recommend to those interested primarily in performing simulation experiments. For this reason, we discuss here only the general types of commonly used tests, not specifics about their implementation. With this information, you should be able to critically evaluate the testing that a particular generator has undergone; if you wish to go further and test a generator yourself, you must look for details in the cited references.

Certain types of test, often called theoretical, concentrate on the "structure" in n-dimensional space of the overlapping n-tuples: $\{U_i, U_{i+1}, \ldots, U_{i+n}\}$, $\{U_{i+1}, U_{i+2}, \ldots, U_{i+n+1}\}$, $\ldots$, constructed from the output $\{U_1, U_2, \ldots\}$ of a given generator. For generators of certain types, including full-period mixed congruential generators and prime-modulus multiplicative congruential generators, the structure is a function of the parameters of the algorithm (e.g., the constants a, c, and m in a congruential generator). Hence, for each dimension n of interest, we can assess the amount of structure without using any actual output sample from the generator. Structure can be thought of in two dimensions by referring back to Figures 4.0.2 and 4.0.3 and asking for numerical measures of the patterns that can be observed in these scatter plots. An early idea, described by Coveyou and MacPherson (1967) and detailed in Knuth (1981) as the spectral test, looks at the distance between the most widely separated family of planes (or hyperplanes in more than two dimensions) going through the points. An algorithm for the spectral test is given by Knuth (1981, Chap. 3.3.4). See, also, Hopkins (1983) for the most recent work on implementation of this test.

Another popular measure of structure, the lattice test (Marsaglia, 1972, including algorithm) constructs cells from each set of four contiguous points and looks at the ratio of the largest side of the cell to the smallest, with values near one being optimal (with analogous extensions to higher than two dimensions). Both these tests are described well in Atkinson (1980) and again, with three other measures of structure, in Fishman and Moore (1985).

These theoretical tests are generally recognized as being the most sensitive tests available for assessing the randomness of a particular generator. Interpretation of the results, especially when multiple tests in many dimensions are performed, is difficult, however, unless the generator is obviously faulty. As an example, Fishman and Moore (1985) present a performance measure (unspecified for our purposes here) that is 0.8411 for their "best" generator (in our Generator 2), 0.000142 for a reference "bad" generator, and 0.3375 for our Generator 1. By this measure, in two-dimensional space, Generator 1 is inferior, but we are still left with the question of whether Generator 1 is unacceptable for simulation work.

Another caution concerning theoretical tests, emphasized by Atkinson (1980), is that they can be applied only to generators of certain types.

The second class of random number generator tests are the empirical or statistical tests. For these measures, we require an output sample of uniforms and must, therefore, worry about the size of the sample and the seed from which it starts. Often, multiple samples from various seeds are tested. These tests vary widely in form, but often can be traced to a common idea: use the uniform sample to mimic a common situation with which we naturally associate randomness, and see if the usual (known) probability laws are followed.

We give several examples: first, think of using the uniforms to generate random permutations (Section 4.4) of the 52 cards in a deck of playing cards and seeing whether we are dealt different hands with the proper frequencies; second, we can convert the uniforms to coupons or lottery tickets and ask how long before we have a full set or win the lottery; finally, we can think of bivariate pairs of uniforms as car-parking locations on a square and ask if approximately equal numbers of cars are parked in equally sized subsquares. Detailed descriptions of various of these tests can be found in Fishman and Moore (1982), in Learmonth (1976), and in most other simulation textbooks. In addition, certain algorithms are available as part of Chapter G in the older IMSL Library (IMSL, 1984) and as Chapter 17 in the latest version of the IMSL Library (IMSL, 1987).

Although these empirical tests are not considered to be as stringent as theoretical tests, they do have the advantage that the type of randomness measured is intuitively more clear and that "failures" are based on precise statistical significance tests. Often, both empirical and theoretical tests are performed, with theoretical tests providing fine differentiation between competing good generators and empirical tests providing evidence of good general behavior and serving to validate the results of the theoretical tests (see, e.g., Atkinson, 1980).

5.6 Conclusions on Generating and Testing Pseudo-Random Number Generators

We have shown in this chapter that one can find very good pseudo-random number generators, and also that the pseudo-random number generators provided in some packages and with some computer systems are defective. This leads to the following guidelines on the choice and testing of pseudo-random number generators, which, when done completely, is a time-consuming and arduous task, not to be undertaken lightly.

1. Testing is *unnecessary* in the sense that very good pseudo-random number generators are available.
2. Testing is *necessary* in the sense that bad pseudo-random number generators exist on many computer systems and are commonly used.
3. It is preferable to substitute a good pseudo-random number generator with documented properties for a pseudo-random number generator you know nothing about.

And even if you follow the last of these rules, it is wise to use some of the graphical testing methods given in this book to make sure that the implementation is correct.

Exercises   **5.1.** Prove part a of Theorem 5.0.1.

5.2. Show that in Theorem 5.0.1 when $U > p$, we can define the new random variable U_2 in an equivalent fashion as $(U - p)/(1 - p)$.

5.3. Show that Theorem 5.0.1 is the same as the following proposition: *Let E be an exponential(λ) random variable and suppose we observe that E is greater than x_0. Then $E_0 = E - x_0$ is an exponentially distributed random variable with parameter λ, which is independent of the event $E > x_0$. (This is known as the memoryless property of exponential random variables.)*

   What is the analogue of part b of Theorem 5.0.1?

5.4. State and prove the recycling theorem for discrete uniform random variables.

5.5. Relate the discrete recycling theorem to the geometric distribution.

c5.6. Continue the example of the mid-square method for generating random integers for 50 steps and plot the output as was done in Figure 4.0.2 for RANDU. Are any patterns evident in the scatter plot? If possible, do this on a computer so that you can go beyond 50 steps and get plots like those in Figure 4.0.2.

c5.7. Continue the sequence generated by the shift register in Example 5.2.1 and plot the results, after conversion to decimal numbers, as in Exercise 5.6.

T5.8. Compute the matrix T for Example 5.2.1.

T5.9. Show that the sum of two independent uniform$(0, 1)$ random variables, mod 1, is again uniform$(0, 1)$.

T5.10. The division simulation algorithm for computing prime-modulus multiplicative congruential generators on a binary machine—that is:

$$U_i = aU_{i-1} \bmod (2^\beta - \gamma)$$

where $m = 2^\beta - \gamma$ is the largest prime number the binary computer can hold—is as follows:

$$\text{Let } U'_{i-1} = aU'_{i-1} \bmod 2^\beta$$
$$= aU_{i-1} - C2^\beta$$

where C is the integer part of $aU_{i-1}/2^\beta$. (Note that this division is not modulus arithmetic and, since the division is a power of 2, it is a simple shift operation.)

   1. Compute U'_i and $U'_i + C\gamma$.
   2. If $U'_i + C\gamma < 2^\beta - \gamma$, then set

$$U_i = U'_i + C\gamma$$

   3. Else set

$$U_i = U'_i + C\gamma - (2^\beta - \gamma)$$

Show that the algorithm generates the desired sequence U_i.

5.11. In the case of $\gamma = 1$, which is true when $\beta = 31$, one needs to add C to U'_i only once in the second step in the algorithm, and test for an overflow in the operation. The computation is then simple and will usually save time over division of aU_i by $2^\beta - 1$. Discuss the implementation and merits of the division simulation algorithm in this case.

References

Abramowitz, M., and Stegun, I. A. (1965). *Handbook of Mathematical Functions*. Dover: New York.

Andrews, D. F., Bickel, P. J., Hampel, F. R., Huber, P. J., Rogers, W. H., and Tukey, J. W. (1972). *Robust Estimates of Location: Survey and Advances*. Princeton University Press: Princeton, NJ.

Atkinson, A. C. (1980). Tests of Pseudo-Random Numbers. *Applied Statistics, 29*(2), 164–171.

Binder, K., and Stauffer, D. (1984). A Simple Introduction to Monte Carlo Simulation and Some Specialized Topics. In *Applications of the Monte Carlo Method in Statistical Physics*, K. Binder, Ed., Springer: Berlin, pp. 1–36.

Birtwistle, G. M., Dahl, O. J., and Myhrhaug, D. (1979). *Simula Begin*. Brookfield: Brookfield, VT.

Bratley, P., Fox, B. L., and Schrage, L. E. (1983). *A Guide to Simulation*. Springer: New York.

Brent, R. P. (1974). Algorithm 488: A Gaussian Pseudo-Random Number Generator. *Communications of the Association for Computing Machinery, 17,* 704–706.

Brown, M., and Solomon, H. (1979). On Combining Pseudorandom Number Generators. *Annals of Statistics, 7*(3), 691–695.

Coveyou, R. R., and MacPherson, R. D. (1967). Fourier Analysis of Uniform Random Number Generators. *Journal of the Association for Computing Machinery, 14,* 100–119.

Fishman, G., and Moore, L. (1982). A Statistical Evaluation of Multiplicative Congruential Random Number Generators with Modulus $2^{31} - 1$. *Journal of the American Statistical Association, 77,* 129–136.

Fishman, G., and Moore, L. (1985). An Exhaustive Analysis of Multiplicative Congruential Random Number Generators with Modulus $2^{31} - 1$. *SIAM Journal of Scientific and Statistical Computing, 7*(1), 24–45.

Gentle, J. E. (1981). Portability Considerations for Random Number Generators. In *Computer Science and Statistics: Proceedings of the 13th Symposium on the Interface*, W. F. Eddy, Ed. Springer: New York, pp. 158–164.

Golomb, S. W. (1967). *Shift Register Sequences*. Holden-Day: San Francisco.

Green, B. F., Jr., Smith, J. E. K., and Klem, L. (1959). Empirical Tests of an Additive Random Number Generator. *Journal of the Association for Computing Machinery, 6,* 527–537.

Hopkins, T. R. (1983). Algorithms AS193: A Revised Algorithm for the Spectral Test. *Applied Statistics, 32*(3), 328–335.

Hutchinson, D. W. (1966). A New Uniform Pseudo-random Number Generator. *Communications of the Association for Computing Machinery, 9*(6), 432–433.

IMSL Library (1984). IMSL, Inc., Houston.

IMSL Stat/Library (1987). IMSL Stationary: User Manual, Version 1.0. IMSL, Inc., Houston.

Jansson, B. (1966). *Random Number Generators*. Almqvist and Wiksell: Stockholm.

Katzen, H., Jr. (1970). *APL Programming and Computer Techniques*. Van Nostrand: New York.

Kendall, M. G., and Babbington-Smith, B. (1938). Randomness and Random Sampling Numbers. *Journal of the Royal Statistical Society, A, 101,* 147–166.

Kendall, M. G., and Babbington-Smith, B. (1939a). Tables of Random Sampling Numbers. *Tracts for Computers*, No. 24. Cambridge University Press: Cambridge.

Kendall, M. G., and Babbington-Smith, B. (1939b). Second Paper on Random Sampling Numbers. *Journal of the Royal Statistical Society, Suppl., 6,* 51–61.

Kennedy, W. J., and Gentle, J. E. (1980). *Statistical Computing*. Dekker: New York.

Kiviat, P. J., Villanueva, R., and Markowitz, H. M. (1983). In *SIMSCRIPTII.5 Programming Language*, A. Mullarney, Ed. CACI, Inc.: Los Angeles.

Knuth, D. E. (1981). *The Art of Computer Programming.* Vol. 2, *Seminumerical Algorithms,* 2nd ed. Addison-Wesley: Reading, MA.

Learmonth, G. P. (1976). Empirical Tests of Multipliers for the Prime Modulus Random Number Generator $X_{i+1} = AX_i(\text{mod } 2^{31} - 1)$. *Proceedings of the Ninth Interface Symposium on Computer Science and Statistics,* D. C. Hoaglin and R. E. Welsch, Eds. Prindle, Weber & Schmidt: Boston, pp. 178–183.

Learmonth, G. P., and Lewis, P. A. W. (1974). Statistical Tests of Some Widely Used and Recently Proposed Uniform Random Number Generators. In *Proceedings of the Seventh Conference on the Computer Science and Statistics Interface,* W. J. Kennedy, Ed. Iowa State University Press: Ames.

Lee, Y. K. (1985). Random Number Generation. *Byte, 10*(13), 426.

Lehmer, D. (1951). Mathematical Methods in Large-Scale Computing Units. *Annals of the Computing Laboratory [Harvard University], 26,* 141–146.

Lewis, P. A. W., Goodman, A. S., and Miller, J. M. (1969). A Pseudo-random Number Generation for the System 360. *IBM Systems Journal, 8,* 136–146.

Lewis, P. A. W., Orav, E. J., and Uribe, L. C. (1988). *Enhanced Simulation and Statistics Package.* Wadsworth & Brooks/Cole: Pacific Grove, CA.

Lewis, P. A. W., and Uribe, L. C. (1981). The New Naval Postgraduate School Random Number Generator Package LLRANDOMII. Naval Postgraduate School Technical Report, PS-55-81-005. Monterey, CA.

Lewis, T. G., and Payne, W. H. (1973). Generalized Feedback Shift Register Pseudorandom Number Algorithm. *Journal of the Association for Computing Machinery, 20,* 456–468.

MacLaren, M. D., and Marsaglia, G. (1965). Uniform Random Number Generators. *Journal of the Association for Computing Machinery, 12*(1), 83–89.

Marsaglia, G. (1972). The Structure of Linear Congruential Sequences. In *Applications of Number Theory to Numerical Analysis,* S. K. Zaremba, Ed. Academic Press: New York.

Marsaglia, G. (1976). Random Number Generation. In *Encyclopedia of Computer Science,* A. Ralston, Ed. Van Nostrand Reinhold: New York.

Marsaglia, G. (1985). A Current View of Random Number Generators. In *Computer Science and Statistics: Proceedings of the Sixteenth Symposium on the Interface,* L. Billard, Ed. North-Holland: Amsterdam, pp. 3–10.

Marshall, A. W., and Olkin, I. (1979). Majorization in Multivariate Distributions. *Annals of Statistics, 2,* 1189–1200.

McLeod, A. I. (1985). A Remark on Algorithm AS183. An Efficient and Portable Pseudo-Random Number Generator. *Applied Statistics, 34,* 198–200.

Payne, W. H., Rabung, J. R., and Bogyo, T. P. (1969). Coding the Lehmer Pseudorandom Number Generator. *Communications of the Association for Computing Machinery, 12,* 85–86.

Rand Corporation (1955). *A Million Random Digits with 100,000 Normal Deviates.* Free Press: Glencoe, IL.

Ripley, B. D. (1983). Computer Generation of Random Variables: A Tutorial. *International Statistical Review, 51,* 301–319.

Rosenblatt, M. (1975). Multiply Schemes and Shuffling. *Mathematics of Computation, 29,* 929–934.

SAS User's Guide (1982). SAS Institute, Inc., Cary, NC.

Schrage, L. (1979). A More Portable FORTRAN Random Number Generator. *Association for Computing Machinery Transactions on Mathematical Software, 5,* 132–139.

Sparks, D. (1983). RND is Fatally Flawed. *CALL-A.P.P.L.E.,* January, pp. 29–32.

SPSSX Statistical Algorithms (1983). M. Norusis and C. Ming Wang, Eds. SPSS, Inc., Chicago, p. 186.

Sylwestrowicz, J. D. (1982). Parallel Processing in Statistics. *COMPSTAT 1982*. Physica: Vienna, pp. 131–136.

Tippett, L. H. C. (1927). Random Sampling Numbers. *Tracts for Computers*, No. 15. Cambridge University Press: Cambridge.

von Neumann, J. (1951). Various Techniques Used in Connection with Random Digits. *National Bureau of Standards Applied Mathematics Series*, *12*, 899–900.

Wichmann, B. A., and Hill, I. D. (1982). An Efficient and Portable Pseudo-Random Number Generator. *Applied Statistics*, *31*(2), 188–190.

Wichmann, B. A., and Hill, I. D. (1984). Correction to AS183: An Efficient and Portable Pseudo-Random Number Generator. *Applied Statistics*, *33*(1), 123.

Wichmann, B. A., and Hill, I. D. (1987). Programming Insight: Building a Random-Number Generator, *Byte*, *12*(3), 127–128.

Part

II

Sophisticated Simulation

Descriptions and Quantifications of Univariate Samples: Numerical Summaries

6.0 Introduction

In Chapters 6 and 7 we consider samples of independent *univariate* random variables whose distribution we wish to characterize, either by numerical summaries or by graphical displays. The emphasis in this chapter is on numerical summaries, which before the advent of cheap and widely available computer graphics, were the primary means of looking at data distributions.

As examples of data, we can think of simulated waiting times in a queue, maximum likelihood estimates of a parameter obtained from simulated samples from a given distribution, or response times at a computer terminal. The characterizations we use depend on the type of data (e.g., discrete or continuous), on the distribution of the data, and on the questions we want to answer. However, they generally fall into two categories: numerical summaries that partially characterize the data, and graphical displays.

Numerical summaries include:

- Mean and median to describe location
- Variance, coefficient of variation, and interquartile range to measure scale or spread
- Normalized third and fourth moments to quantify skewness and kurtosis (i.e., heavy-tailedness), respectively
- Quantiles, percentiles, and more generally, the cumulative distribution function (cdf) to describe distributional shape

For each of these summaries, we use the observed sample values to estimate the corresponding population parameter (e.g., the sample average estimates the population mean). Formal techniques exist to measure the precision of the numerical summaries and compare them to hypothesized values. Some of these procedures are discussed, but most rely on distributional assumptions and are best left to a study of mathematical statistics (Lehmann, 1959). Alternative, essentially nonparametric, methods for assessing the variability of these numerical summaries are given in Chapter 9. The methods are sectioning, jackknifing, and bootstrapping. They are generalizations of the important method given in Chapter 4 on crude simulation (Equation 4.1.2) to estimate the precision of a sample average: that is $S_{\bar{X}} = S_X/(m^{1/2})$.

Graphical displays include:

- A plot of the empirical cumulative distribution function (ecdf) to display an estimate of the entire cdf
- Histograms, density estimates, and boxplots to look at and summarize the shape of the probability density function (pdf)
- Probability (P–P) plots and quantile (Q–Q) plots to visually compare the distribution of a data set to standard theoretical distributions

Such comparisons of distributions can be made formal through the Kolmogorov–Smirnov and Shapiro–Wilk type tests discussed in Section 6.2.

The ecdf is a natural outgrowth of percentile estimates and, as such, is presented in Section 6.2; the other graphical displays are covered separately in Chapter 7.

Before proceeding, it should be noted again that we will deal explicitly *only with samples of independent, identically distributed univariate random variables.* Independent samples will always be the output of crude simulations where we use replication to produce independent values of the variable of interest. Even in some sophisticated simulations—for instance, the regenerative simulations considered in Volume II—the output *may* satisfy iid assumptions and be amenable to the types of descriptive analysis considered here. In many cases, however, simulation output is correlated and the distributions are time dependent.

If we remove the assumption of independence, the ideas and procedures that follow are still applicable, but the distributional results are not. Some of the analogous theoretical results that exist for dependent sequences are presented in Volume II during discussion of systems simulation. When the data are *not identically distributed*, the quantities we will be discussing (the mean, distribution function, quantiles, etc.) have no meaning. Methods to deal with the most commonly occurring nonstationary sequence in a simulation setting, namely, the transient preceding convergence to steady state, are again presented in the second volume.

To illustrate the statistical ideas that appear in this chapter and the next, we analyze two typical simulation outputs. The first concerns waiting times from a GI|GI|3 queue, described in the example presented in Section 3.3; the details are specified exactly in Example 6.2.2. The second data set comes from a simulation problem that arises in physics: comparison of simulated maximum likelihood and moment estimates of the polarization parameter in a muon decay distribution.

EXAMPLE   ### 6.0.1 An Estimation Problem from Physics

The decay distribution of electrons from muon decay is given by the probability density function

$$f(x) = \frac{1 + \delta x}{2} \qquad -1 \leqslant x \leqslant 1; -1 \leqslant \delta \leqslant 1 \qquad (6.0.1)$$

where $x = \cos \theta$ and θ is the angle with respect to the direction of polarization of the muon. Here "δ" is a parameter that takes the value $-\frac{1}{3}$ for a completely polarized muon and is $-(\frac{1}{3})P$ for polarization P. Since P can range only between -1 and 1, values of δ in $(-\frac{1}{3}, \frac{1}{3})$ are of interest in this context, even though the density given by Equation 6.0.1 is valid for all δ in $[-1, 1]$.

We wish to estimate δ from n observations on electron decay; the simulation problem is to determine whether maximum likelihood or moment estimation of δ is better in relatively small samples.

The density as given in Equation 6.0.1 was discussed in Exercise 4.24. It is a valid density for $-1 \leqslant \delta \leqslant 1$, the value $\delta = 0$ giving a uniform density on $[-1, 1]$ and the extremes giving wedges that are the densities of the maxima and minima, respectively, of two uniform $(-1, 1)$ random variables (see Exercise 4.3). Thus, in general, all values of δ in $[-1, 1]$ could be of interest, although the values of interest in the physics problem are restricted.

Since for the density above (Equation 6.0.1)

$$E(X) = \int_{-1}^{1} \frac{x(1 + \delta x)\,dx}{2} = \frac{\delta}{3}$$

and

$$\text{var}(X) = \int_{-1}^{1} \frac{\{x - E(X)\}^2(1 + \delta x)\,dx}{2} = \frac{3 - \delta^2}{9} \tag{6.0.2}$$

the moment estimator of δ is

$$\tilde{\delta} = 3\bar{X} = \frac{3}{n} \sum_{i=1}^{n} X_i \tag{6.0.3}$$

with

$$E(\tilde{\delta}) = \delta \quad \text{and} \quad \text{var}(\tilde{\delta}) = \frac{3 - \delta^2}{n} \tag{6.0.4}$$

In addition, by the central limit theorem, $\tilde{\delta}$ will have a Normal distribution for reasonably large n.

The likelihood function for the sample is

$$L(\delta) = \prod_{i=1}^{n} \frac{1 + \delta x_i}{2} \tag{6.0.5}$$

with

$$\ln\{L(\delta)\} = \sum_{i=1}^{n} \{\ln(1 + \delta x_i) - \log 2\}$$

so that the maximum likelihood estimate (mle) is the solution of the equation

$$\frac{d\ln\{L(\delta)\}}{d\delta} = \sum_{i=1}^{n} \frac{x_i}{1 + \delta x_i} = 0 \tag{6.0.6}$$

The solution $\hat{\delta}$ of this equation for δ in $[-1, 1]$ is unique, since $d^2 \ln L(\delta)/d\delta^2$ is always negative, so that it is easy to find $\hat{\delta}$ numerically for a given set of data.

Now maximum likelihood theory says that the estimate $\hat{\delta}$ is asymptotically consistent and efficient, with

$$\text{var}(\hat{\delta}) \sim \frac{2\delta^2}{\left[\left\{\log\left(\frac{1+\delta}{1-\delta}\right) - 2\delta\right\}n\right]} \tag{6.0.7}$$

Thus the asymptotic relative efficiency (ARE) of the estimate $\hat{\delta}$ relative to the moment estimate $\tilde{\delta}$ is

$$\text{ARE} = \frac{2\delta^2}{3 - \delta^2} \times \frac{1}{\log\left(\frac{1+\delta}{1-\delta}\right) - 2\delta} \tag{6.0.8}$$

It can be shown that the ARE is one at $\delta = 0$ (uniform case), since there the moment and mle estimators are approximately the same. The mle estimator will have smaller asymptotic variance than the moment estimator when $\delta \neq 0$, but, since we are interested in small-sample estimates, the question of small-sample variance, bias, and distribution of $\hat{\delta}$ is important.

The solution to this rather typical statistical problem will be pursued by simulation throughout this and the next chapter.

6.1   Sample Moments

Sample moments (and numerical summaries in general) compress an entire sample of data into one, or possibly a few, representative quantities. In this section we specifically consider the mean, variance, skewness, kurtosis, and coefficient of variation, all of which are functions of the sample moments. From such statistics we can picture and summarize the distribution of any sample—its *location*, *spread*, or *shape*.

In simulation, that summary information can be used to model the input or output of the physical system we want to simulate or to analyze the output of the simulation. Since our simulation is presumably being run to answer some question, we may be able to base our decision (e.g., whether four tellers in a bank are enough) on such numerical summaries of the simulation output (e.g., the finding that the *average waiting time* of the 100th customer is 3.5 minutes). Finally, when validating a simulation program, all that may be known for the special cases used in the validation is moments of distributions. We will want to match those theoretical moments to the estimated moments from the simulation output. This is often the case in queueing theory.

The price we pay for data compression is the possibility that we are ignoring important details (for instance, that the average waiting time is indeed 3.5 minutes but many people wait more than 10 minutes), or we may be using the wrong estimator—maybe the median or trimmed mean would give a better indication of the "center" of the waiting time distribution. The material in this section, therefore, should be used in conjunction with, not in lieu of, the other quantizations in Sections 6.2 and 6.3 and the graphical displays in Sections 7.1 and 7.2. Two prepared programs for output analysis, HISTPC and SMTBPC, discussed in Section 7.1 and Chapter 8 respectively, illustrate the merger of graphical techniques and multiple numerical summaries for the analysis of simulation output. (The programs HISTPC and SMTBPC are described fully in the *Enhanced Simulation and Statistics Package* available as a supplement to this book.)

6.1.1 Some Common Quantizations

i. Location

One of the most basic attributes of the distribution of a random variable X is its location or center. Often we will be considering *symmetric*, unimodal distributions, examples of which are the Normal, logistic, and Laplace distributions. (See Table 6.1.2 for details on these distributions.) A distribution $F_X(x)$ (or by convention a random variable X) is said to be *symmetric* about a point c if $F_X(c - x) = P(X \leqslant c - x) = P(X > c + x) = 1 - F_X(c + x)$.

In the symmetric case the three classical quantizations, the mean (the integral average), median (the 50% point), and mode (the "most probable" point), agree: that is,

$$E(X) = \text{median}(X) = \text{mode}(X) = c$$

The center of a symmetric, unimodal distribution can be estimated by the sample average, which is discussed below, by the sample median, to be considered in Section 6.3, or by more recently introduced robust measures such as the *trimean*, *trimmed mean*, or *biweight mean* (Hoaglin, Mosteller, and Tukey, 1983; see also the histogram program HISTPC in the software supplement to this book). Choosing the "best" estimator among these is a complicated problem that depends on the unknown distribution of the data and cannot be dealt with in this overview. An example in Section 8.2 does illustrate how long-tailed distributions require robust estimators like the median, and further reading should be done in Hoaglin, Mosteller, and Tukey (1983).

When the distribution is not symmetric or unimodal, one must first consider the question that is to be answered and then decide which measure of location—for instance, the integral average $E(X)$, the 50% point median (X), or the most probable value mode (X)—fits the situation. Estimation will depend on the choice of quantization. It may not be sensible, of course, to talk about a unique center for the distribution. This would be the case if the distribution were clearly a mixture of two quite disparate distributions.

Frequently, sometimes by default rather than by choice, the mean is taken to represent the center of the distribution and the sample average,

$$\bar{X} = \sum_{i=1}^{m} \frac{X_i}{m} \tag{6.1.1}$$

where $X_1, X_2, \ldots, X_m$ are the data, is used as the estimator. The mean and sample average are often used because:

- $E(X)$ or some simple function of it is the natural location parameter incorporated into many standard families of distributions [e.g., $E(X) = \mu$ if $X \sim$ Normal(μ, σ) and $E(X) = 1/\lambda$ if $X \sim$ exponential(λ)]
- $E(X)$ appears in the classical limit theorems such as the strong law of large numbers and the central limit theorems
- As noted in Section 4.2.3, a simple method exists for estimating the variance of the estimator $\bar{X}$, namely, S^2/m
- The sample average is the least-squares estimator of $E(X)$. It is unbiased and consistent and has minimum variance among all unbiased estimators of $E(X)$ that are linear combinations of the data points (i.e., it is a "best, linear, unbiased estimator").

These properties plus the widespread use of $E(X)$ and $\bar{X}$ are the reasons for their prominence in this section. At the same time, there are situations of (a) "long-tailed" or other non-Normal distributions and (b) nonsymmetric distributions, where $E(X)$ and $\bar{X}$ may not be appropriate. The classic example is the symmetric Laplace distribution (Problem 4.23, Table 6.1.2), for which the median is the mle estimator and has smaller variance than the sample average for all sample sizes. It is always best to compute more than one estimate of location (i.e., at least the mean and the median). As an illustration, six estimates of location are produced in the numerical summary in HISTPC (see Section 7.1).

ii. Scale

The dispersion, spread, or scale of a random variable X is a concept that has many definitions (quantizations) but usually reflects how far realizations of X can be from the "center" of X's distribution. The variance of X,

$$\text{var}(X) = E[\{X - E(X)\}^2] = \sigma^2 \tag{6.1.2}$$

or its square root, the standard deviation of X, has been adopted as the most common quantization of spread or scale for two reasons.

- A close relationship to the "scale" parameter that has been incorporated into many common distributions. This includes $\sigma^2 = \text{var}(X)$ in the Normal(μ, σ) case and $1/\lambda^2 = \text{var}(X)$ in the exponential(λ) case.
- The appearance of $\text{var}(X)$ as the scaling factor in the central limit theorems and in Chebycheff's inequality.

Often, interpretation of scaling is done by mentally converting to confidence intervals. In the case of $\text{var}(X)$, attention goes to the standard deviation,

$$\text{s.d.}(X) = \{\text{var}(X)\}^{1/2} = \sigma \tag{6.1.3}$$

and the rule of thumb about the distribution of the population: the mean, plus or minus 2 standard deviations, will include, on the average, 95% of the data. Be aware that this rule is based on Normality assumptions for X and, as Example 6.1.1 shows, it can be misleading when X comes from a highly non-Normal distribution.

EXAMPLE   **6.1.1 Interpretation of the Variance and Standard Deviation**

Table 6.1.1 shows the probability associated with each tail of various distributions, using as a reference the usual (Normal theory) rule of thumb: any value more than 2 standard deviations greater (less) than the mean has only a 0.025 (0.025) chance of occurring.

As noted in connection with the discussion of Equation 4.2.7, the logistic(β) distribution has a shape close to that of the Normal distribution, whereas the other two distributions given in Table 6.1.1 are more non-Normal.

An alternate quantization of scale, the interquartile range or IQR, is discussed in Section 6.3 and has an interpretation that is consistent for all distributions. If we let $x_{0.25}$ and $x_{0.75}$ be the 0.25 and 0.75 quantiles (see Equation 3.5.5 and the definition of quantiles in Section 6.3), then $IQR = x_{0.75} - x_{0.25}$. Furthermore, let

$$C = \frac{x_{0.75} + x_{0.25}}{2}$$

represent the center of the distribution. Then we know two things:

- There is a 25% chance of finding data less than $C - (IQR/2)$.
- There is a 25% chance of finding data greater than $C + (IQR/2)$.

Notice that for symmetric distributions

$$C = E(X) = \text{median}(X)$$

The IQR ignores the far tails of the distribution and gives a quantization of scale that does not reflect possible realizations of X that are rare, but whose magnitude could influence var(X) greatly. If such large but improbable values do not enter into your concept of scale in a particular situation, then the IQR rather than s.d.(X) should be used.

Table 6.1.1   Tail probabilities for common distributions

			Probability	
Distribution	Mean	S.D.	$< \mu - 2(\text{s.d.})$	$> \mu + 2(\text{s.d.})$
Normal(μ, σ)	μ	σ	.025	.025
Logistic(μ, β)	μ	$\beta\pi/(3)^{1/2}$	.0259	.0259
Laplace(μ, β)	μ	$\sqrt{2}\beta$	.0296	.0296
Exponential(λ)	$1/\lambda$	$1/\lambda$	0	.0498

EXAMPLE   **6.1.2 A Case for Using the Interquartile Range (IQR) for Scale**
The Cauchy density (Exercise 4.9; Table 6.1.2) can be written to include a scale parameter b as follows:

$$f(x) = \frac{1}{\pi b[1 + \{(x - a)/b\}^2]} \qquad -\infty < x < \infty, b > 0, -\infty < a < \infty$$
(6.1.4)

$$F(x) = \frac{1}{2} + \frac{1}{\pi}\tan^{-1}\left(\frac{x - a}{b}\right)$$
(6.1.5)

Regardless of b, the variance associated with this density is infinite and therefore gives you no indication of scale. The interquartile range, however, is

$$x_{0.75} - x_{0.25} = (a + b) - (a - b) = 2b$$

and does quantify the relative spread of the distribution.

Other robust quantizations of scale such as the mean absolute deviation (estimated in HISTPC)

$$E(|X - \text{median}(X)|)$$
(6.1.6)

and the median absolute deviation,

$$\text{median}(|X - \text{median}(X)|)$$
(6.1.7)

offer different definitions of distance and distributional center that may be more relevant to non-Normal distributions. Discussion of these is left to outside reading (Hoaglin, Mosteller, and Tukey, 1983, Chap. 12).

Estimation of the IQR is presented in Section 6.3, while $\text{var}(X)$ is frequently estimated as

$$S^2 = \sum_{i=1}^{m} \frac{(X_i - \bar{X})^2}{m - 1}$$
(6.1.8)

where $X_1, X_2, \ldots, X_m$ is the random sample. The unbiased estimator S^2 is usually called the *sample variance* and its properties are presented in Section 6.1.2. Alternative estimators include

$$S^{*2} = \sum_{i=1}^{m} \frac{(X_i - \bar{X})^2}{m}$$
(6.1.9)

which is biased but has a smaller variance than S^2, and also some robust estimators that may work better for non-Normal data (Hoaglin, Mosteller, and Tukey, 1983, Chap. 12).

Finally, the sample standard deviation

$$S = (S^2)^{1/2} \tag{6.1.10}$$

is used to estimate the population standard deviation σ.

iii. Symmetry

A distribution $F_X(x)$ is symmetric if there is some point c such that the probability of observing a value X greater than or equal to $c + x$ is equal to the probability of observing a value of X less than or equal to $c - x$, for every x on the real line. Figure 6.1.1 shows the density of a symmetric distribution, in this case the Normal$(0, 1)$. Note that in the continuous case, symmetry in a probabilistic sense corresponds to functional symmetry of the density function (pdf). Other well-known symmetric distributions are the uniform$(0, 1)$ and the Cauchy distributions.

If the distribution is not symmetric, the imbalance in the distribution can take many forms, but we hope for one of two simple cases:

  a. Extreme high values are consistently more probable than extreme low values.
  b. Extreme low values are consistently more probable than extreme high values.

Case a is illustrated in Figure 6.1.2 by the exponential(1) density and Case b is illustrated in Figure 6.1.3 by the Weibull$(10, 0.15)$ density (Exercise 4.20 and Table 6.1.2).

The three alternatives above, namely symmetry or Case a or Case b, can be distinguished by looking at the *skewness* (third central moment)

$$\mu_3 = E[\{X - E(X)\}^3] \tag{6.1.11}$$

of the distribution, although other forms of nonsymmetry may not be revealed. If

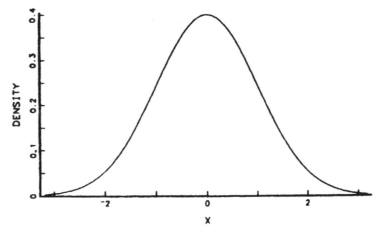

Figure 6.1.1 Normal$(0, 1)$ density function. Illustration of a symmetric probability density function. Coefficient of skewness = $\gamma_1 = 0$; coefficient of kurtosis = $\gamma_2 = 0$.

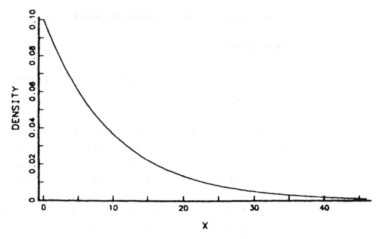

Figure 6.1.2 Positively (right) skewed exponential(1) probability density function. Coefficient of skewness = γ_1 = 2; coefficient of kurtosis = γ_2 = 6.

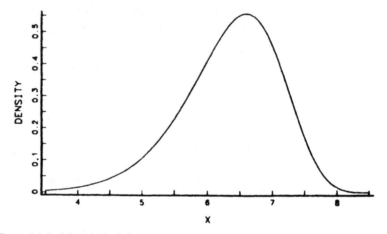

Figure 6.1.3 Negatively (left) skewed Weibull(10, 0.15) probability density function. Coefficient of skewness = γ_1 = −0.638; coefficient of kurtosis = γ_2 = 0.570.

the distribution is symmetric, the skewness must be zero; if Case a is true then the skewness is positive and we say the distribution has a right skew; if Case b is true then the skewness is negative and we say the distribution has a left skew.

From the definition of skewness, it can be seen that scaling the random variable X to be βX changes the skewness by the factor β^3. To avoid this problem, the *coefficient of skewness*

$$\gamma_1 = \frac{\mu_3}{\sigma^3} = \frac{E[\{X - E(X)\}^3]}{\{\mathrm{var}(X)\}^{3/2}}$$

is often used instead of μ_3. *It is the skewness of the standardized variable* $\{X - E(X)\}/\sigma$ and thus is a location- and scale-invariant measure.

Other common measures of symmetry are based on quantiles and are discussed in Section 6.3.

An unbiased estimator of skewness for the random sample $X_1, X_2, \ldots, X_m$ is given by

$$\hat{\mu}_3 = m \sum_{i=1}^{m} \frac{(X_i - \bar{X})^3}{(m-1)(m-2)} \tag{6.1.12}$$

while the corresponding (but generally biased) estimator of the coefficient of skewness is

$$\hat{\gamma}_1 = \frac{\hat{\mu}_3}{S^3} \tag{6.1.13}$$

Properties of each of these estimators are presented in Section 6.1.2.

iv. Heavy-Tailedness

Another way in which we can describe the shape of a distribution is by how quickly the tails of the density fall off to zero or, equivalently, the relative likelihood of extreme values. If the tails fall off quickly, we say that the tails are "light" and we can usually infer that the body of the distribution is rather flat. If the tails fall off slowly, then we say that the tails are "heavy" and that the body must be peaked.

Note too that "heavy-tailedness" is a relative idea and is usually defined with the Normal distribution as reference. (In fact a Normal distribution is heavy-tailed with respect to a uniform distribution, or, if we use the Normal as reference, the uniform distribution is light-tailed relative to the Normal distribution.)

Two problems arise immediately from these ideas:

- By changing the scale of the distribution, we can enlarge the tails without changing the shape. Hence, our quantization must not react to changes in scale.
- Again, "heavy" and "light" are relative terms, so our quantization should allow us to use some common distribution as a reference, usually the Normal distribution.

To satisfy these requirements, we start with the kurtosis, the fourth central moment,

$$\mu_4 = E[\{X - E(X)\}^4] \tag{6.1.14}$$

divide by the variance squared to make the quantity unitless, and then subtract 3 so that the final result, the *coefficient of kurtosis*,

$$\gamma_2 = \frac{\mu_4}{\sigma^4} - 3 \tag{6.1.15}$$

takes the value 0 for a Normal(μ, σ) random variable. Note that μ_4/σ^4 is the *kurtosis for the standardized variable* $\{X - E(X)\}/\sigma$ and thus is a location- and scale-invariant measure.

Likewise, for estimation we start with an unbiased estimator of the kurtosis,

$$
\hat{\mu}_4 = \frac{\sum_{i=1}^{m} (X_i - \bar{X})^4(m^2 - 2m + 3)}{(m - 1)(m - 2)(m - 3)} - \frac{3(S^2)^2(m - 1)(2m - 3)}{m(m - 2)(m - 3)}
\tag{6.1.16}
$$

which is transformed into a (generally biased) estimator of the coefficient of kurtosis,

$$
\hat{\gamma}_2 = \frac{\hat{u}_4}{S^4} - 3
\tag{6.1.17}
$$

In general, large values of $\hat{\gamma}_2$ correspond to heavy tails and small values to light tails, with the uniform distribution having the smallest possible value of -1.2. However, the coefficient of kurtosis may not always be measuring heavy and light tails (Mood, Graybill, and Boes, 1974, p. 76) and the estimator of γ_2 (Equation 6.1.17) itself has a very high variance. For these reasons, the coefficient of kurtosis (and also the coefficient of skewness) should be supplemented with graphical views of the density (Section 7.1).

Some of these ideas are expressed and measured much more simply in terms of the quantiles of the distribution and we return to this topic in subsection iv of Section 6.3.3.

v. Relative Dispersion

One final quantization applies to one-sided distributions—usually distributions whose densities are zero for negative values. In such cases, the scale often increases "naturally" as the center moves away from zero so that dispersion can only be thought of as relative to location. In fact, the mean of the distribution may be a scale parameter rather than a shift parameter, as in the case of the exponential distribution. This relationship is natural mathematically because there can be no negative values to balance large positive values. Hence, if the mean is close to zero, most values must be close to zero. It can also be natural physically because we expect more "play" in the data when we measure large quantities or measure very finely. For example:

• We would not be surprised if a switchboard that handled an average of 100,000 calls a day had fluctuations (standard deviation) of $(100,000)^{1/2} \approx 316$ calls around that average. We would be surprised, however, if a smaller switchboard that handled only about 2000 calls a day (on average) had similar fluctuations (standard deviation) of 316 calls around that lower average. This judgment comes in part from the standard, implicit assumption that counts have a Poisson(λ) distribution for which the mean is λ and the variance is also λ. Consequently the ratio of the mean to the standard deviation is important. Thus, with this implicit assumption, an average 2000 calls per day should correspond to a standard deviation of 45.

• The average height of a man may be 186 centimeters with a standard deviation of 15 centimeters but, if we change to millimeters, the average is recorded as 1860 millimeters and the standard deviation, for the same data, increases to 150 millimeters.

To adjust for this relationship, it is useful to report, in addition to the mean and standard deviation of the distribution, the *coefficient* of *variation*,

$$C(X) = \frac{\sigma}{\mu} = \left[E\left\{ \frac{(x - \mu)}{\mu} \right\}^2 \right]^{1/2} \tag{6.1.18}$$

which is the standard deviation of the standardized random variable $(X - \mu)/\mu$. Dividing by the mean is one way, though not the only way, to adjust for proportional changes in mean and standard deviation. In fact, since $C(X)$ is the standard deviation of X adjusted for scale (measured by the mean), $C(X)$ is a scale-free or standardized measure of relative dispersion.

Notice that for the first example above, a high coefficient of variation obtained with data from the small switchboard would signal the anomalous situation at the small switchboard, whereas in the second example the coefficient of variation would be the same (0.081) in both cases, which is reasonable, since the data are essentially the same, i.e., have just undergone a change of scale.

The coefficient of variation takes the form that it does with the implicit proportionality between mean and standard deviation because, for the most common one-sided, continuous distributions, the exponential and Gamma families (Section 3.5, and Table 6.1.2), the mean and the standard deviation are proportional. In fact, these families serve as reference standards with the coefficient of variation equal to one for an exponential(λ) random variable ($\lambda > 0$). In addition, the coefficient of variation squared equals the reciprocal of the shape parameter k, when the random variable is distributed Gamma(k, μ). For discrete random variables, the Poisson(λ) distribution serves as reference with $C(X) = \lambda^{-1/2}$.

Estimation of the coefficient of variation is done in the obvious way,

$$\tilde{C}(X) = \frac{S}{\overline{X}} \tag{6.1.19}$$

with the properties of this estimator left to the general discussion in Section 6.1.2.

vi. Summary Tables for Common Distributions

All the theoretical moments mentioned in this section are displayed in Tables 6.1.2 and 6.1.3 for the common continuous distributions and for discrete distributions, respectively. The tables also summarize many other properties that we have used or will be using. The inverse transforms, for instance, are used for random variable generation (Chapter 4), computation of quantiles (Section 6.3), and probability plots (Section 7.2). We will be returning to these tables in later chapters, as needed (e.g., during sophisticated random variate generation in Volume II). Similar but more

Table 6.1.2 Continuous distributions

Parametric family of distributions	Reference	Range	Parameters	Cumulative distribution function (cdf)
Normal(μ, σ)	Equations 3.4.4, 3.4.5, 4.2.5, and 4.2.6a	$-\infty < x < \infty$	$-\infty < \mu < \infty$ $\sigma > 0$	$F(x) = \dfrac{1}{(2\pi)^{1/2}} \displaystyle\int_{-\infty}^{(x-\mu)/\sigma} \left\{\exp\dfrac{-v^2}{2}\right\} dv$
Logistic(μ, β)	Equations 4.2.7 and 4.2.8	$-\infty < x < \infty$	$-\infty < \mu < \infty$ $\beta > 0$	$F(x) = \left\{1 + e^{-(x-\mu)\beta}\right\}^{-1}$
Laplace(μ, β)	Exercise 4.23	$-\infty < x < \infty$	$-\infty < \mu < \infty$ $\beta > 0$	$F(x) = \tfrac{1}{2}\exp\left(\dfrac{x-\mu}{\beta}\right) \qquad\qquad x \leqslant \mu$ $\qquad = \tfrac{1}{2} + \dfrac{1}{2}\left[1 - \exp\left\{-\dfrac{(x-\mu)}{\beta}\right\}\right] \quad x > \mu$
Cauchy(a, b)	Exercise 4.9	$-\infty < x < \infty$	$-\infty < a < \infty$ $b > 0$	$F(x) = \dfrac{1}{2} + \dfrac{1}{\pi}\tan^{-1}\left(\dfrac{x-a}{b}\right)$
Exponential(β)	Equation 3.2.3	$0 \leqslant x < \infty$	$\beta > 0$	$F(x) = 1 - e^{-\beta x}$
Gamma(k, β)	Equations 3.5.1 and 3.5.2	$0 \leqslant x < \infty$	$\beta > 0$ (scale) $k > 0$ (shape)	$F(x) = \dfrac{1}{\Gamma(k)} \displaystyle\int_{0}^{\beta x} v^{k-1}e^{-v}dv$
Weibull(k, β)	Exercise 4.20	$0 \leqslant x < \infty$	$\beta > 0$ (scale) $k > 0$ (shape)	$F(x) = \left[1 - \exp\left\{-(\beta x)^k\right\}\right]$
Lognormal(μ, σ)		$0 \leqslant x < \infty$	$-\infty < \mu < \infty$ $\sigma > 0$	$F(x) = \dfrac{1}{\sqrt{2\pi}\sigma} \displaystyle\int_{0}^{x}\left[\exp\left\{\dfrac{-\dfrac{(lnv - \mu)^2}{2\sigma^2}}{v}\right\}\right] dv$
Uniform(a, b)	Equation 4.0.2	$a \leqslant x \leqslant b$	$-\infty < a < b < \infty$	$F(x) = \dfrac{x-a}{b-a}$
Beta(p, q)		$0 \leqslant x \leqslant 1$	$p > 0$ $q > 0$	$F(x) = \dfrac{1}{B(p, q)} \displaystyle\int_{0}^{x} u^{p-1}(1-u)^{q-1}du$

Density (pdf)	Inverse distribution function, $F^{-1}(p)$	Generation	Mean, μ
$f(x) = \dfrac{1}{\sqrt{2\pi}\sigma}\exp\left\{-\dfrac{(x-\mu)^2}{2\sigma^2}\right\}$	$\Phi^{-1}(x)$	Sections 4.2.5i and 4.2.6	μ
$f(x) = \dfrac{e^{-(x-\mu)/\beta}}{[\beta\{1 + e^{-(x-\mu)/\beta}\}^2]}$	$x = \mu + \beta\left\{\ln\left[\dfrac{p}{1-p}\right]\right\}$	Equation 4.2.8	μ
$f(x) = \dfrac{1}{2\beta}\exp\left(-\dfrac{|x-\mu|}{\beta}\right)$	$x = +\beta\ln(2p) \qquad p \leqslant \frac{1}{2}$ $x = -\beta\ln(2(1-p)) \qquad p > \frac{1}{2}$	Exercise 4.23	μ
$f(x) = \dfrac{1}{\pi b}\dfrac{1}{[1 + \{(x-a)/b\}^2]}$	$x = a + b\tan[\pi(p-\frac{1}{2})]$	Exercise 4.9	Undefined
$f(x) = \beta e^{-\beta x}$	$x = \dfrac{-\ln(1-p)}{\beta}$	Section 4.2.4i	$1/\beta$
$f(x) = \dfrac{\beta^k x^{k-1} e^{-\beta x}}{\Gamma(k)}$	No closed form	Volume II	k/β
$f(x) = k\beta^k x^{k-1}\exp\{-(\beta x)^k\}$	$x = \dfrac{[-\ln(1-p)]^{1/k}}{\beta}$	Exercise 4.20	$\Gamma\left(\dfrac{1}{k}+1\right)\Big/\beta$
$f(x) = \dfrac{1}{x\sqrt{2\pi}\sigma}\exp\left[-\dfrac{(\ln x-\mu)^2}{2\sigma^2}\right]$	No closed form	Exponential of a Normal (μ, σ) variable	$e^{\mu+\sigma^2/2}$
$f(x) = \dfrac{1}{b-a}$	$x = a + (b-a)p$	Chapter 5	$(a+b)/2$
$f(x) = \dfrac{1}{B(p,q)}x^{p-1}(1-x)^{q-1}$	No closed form	Volume II	$p/(p+q)$

Table 6.1.2 (continued)

Parametric family of distributions	Standard deviation, σ	Coefficient of variation, $C(X)$	Coefficient of skewness, γ_1	Coefficient of kurtosis, γ_2	Interquartile range, IQR
Normal(μ, σ)	σ	NA	0	0	1.35σ
Logistic(μ, β)	$\dfrac{\pi\beta}{\sqrt{3}}$	NA	0	1.2	2.197β
Laplace(μ, β)	$\sqrt{2}\beta$	NA	0	3	1.386β
Cauchy(a, b)	Undefined	Undefined	Undefined	Undefined	$2b$
Exponential(β)	$1/\beta$	1	2	6	$1.099/\beta$
Gamma(k, β)	$k^{1/2}/\beta$	$k^{-1/2}$	$2/k^{1/2}$	$6/k$	No closed form
Weibull(k, β)	$\left\{\Gamma\left(\dfrac{2}{k} + 1\right) - \Gamma^2\left(\dfrac{1}{k} + 1\right)\right\}$	σ/μ	*	†	Not simple
Lognormal(μ, σ)	$e^{2\mu+\sigma^2}(e^{\sigma^2} - 1)$	$e^{\mu}e^{\sigma^2/2}(e^{\sigma^2} - 1)$			No closed form
Uniform(a, b)	$\dfrac{b - a}{12^{1/2}}$	NA	0	-1.2	$0.5(b - a)$
Beta(p, q)	$\left\{\dfrac{pq}{(p + q + 1)(p + q)^2}\right\}^{1/2}$	$\dfrac{\sigma}{\mu}$	‡	§	No closed form

$$* = \frac{\Gamma(1 + 3/k) - 3\Gamma(1 + 1/k)\Gamma(1 + 2/k) + 2\Gamma^3(1 + 1/k)}{\{\Gamma(1 + 2/k) - \Gamma^2(1 + 1/k)\}^2}$$

$$\dagger = \frac{\Gamma(1 + 4/k) - 4\Gamma(1 + 1/k)\Gamma(1 + 3/k) + 6\Gamma^2(1 + 1/k)\Gamma(1 + 2/k) - 3\Gamma^4(1 + 1/k)}{\{\Gamma(1 + 2/k) - \Gamma^2(1 + 1/k)\}^{3/2}} - 3$$

$$\ddagger = 2(q - p)\frac{\{p^{-1} + q^{-1} + (pq)^{-1}\}^{1/2}}{p + q + 2}$$

$$\S = 3(p + q + 1)\frac{2(p + q)^2 + pq(p + q - 6)}{pq(p + q + 2)(p + q + 3)}$$

Table 6.1.3   Discrete distributions

Parametric family of distributions	Reference	Range	Parameters	Probability function, $p_k = P\{X = k\}$
Discrete uniform	Part b of subsection ii, Section 4.2.5	$a \leqslant k \leqslant (a+b)$	$-\infty < a \leqslant b < \infty$	$p_k = \dfrac{1}{b - a + 1}$ $k = a, \ldots, (a+b)$ $= 0$ otherwise
Binomial		$k = 0, 1, \ldots, N$	$0 \leqslant p \leqslant 1$ $N \geqslant 1$	$p_k = \dbinom{N}{k} p^k (1-p)^{N-k}$ $k = 0, \ldots, N$ $= 0$ otherwise
Poisson	Equation 3.2.3 Exercise 4.4	$k = 0, 1, \ldots$	$\lambda > 0$	$p_k = \dfrac{e^{-\lambda} \lambda^k}{k!}$ $k = 0, 1, \ldots$ $= 0$ otherwise
Geometric	Equation 4.2.14 Part c of subsection ii of Section 4.2.5	$k = 0, 1, \ldots$	$0 \leqslant p < 1$ $(q = 1 - p)$	$p_k = (1-p)p^k$ $k = 0, 1, \ldots$ $= 0$ otherwise
Negative binomial		$k = 0, 1, \ldots$	$0 \leqslant p < 1$ $r > 0$ $(q = 1 - p)$	$p_k = \dbinom{r + k - 1}{k} q^r p^k$ $k = 0, 1, \ldots$ $= 0$ otherwise
Pareto	Exercise 4.16	$k = A, A+1, \ldots$	$\alpha > 0$ $A > -\infty$	$p_k = \left(\dfrac{1}{k - A}\right)^{\alpha} - \left(\dfrac{1}{k + 1 - A}\right)^{\alpha}$ $k = A, A+1$ $= 0$ otherwise

extensive tables are given in Mood, Graybill, and Boes (1974, Appendix B, p. 537). Much more detailed information about parametric families of distributions can be found in the three volumes by Johnson and Kotz (1970).

6.1.2   Properties of Moment Estimators

Even if we can specify the appropriate quantization of a distribution, we must still decide on the best way to estimate that quantization. As mentioned before, "best" will depend on the distribution of the data but, in general, it is desirable to have an estimator that is accurate (small bias or no bias) and precise (small spread) and has known distribution (to perform hypothesis tests).

Here we look at the properties for the estimators emphasized in Section 6.1.1. All those estimators share many theoretical properties, since they are all functions (in fact either powers or ratios) of the centralized sample moments:

Table 6.1.3 (continued)

Parametric family of distributions	Cumulative distribution function $P\{X \leqslant k\}$	Mean, μ	Standard deviation, σ	Coefficient of variation, $C(X)$
Discrete uniform	$\dfrac{k - a + 1}{b - a + 1}$	$a + \dfrac{b - a + 1}{2}$	$\dfrac{(b - a + 1)^2 - 1}{12}$	NA
Binomial	$\displaystyle\sum_{i=0}^{k} \binom{n}{i} p^i (1 - p)^{n-i}$	Np	$(Npq)^{1/2}$	$\left\{\dfrac{q}{(Np)}\right\}^{1/2}$
Poisson	$e^{-\lambda} \displaystyle\sum_{i=0}^{k} \dfrac{\lambda^i}{i!}$	λ	$\lambda^{1/2}$	$\lambda^{-1/2}$
Geometric	$1 - p^{k+1}$	$\dfrac{p}{q}$	$\dfrac{p^{1/2}}{q}$	$p^{-1/2}$
Negative binomial	$\displaystyle\sum_{i=0}^{k} \binom{r + i - 1}{i} q^r p^i$	$\dfrac{rp}{q}$	$\dfrac{(rp)^{1/2}}{q}$	$(rp)^{-1/2}$
Pareto	$1 - \left(\dfrac{1}{k + 1 - A}\right)^\alpha$	No closed form; finite if $\alpha \geqslant 2$	No closed form; finite if $\alpha \geqslant 2$	No closed form; finite if $\alpha \geqslant 2$

$$\eta_1 = \sum_{i=1}^{m} \frac{X_i}{m} \tag{6.1.20}$$

$$\eta_k = \sum_{i=1}^{m} \frac{(X_i - \eta_1)^k}{m} \qquad k = 2, 3, \ldots \tag{6.1.21}$$

(Notice that the sample size, m, has been suppressed in the notation.)

Explicitly, the relationships between the estimators in Section 6.1.1 and the centralized sample moments are:

$$\bar{X} = \eta_1 \tag{6.1.22}$$

$$S^2 = \left\{\frac{m}{m - 1}\right\} \eta_2 \tag{6.1.23}$$

$$S = (S^2)^{1/2} = \left\{\frac{m\eta_2}{m - 1}\right\}^{1/2} \tag{6.1.24}$$

Coefficient of skewness, γ	Coefficient of kurtosis	Generation
0 (symmetric)	$\dfrac{3\{3(b-a+1)^2-7\}\{(b-a+1)^2-1\}}{5\{(b-a+1)^2-1\}^2}-3$	Part b of subsection ii, Section 4.2.5
$(q-p)(Npq)^{-1/2}$	$(1-6pq)(Npq)^{-1}$	Sum of n binary (p) random variables
$\lambda^{-1/2}$	λ^{-1}	Exercise 4.4
$(1+p)p^{-1/2}$	$\dfrac{1+7q+q^2}{q}-3$	Part c of subsection ii, Section 4.2.5
$(1+p)(rp)^{-1/2}$	$\dfrac{\{1+(3r+4)q+q^2\}}{q}-3$	Volume II
No closed form; finite if $\alpha \geqslant 3$	No closed form; finite if $\alpha \geqslant 4$	Exercise 4.16

$$\hat{\mu}_3 = \left\{\frac{m^2}{(m-1)(m-2)}\right\}\eta_3 \tag{6.1.25}$$

$$\hat{\gamma}_1 = \frac{\hat{\mu}_3}{S^3} \tag{6.1.26}$$

$$\hat{\mu}_4 = \frac{[\{m(m^2-2m+3)\eta_4\}-\{3m(2m-3)\eta_2^2\}]}{\{(m-1)(m-2)(m-3)\}} \tag{6.1.27}$$

$$\hat{\gamma}_2 = \frac{\hat{\mu}_4}{S^4}-3 \tag{6.1.28}$$

$$\tilde{C}(X) = \frac{[\{m/(m-1)\}\eta_2]^{1/2}}{\eta_1} \tag{6.1.29}$$

We will assume throughout that the population moments we are trying to estimate are finite, in order for our quantizations to make sense.

i. Accuracy

An estimator can be considered accurate if it is unbiased or, equivalently, if the expected value of the estimator equals the quantity we are trying to estimate. Our estimates for the mean, variance, skewness, and kurtosis were all chosen to be unbiased for all sample sizes:

$$E(\bar{X}) = \mu \tag{6.1.30}$$

$$E(S^2) = \sigma^2 \tag{6.1.31}$$

$$E(\hat{\mu}_3) = \mu_3 \tag{6.1.32}$$

$$E(\hat{\mu}_4) = \mu_4 \tag{6.1.33}$$

while the standard deviation, coefficient of skewness, coefficient of kurtosis, and coefficient of variation are asymptotically unbiased (Cramér, 1946, Chap. 27). Thus if $O(1/m)$ indicates a function of m (e.g., K/m) that when divided by $1/m$ goes to a constant as $m \rightarrow \infty$, we have

$$E(S) = \sigma + O\left(\frac{1}{m}\right) \tag{6.1.34}$$

$$E(\hat{\gamma}_1) = \frac{\mu_3}{\sigma^3} + O\left(\frac{1}{m}\right) = \gamma_1 + O\left(\frac{1}{m}\right) \tag{6.1.35}$$

$$E(\hat{\gamma}_2) = \frac{\mu_4}{\sigma^4} - 3 + O\left(\frac{1}{m}\right) = \gamma_2 + O\left(\frac{1}{m}\right) \tag{6.1.36}$$

$$E\{C(X)\} = \frac{\sigma}{\mu} + O\left(\frac{1}{m}\right) = C(X) + O\left(\frac{1}{m}\right) \tag{6.1.37}$$

ii. Precision

An estimator is precise if it "usually" takes values "close" to the quantity we are trying to estimate. This property can be interpreted as the estimator having a small variance or perhaps a small mean-squared error, but this requires knowing the distribution of the data. Instead, we consider asymptotic closeness and note that *all the estimators here are consistent*, so that as the sample size gets large, the probability that our estimator deviates from the quantity estimated goes to zero (Cramér, 1946, Chaps. 20 and 27).

iii. Distribution

Knowing the exact distribution of any estimator requires information about the underlying distribution of the data. Hence, we are again forced to consider asymptotic results that state that *usually all the estimators here are asymptotically Normally distributed* (Cramér, 1946, Chap. 28).

The word "usually" is included because of two assumptions:

1. For an estimator that is a function of only one centralized sample moment η_k, we require the existence of $E(X^{2k})$. This will not be a problem if the data come from any of the common distributions other than the Cauchy distribution.

2. For an estimator that is a function of two centralized sample moments, and in particular a ratio, there are continuity conditions on the function and its first- and second-order partial derivatives (Cramér, 1946, p. 336). For the estimators considered here, this assumption reduces to the finiteness of the mean of the estimator and the finiteness of the variance term of order $O(1/m)$. For example, the coefficient of variation σ/μ is not asymptotically Normal if the data are Normal(0, 1), since $\sigma/\mu = \infty$. The sample coefficient of variation will generally be finite but will grow in an unstable manner as $m \to \infty$.

For completeness, the variances for the estimators considered here are given in Table 6.1.4. The lead terms in Table 6.1.4 are the variances that appear in the limiting Normal distributions of the estimators. These terms are $O(1/m)$ for all estimators. If the data are Normal(μ, σ), then the means and variances for the most common estimators simplify and can be found in Table 6.1.5. *Note that the variances of $\hat{\gamma}_1$ and $\hat{\gamma}_2$ for Normal populations are, from Table 6.1.5, approximately 6/m and 24/m for large m.* These simple results are useful in simulations to give general rules of thumb about precision of estimates of γ_1 and γ_2.

Table 6.1.4 Variances for the moment-based estimators

Estimator	Variance
$\bar{X}$	$\dfrac{\sigma^2}{m}$
S^2	$\dfrac{m(\mu_4 - \sigma^4)}{(m-1)^2} - \dfrac{2(\mu_4 - 2\sigma^4)}{(m-1)^2} + \dfrac{\mu_4 - 3\sigma^4}{m(m-1)^2}$
S	$\dfrac{\mu_4 - \sigma^4}{4(m-1)\sigma^2} + O\!\left(\dfrac{1}{m^2}\right)$
$\hat{\mu}_3$	$\dfrac{m^3(\mu_6 - 6\mu_4\sigma^2 - \mu_3^2 + 9\sigma^6)}{(m-1)^2(m-2)^2} + O\!\left(\dfrac{1}{m^2}\right)$
$\hat{\gamma}_1$	$\dfrac{(m-1)K(\gamma_1)}{4(m-2)^2\sigma^{10}} + O\!\left(\dfrac{1}{m^{3/2}}\right)$
	where
	$K(\gamma_1) = 4\sigma^4\mu_6 - 12\sigma^2\mu_3\mu_5 - 24\sigma^6\mu_4 + 9\mu_3^2\mu_4 + 35\sigma^4\mu_3^2 + 36\sigma^{10}$
$\hat{\mu}_4$	$\dfrac{m(m^2 - 2m + 3)^2 K(\mu_4)}{(m-1)^2(m-2)^2(m-3)^2} + O\!\left(\dfrac{1}{m^{3/2}}\right)$
	where
	$K(\mu_4) = \mu_8 - 8\mu_3\mu_5 - \mu_4^2 + 16\sigma^2\mu_3^2$
$\hat{\gamma}_2$	$\dfrac{(m-1)^2(m^2 - 2m + 3)^2 K(\gamma_2)}{(m-2)^2(m-3)^2 m^3\sigma^{12}} + O\!\left(\dfrac{1}{m^{3/2}}\right)$
	where
	$K(\gamma_2) = \sigma^4\mu_8 - 4\sigma^2\mu_4\mu_6 - 8\sigma^4\mu_3\mu_5 + 4\mu_4^3 - \sigma^4\mu_4^2 + 16\sigma^2\mu_3^2\mu_4 + 16\sigma^6\mu_3^2$
$\bar{C}(X)$	$\dfrac{\mu^2(\mu_4 - \sigma^4) - 4\mu\sigma^2\mu_3 + 4\sigma^6}{4\mu^4\sigma^2(m-1)} + O\!\left(\dfrac{1}{m^{3/2}}\right)$

Table 6.1.5 Means and variances for the common moment-based estimators, assuming the underlying data are Normal(μ, σ)

Estimator	Mean	Variance
$\bar{X}$	μ	$\dfrac{\sigma^2}{m}$
S^2	σ^2	$\dfrac{2\sigma^4}{m-1}$
S	$\dfrac{\Gamma(m/2)}{\Gamma\{(m-1)/2\}} \left\{\dfrac{2}{m-1}\right\}^{1/2} \sigma$ $= \sigma + O\left(\dfrac{1}{m}\right)$	$\left(1 - \dfrac{2\Gamma^2(m/2)}{(m-1)\Gamma^2(m-1)/2)}\right)\sigma^2$ $= \dfrac{\sigma^2}{2(m-1)} + O\left(\dfrac{1}{m^2}\right)$
$\hat{\gamma}_1$	0	$\dfrac{6m(m-1)}{\{(m-2)(m+1)(m+3)\}}$
$\hat{\gamma}_2$	$-\dfrac{6}{m+1}$	$\dfrac{24(m-1)^2(m^2-2m+3)^2}{m(m+1)^2(m+3)(m+5)(m-2)(m-3)}$

We could get rough confidence intervals for our estimator by using the asymptotic Normality along with either (a) estimates of the asymptotic variance or (b) estimates of the variance assuming Normal data. However, in the first case we would need to estimate terms such as $E(X^{2k})$ to find the variance of an estimator based on η_k, and such estimation is prone to large errors.

In the second case, (b), non-Normality of the data could cause wide discrepancies in the confidence interval. Hence neither method is recommended as anything more than an informal procedure. Using asymptotic Normality for testing and confidence intervals is considered in Section 6.1.4 and again, from a nonparametic viewpoint, in Chapter 9.

6.1.3   Asymptotic Expansions for Functions of Central Moment Estimators

We describe now briefly the general method used to obtain asymptotic expansions and, for example, the results in Tables 6.1.4 and 6.1.5. This method, which comes from the geometric expansion of a function of central moment estimators, is the following (it is sometimes referred to as the delta method).

If $\tilde{\theta}_m$ represents our estimator for the quantization θ, then the general form of the asymptotic expansions are

$$E(\tilde{\theta}_m) = \theta + \frac{a_1}{m} + \frac{a_2}{m^2} + \frac{a_3}{m^3} + \cdots \tag{6.1.38}$$

$$\text{var}(\tilde{\theta}_m) = \frac{b_1}{m} + \frac{b_2}{m^{3/2}} + \frac{b_3}{m^2} + \frac{b_4}{m^{5/2}} + \cdots \tag{6.1.39}$$

The constants $a_1, a_2, \ldots$ and $b_1, b_2, \ldots$ will depend on the distribution of the data and on the estimator $\tilde{\theta}_m$. These results can be proven directly when $\tilde{\theta}_m = \eta_k$ for some k (see Cramér, 1946, Chap. 27, and Example 6.1.3 below for an indication of how to start the calculation). What is remarkable is that the same two expansions hold for many common estimators, even those not based on moments. We will make use of these facts in Section 8.2, and again when we study ratio estimators in the context of the regenerative method of systems simulation in the second volume of this book.

EXAMPLE   **6.1.3   Asymptotic Expansion for the Mean of the Reciprocal of the Sample Mean**
The following is perhaps the simplest example one can find of how the asymptotic expansions of Equations 6.1.38 and 6.1.39 are derived for ratio estimators.

In many contexts one is interested in estimating the reciprocal of the mean of a distribution $1/\mu$: in life-testing, for example, $\lambda = 1/\mu$ is the rate of failures per unit time. The usual ad hoc estimator $1/\bar{X}$ has a mean $E(1/\bar{X})$, whose expansion is obtained as follows:

$$E\left(\frac{1}{\bar{X}}\right) = E\left[\frac{1}{\mu + (\bar{X} - \mu)}\right] = \left(\frac{1}{\mu}\right)E\left\{\frac{1}{\left[1 - \left(\frac{\mu - \bar{X}}{\mu}\right)\right]}\right\}$$

$$= \frac{1}{\mu}E\left\{1 - \frac{(\mu - \bar{X})}{\mu} + \frac{(\mu - \bar{X})^2}{\mu^2} - \cdots\right\}$$

where the geometric expansion is justified on the basis of the fact that $(\mu - \bar{X})/\mu$ is, with high probability, less than one as $m \to \infty$. Then

$$E\left(\frac{1}{\bar{X}}\right) = \frac{1}{\mu}\left\{1 - \frac{E(\mu - \bar{X})}{\mu} + \frac{E\{(\mu - \bar{X})^2\}}{\mu^2}\right\} + O\left(\frac{1}{m^2}\right)$$

$$= \frac{1}{\mu}\left\{1 + \frac{C^2(X)}{m}\right\} + O\left(\frac{1}{m^2}\right)$$

where $C(X)$ is the coefficient of variation defined at Equation 6.1.18 and we have used the fact that $E(X) = \mu$ and, from Section 4.2, that $E(\mu - \bar{X})^2 = \text{var}(\bar{X}) = \sigma^2/m$. The fact that higher order terms in the expansion are $O(1/m^2)$ is harder to prove.

The result says that the bias in $1/\bar{X}$ is severe for small m if $C(X)$ is large, so that the standard deviation is large relative to the mean; that is, the random variable is highly dispersed.

6.1.4   Tests and Confidence Intervals

It was pointed out in Section 6.1.2 that the exact distribution of any estimator will depend on the unknown distribution of the data. To proceed to hypothesis testing, therefore, we must be willing to accept results that are only approximately true and, even then, we will find useful procedures only when testing for the mean. Non-parametric procedures for estimating variability and obtaining confidence intervals are given in Chapter 9, and these use the results for the mean.

As discussed in Chapter 4, the simplest estimator, the sample mean $\bar{X}$, has a variance,

$$\text{var}(\bar{X}) = \frac{\sigma^2}{m}$$

which we can estimate from the data by

$$S_{\bar{X}}^2 = \frac{S^2}{m}$$

Hence we can form the t statistic (see Section 3.4)

$$T_m = \frac{\bar{X} - E(X)}{(S_{\bar{X}})^{1/2}}$$

where $(S_{\bar{X}}^2)^{1/2}$ is often called the *standard error* (of $\bar{X}$). As long as S^2 is a consistent estimator of σ^2 [which assumes that $E(X^2)$ and $E(X^4)$ are finite] and m is "large enough," we can insert a hypothetical value μ_0 for $\mu = E(X)$ and test the hypothesis

$$H_0: \mu = \mu_0$$

according to whether

$$|T_m| = \left| \frac{(\bar{X} - \mu_0)}{(S_{\bar{X}}^2)^{1/2}} \right|$$

is greater than (reject H_0) or less than (do not reject H_0) a Normal critical value $z_{\alpha/2}$. For an α-level test, $z_{\alpha/2}$ should be the $(1 - \alpha/2)$ quantile of the standard Normal distribution.

This procedure can also be inverted to give a $100(1 - \alpha)\%$ confidence interval,

$$(\bar{X} - z_{\alpha/2}(S_{\bar{X}}^2)^{1/2}, \bar{X} + z_{\alpha/2}(S_{\bar{X}}^2)^{1/2}) \tag{6.1.40}$$

If the distribution of the X_i is known to be Normal(μ, σ), then the exact distribution of T is known to be the t distribution with $(m - 1)$ degrees of freedom. In this case the t, rather than the Normal, cutoff points should be used. If m is small (say $m \leqslant 25$), it is also usually better to use the more conservative t-distribution cutoff points, even if the data are non-Normal.

This test and confidence interval work because:

- The variance of $\bar{X}$ has a simple form, which can be estimated easily from the data.
- The test statistic T_m has a known small sample distribution for Normal data that holds, approximately, for many other data distributions. (See subsection

i of Section 8.2.2, however, for counterexamples and an exploration by simulation, of the robustness of the t statistic.)

The test statistic T_m is asymptotically Normal if $E(X^2)$ is less than infinity and, unless the data are highly non-Normal, m does not have to be very large (usually around 60) before the asymptotics take hold (again, see Section 8.2.2).

This "t test" is therefore recommended whenever you are interested in an average of random variables (that are not highly non-Normal).

It may seem that the same type of procedure can be extended to the other estimators considered in Table 6.1.4, for example, since

$$\text{var}(S^2) = \frac{m(\mu_4 - \mu_2^2)}{(m-1)^2} - \frac{2(\mu_4^2 - 2\mu_2^2)}{(m-1)^2} + \frac{(\mu_4 - 3\mu_2^2)}{m(m-1)^2} \tag{6.1.41}$$

where

$$\mu_k = E[\{X - E(X)\}^k] \qquad k = 2, 4$$

we can estimate $\text{var}(S^2)$ by inserting the central sample moments $\hat{\mu}_2$ and $\hat{\mu}_4$ for μ_2 and μ_4 and call the estimator $\hat{\text{var}}(S^2)$. Then, as long as $E(X^4)$ is less than infinity,

$$\frac{(S^2 - \sigma^2)}{\{\hat{\text{var}}(S^2)\}^{1/2}}$$

will be asymptotically Normal and can be used for testing or confidence intervals.

Unfortunately, the convergence to Normality can be shown to be very slow and, since the small sample distribution is unknown and possibly highly non-Normal, it is best not to use this procedure.

If the sample size is large enough, a different technique called *sectioning* (Chapter 9) makes use of both the asymptotic Normality of most estimators and the robustness of the t statistic to construct hypothesis tests and confidence intervals for statistics such as S^2. For small samples it is better to use the numerical summaries and the graphics that are introduced in the next chapter to get a feel for the data, and not to try to perform formal tests. If necessary, jackknifing and bootstrapping (Chapter 9) can be used to give approximate confidence intervals in small samples.

6.1.5 An Application of Moment Estimation

The ideas previously expressed concerning moment estimates are now used to summarize the distributions of one of the two simulated data sets mentioned in Section 6.0.

EXAMPLE **6.1.4 Moment and Maximum Likelihood Estimators for δ**

In Example 6.0.1 we outlined the problem of estimating the parameter δ in the decay distribution (Equation 6.0.1) of electrons from muon decay. The two competing

Table 6.1.6   Simulated results for estimates of δ for $n = 20$

Sample moment	$\delta = -0.4$		$\delta = 0.8$	
	Moment estimator, $\tilde{\delta}$	mle, $\hat{\delta}$	Moment estimator, $\tilde{\delta}$	mle, $\hat{\delta}$
$\bar{X}(S/m^{1/2})$	$-0.401(0.012)$	$-0.399(0.012)$	$0.797(0.011)$	$0.762(0.008)$
S	0.386	0.377	0.343	0.263
$\hat{\gamma}_1$	0.068	0.322	-0.207	-1.134
$\hat{\gamma}_2$	-0.111	-0.327	-0.235	0.907

estimators are the moment estimator $\tilde{\delta}$ given at Equation 6.0.3 and the maximum likelihood estimator $\hat{\delta}$ given implicitly at Equation 6.0.6.

A simulation was performed for sample size $n = 20$ at values of δ equal to -0.4 and 0.8 using $m = 1000$ replications in the simulation. These two values of δ were chosen to be representative of values of δ that might occur in practice.

Table 6.1.6 shows sample moments of the simulated distributions of $\tilde{\delta}$ and $\hat{\delta}$.

The quantities in parentheses next to the mean values are the estimated standard deviations $(S/m^{1/2})$ of the estimated means of $\tilde{\delta}$ and $\hat{\delta}$ (see Section 4.2.4). Recall that $\tilde{\delta}$ is an unbiased estimator of δ and note that for $\delta = -0.4$ and $\delta = 0.8$, the means $(-0.401$ and $0.797)$ are within 2 standard deviations of their true values. This should give confidence that the simulation is running correctly. The mle estimate, $\hat{\delta}$, at $\delta = 0.8$, however, appears to be biased, since its estimated mean of 0.762, is more than 4 (estimated) standard deviations $(4 \times .008)$ away from 0.800.

In the second row of Table 6.1.6, the estimated standard deviations of the mle estimators $\hat{\delta}$ are smaller than the respective estimated standard deviations of the moment estimators. This effect is more marked at $\delta = 0.8$. The standard deviation of $\tilde{\delta}$ is known through Equation 6.0.4 to be 0.344 at $\delta = 0.8$ and 0.377 at $\delta = -0.4$, which compares well with the estimated values in Table 6.1.6 of 0.343 and 0.386. The small differences are of course due to sampling error in the simulation with only $m = 1000$ replications.

Although we would usually want to use the estimator with smaller variance, namely $\hat{\delta}$, the bias of $\hat{\delta}$ and non-Normality of its distribution (discussed below in Example 6.2.3) must also be factored into our decision.

The overall departures from Normality of the distributions of $\tilde{\delta}$ and $\hat{\delta}$ can be measured by $\hat{\gamma}_1$ (skewness) and $\hat{\gamma}_2$ (kurtosis). These departures appear to be slightly greater for the mle estimates, especially at $\delta = 0.8$. Since the moment estimator is a sample average, and the central limit theorem holds, this Normality for $\tilde{\delta}$ is not surprising. In assessing this result, it should be recalled that *if* the distributions of $\hat{\delta}$ and $\tilde{\delta}$ are Normal (which we are testing!) then from Table 6.1.5, $\sigma(\hat{\gamma}_1) \sim (6/1000)^{1/2} = 0.077$ and $\sigma(\hat{\gamma}_2) \sim (24/1000)^{1/2} = 0.154$.

Thus departures from Normality for the distribution of $\hat{\delta}$ at $\delta = 0.8$ are probably real. In fact $\hat{\gamma}_2$ is $0.907/0.154 \sim 6$ standard deviations from zero.

Further details on the distributions of $\hat{\delta}$ and $\tilde{\delta}$ will become apparent when graphical presentations of the simulated distributions of $\hat{\delta}$ and $\tilde{\delta}$ are given in later sections of this chapter and in Chapter 7.

6.2   Percentiles, the Empirical Cumulative Distribution Function, and Goodness-of-Fit Tests

Because one statistic—for instance, the mean of a random variable—is seldom sufficient as a summary of an entire distribution, percentiles are often used to give a more complete description of the distribution. Furthermore, the empirical cumulative distribution function (ecdf) is a portmanteau or summary of all the percentiles, usually presented graphically, that can be used to estimate the entire cumulative distribution function. Use of the ecdf is not restricted to continuous random variables, although for a discrete random variable, discrete probabilities at all possible outcomes of the random variable are usually estimated, provided that the number of possible outcomes is not too large.

For a given x and random variable X with distribution function $F_X(x)$ the percentile p, illustrated in Figure 6.2.1, is defined as

$$p = F_X(x) = P(X \leqslant x) \qquad -\infty < x < \infty \qquad (6.2.1)$$

We recall that $F_X(x)$ is a monotonically nondecreasing function of x with $F_X(-\infty) = 0$ and $F_X(\infty) = 1$. In the sections that follow, we will look at three equivalent ways to estimate p by using m independent, simulated, realizations, $X_1, X_2, \ldots, X_m$, of X. Since p can be estimated for any value of x, we can also estimate the function $F_X(x)$ and compare it, graphically and statistically, to various theoretical distribution functions. Strictly speaking, p is a function of x and can be considered a shorthand notation for $F_X(x)$. When it is important to note the dependence on x, we write $p(x)$. We now give details of the three equivalent ways of estimating a percentile and, consequently, the empirical cumulative distribution function (ecdf).

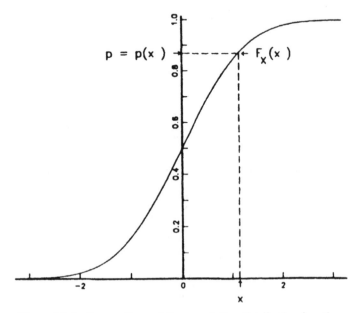

Figure 6.2.1   Percentiles and the cumulative distribution function.

6.2.1   Percentile Estimation: Indicator Functions and Order Statistics

Our first estimate of the percentile p is based on the fact that if $I(\cdot)$ is the usual indicator function so that

$$I(X;x) = \begin{cases} 1 & \text{if } X \leqslant x \\ 0 & \text{if } X > x \end{cases} \tag{6.2.2}$$

then $I(X;x)$ is a Bernoulli random variable with parameter $p = F_X(x)$. This follows because $I(X;x)$ is one if and only if $X \leqslant x$, and this has probability $p = F_X(x)$. The only other possible value for $I(x;X)$ is zero, and this has probability $1 - p = 1 - F_x(X)$. We can calculate that

$$\begin{aligned} E\{I(X;x)\} &= P\{I(X;x) = 1\} \\ &= P(X \leqslant x) = F_X(x) = p \end{aligned} \tag{6.2.3}$$

and

$$\begin{aligned} \text{var}\{I(X;x)\} &= E[\{I(X;x)\}^2] - [E\{I(X;x)\}]^2 \\ &= E\{I(X;x)\} - [E\{I(X;x)\}]^2 \\ &= F_X(x) - \{F_X(x)\}^2 \\ &= F_X(x)\{1 - F_X(x)\} = p(1 - p) \end{aligned} \tag{6.2.4}$$

Hence, $I(X_1;x)$ through $I(X_m;x)$ are iid Bernoulli(p) random variables with mean p and variance $p(1 - p)$. This fact leads to our estimator of p,

$$\hat{p} = \frac{\displaystyle\sum_{i=1}^{\infty} I(X_i;x)}{m} \tag{6.2.5}$$

that is, the sample average of the m indicator functions. Using Equation 6.2.3, we can write

$$E(\hat{p}) = E\left\{\sum_{i=1}^{m} \frac{I(X_i;x)}{m}\right\} = \sum_{i=1}^{m} \frac{p}{m} = p \tag{6.2.6}$$

thus our estimator is unbiased and has the following variance (using Equation 6.2.4):

$$\begin{aligned} \text{var}(\hat{p}) &= \text{var}\left\{\sum_{i=1}^{m} \frac{I(X_i;x)}{m}\right\} \\ &= \sum_{i=1}^{m} \text{var}\left\{\frac{I(X_i;x)}{m}\right\} \\ &= \sum_{i=1}^{m} \frac{p(1 - p)}{m^2} \\ &= \frac{p(1 - p)}{m} \end{aligned} \tag{6.2.7}$$

The result (Equation 6.2.7) is just another example of the fact, discussed in Chapter 4 at Equation 4.2.1, that the variance of a sample average is the variance of the individual components of that sample average divided by m.

An alternative but equivalent way to define $\hat{p}$ is by counting the number of observed X_i that are less than or equal to the given cutoff level x, and dividing by m: that is,

$$\hat{p} = \frac{\text{number of } X_i \leqslant x}{m} \tag{6.2.8}$$

The corresponding random variable

$$Z_m = \text{number of } X_i \leqslant x = \sum_{i=1}^{m} I(X_i; x)$$

is the sum of independent Bernoulli(p) random variables and therefore has a binomial(m, p) distribution. It follows that

$$E(Z_m) = mp \qquad \text{and} \qquad \text{var}(Z_m) = mp(1 - p)$$

so that

$$E(\hat{p}) = E\left(\frac{Z_m}{m}\right) = p$$

and

$$\text{var}(\hat{p}) = \text{var}\left(\frac{Z_m}{m}\right) = \frac{p(1 - p)}{m}$$

as before. Note that Equation 6.2.5 is merely a mathematical formalization of Equation 6.2.8.

The final representation for the percentile estimate $\hat{p}$ requires working with the *order statistics* $X_{(1)}, X_{(2)}, \ldots, X_{(m)}$ of the sample $X_1, X_2, \ldots, X_m$ (see Section 3.4). Intuitively, we require $X_{(1)} \leqslant X_{(2)} \leqslant \cdots \leqslant X_{(m)}$, so that the $X_{(i)}$ are the X_i arranged in increasing order. Formally, we define

$$X_{(i)} = (X_{\pi_i} | \pi \text{ is a permutation of the integers 1 through } m \text{ such that}$$
$$\text{the } X_{\pi_i} \text{ satisfy } X_{\pi_1} \leqslant X_{\pi_2} \leqslant \cdots \leqslant X_{\pi_{m-1}} \leqslant X_{\pi_m}) \tag{6.2.9}$$

The percentile estimator can now be represented by

$$\hat{p} = \frac{\max\{i \,|\, X_{(i)} \leqslant x\}}{m} \tag{6.2.10}$$

since

$$Z_m = \max\{i \mid X_{(i)} \leqslant x\}$$

This form of $\hat{p}$ is mathematically equivalent to the previous two but may be computationally more efficient if the sample is already ordered and/or a large number of percentiles (for various values of x) are required—for instance, for the ecdf. Note that

$$\hat{p} = \sum_{i=1}^{m} \frac{I(X_i; x)}{m} \equiv \sum_{i=1}^{m} \frac{I\{X_{(i)}; x\}}{m} \tag{6.2.11}$$

The definition of Equation 6.2.9 of order statistics also covers the case of data that are not distinct. [In theory, if the random variables are continuous, the order statistics satisfy the inequalities $X_{(1)} < X_{(2)} < \cdots < X_{(m)}$.]

EXAMPLE   **6.2.1 Percentile Estimation**

The 10 numbers given in Table 6.2.1 were generated by a Normal(0, 1) random number generator. Below the simulated data are the corresponding indicator functions and order statistics, and we will be interested in finding the percentile corresponding to $x = 1.28$. We can compute the following:

$$\sum_{i=1}^{10} I(X_i \leqslant 1.28) = 8$$

$$Z_{10} = \text{number of } X_i \leqslant x = 8$$

$$\max\{i \mid X_{(i)} \leqslant x\} = 8$$

so that in each case we can divide by $m = 10$ to get $\hat{p} = 0.8$. The corresponding estimate of the variance of the estimate $\hat{p}$ will be

$$\text{vâr}(\hat{p}) = \frac{\hat{p}(1 - \hat{p})}{10} = 0.8 \times \frac{0.2}{10} = 0.016$$

$$\hat{\sigma}(\hat{p}) = 0.126$$

For comparison, the true value of p can be found from a table of the distribution of Normal(0, 1) random variables to be

Table 6.2.1   An example of order statistics

i	1	2	3	4	5	6	7	8	9	10
X_i	$+0.31$	-2.11	-0.21	$+0.65$	$+1.91$	-1.61	$+1.13$	-0.91	$+1.35$	-0.98
$I(X_i; 1.28)$	1	1	1	1	0	1	1	1	0	1
$X_{(i)}$	-2.11	-1.61	-0.98	-0.91	-0.21	$+0.31$	$+0.65$	$+1.13$	$+1.35$	$+1.91$

$$p = F_X(1.28) = P(N(0, 1) \leqslant 1.28) = 0.9$$

where by the assumptions of our simulations, $E(X) = 0$ and $\sigma(X) = 1$. Note that $\bar{X} = -0.047$, $S = 1.338$, and $S(\bar{X}) = 0.423$. Since a Normal$(0, 1)$ random variable is a continuous random variable, there are no ties in the 10 values given in Table 6.2.1. This could occur if fewer than 2 decimal digits had been used.

6.2.2 Confidence Intervals for Percentiles

To produce confidence intervals for p, we work from the definition of Equation 6.2.8 and note that $\hat{p}$ has a known distribution: that of a binomial(m, p) random variable, divided by m. This fact can be exploited to produce a nonparametric, $100(1 - \alpha)\%$ confidence interval for the percentile p:

$$(P_l, P_u) \tag{6.2.12}$$

where P_l is that value of p such that

$$\sum_{i=Z_m}^{m} \binom{m}{i} p^i (1 - p)^{m-i} = \frac{\alpha}{2}$$

and P_u is that value of p such that

$$\sum_{i=0}^{Z_m} \binom{m}{i} p^i (1 - p)^{m-i} = \frac{\alpha}{2}$$

[Recall that $Z_m = \sum_{i=1}^{m} I(X_i; x)$.] Theoretical justification for the validity of this interval is given in Lehmann (1959).

The bounds P_l and P_u can be found approximately by using tables of the binomial distribution (Mood, Graybill, and Boes, 1974; Conover, 1980, p. 100) or by a numerical computer search. Graphical charts are also available (Conover, 1980, Table A4). Either way, the task is impractical for large m, and it is easier to use a $100(1 - \alpha)\%$ confidence interval based on the asymptotic Normal$[p, \{p(1 - p)/m\}^{1/2}]$ distribution of $\hat{p}$:

$$\left(\hat{p} \pm \frac{z_{\alpha/2} \{\hat{p}(1 - \hat{p})\}^{1/2}}{m^{1/2}} \right) \tag{6.2.13}$$

where $z_{\alpha/2}$ denotes the $\alpha/2$ quantile of a standard Normal$(0, 1)$ distribution. Because the indicator random variables that comprise $\hat{p}$ have finite mean and variance, the central limit theorem for $\hat{p}$ and consistency results for $\{\hat{p}(1 - \hat{p})/m\}^{1/2}$ hold, regardless of the underlying distribution of the X_i.

The value of m necessary for the asymptotic interval above to be approximately true will vary with p: extreme values of p (i.e., $p \sim 0$ or $p \sim 1$) require large values of m.

6.2.3   Estimating the Cumulative Distribution Function: The Empirical Cumulative Distribution Function

In many situations it may be desirable to estimate (or see) the entire cumulative distribution function of the sample: for instance, if we are worried about large values that may come from the tail of a highly skewed distribution, if we need to model the distribution of the input to a simulation, or if we are thinking of using statistical tests that depend on the approximate Normality of the sample. In such cases, we need to estimate and plot the cumulative distribution function (or possibly the probability density function described in Section 7.1). Note that what we are estimating here is a *curve*, not simply a parameter value.

Estimating the cdf requires estimating the percentile for each value of x; hence, in the notation in this section, we will make the dependence of p on x explicit by using $\hat{p} = \hat{p}(x)$. The ecdf for the random variable X is defined by

$$\hat{F}_X(x) = \hat{p}(x) \qquad \text{for all } x \text{ on the real line} \tag{6.2.14}$$

where $\hat{p}(x)$ is defined in three equivalent ways in Section 6.2.1.

The third definition of $\hat{p}(x)$, given in Equation 6.2.10,

$$\hat{p}(x) = \frac{\max\{i \,|\, X_{(i)} \leq x\}}{m}$$

makes it clear that $\hat{F}_X(x)$ is constant except when x goes from less than or equal to $X_{(i)}$ to greater than or equal to $X_{(i)}$ for some i.

Thus for continuous random variables, $\hat{F}_X(x)$ is right continuous, has jumps of height $1/m$ at the points $X_{(i)}$ for $i = 1, \ldots, m$, and is constant otherwise.

From the calculations in Section 6.2.1, we can see that the ecdf is an unbiased function of x

$$E\{\hat{F}_X(x)\} = E\{\hat{p}(x)\} = F_X(x) \tag{6.2.15}$$

and has variance function

$$\mathrm{var}\{\hat{F}_X(x)\} = \mathrm{var}\{\hat{p}(x)\} = \frac{p(x)\{1 - p(x)\}}{m}$$

$$= \frac{F_X(x)\{1 - F_X(x)\}}{m} \tag{6.2.16}$$

This expression has a maximum value when $F_X(x) = 1 - F_X(x) = 0.5$. Thus since $\mathrm{var}\{\hat{F}_X(x)\} \leq 0.25m^{-1} \to 0$ as $m \uparrow \infty$, the ecdf is a (weakly) consistent estimate of $F_X(x)$ at each x.

As noted in Section 6.2.2, $\hat{p}(x)$ or $\hat{F}(x)$ has a known small sample distribution that leads to nonparametric confidence intervals for $F_X(x)$. The ecdf $\hat{F}(x)$ was also seen to have an asymptotic Normal$(F_X(x), [F_X(x)\{1 - F_X(x)\}/m]^{1/2})$ distribution, which yields large sample confidence intervals for $F_X(x)$.

Note however, that since the same data are used to estimate $F_X(x)$ at each x, the estimates $\hat{F}_X(x_1)$ and $\hat{F}_X(x_2)$ are correlated. This can be demonstrated by using the indicator function representation to calculate, for $x_1 \leqslant x_2$,

$$\text{cov}\{\hat{F}_X(x_1), \hat{F}_X(x_2)\} = \frac{1}{m} F_X(x_1)\{1 - F_X(x_2)\} \tag{6.2.17}$$

Hence we cannot immediately say anything statistical about $\hat{F}_X(x)$ for all values of x simultaneously. Such formal statements are left to Section 6.2.4, when we look at goodness-of-fit tests. Instead, we will conclude by considering how much we can do by eye. Figures 6.2.2, 6.2.3, and 6.2.4 show the actual ecdf's for both 100 and 1000 simulated random variables from each of the three distributions: Normal(0, 1), Cauchy(0, 1.313), and Laplace(0, 0.654). In all three cases the scale constants were chosen to make $P(-1.96 \leqslant X \leqslant 1.96) = 0.95$. The x-axis scale in the three figures was chosen to span the entire range of observed values. Figure 6.2.5 compares the bodies of the ecdf's from 1000 simulated Normal(0, 1), Cauchy(0, 1.313), and Laplace(0, 0.654) random variables by considering only the portions of the ecdf's that fall between -3 and 3. The larger spread of the Cauchy random variables is quite clear.

The following comments are relevant:

1. We can see distinct differences between the three curves because of the relative lengths of their "tails." The bodies of the distributions, however, are similar,

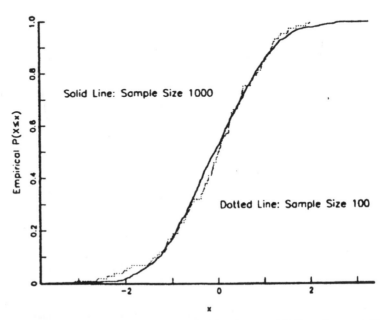

Figure 6.2.2 Empirical distribution for simulated Normal(0, 1) random variables.

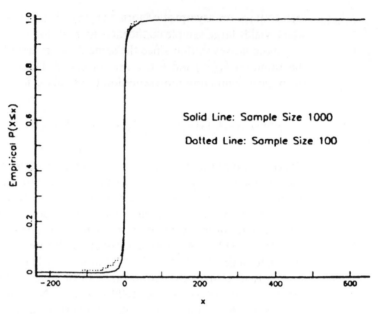

Figure 6.2.3 Empirical distribution for simulated Cauchy(0, 1.313) random variables.

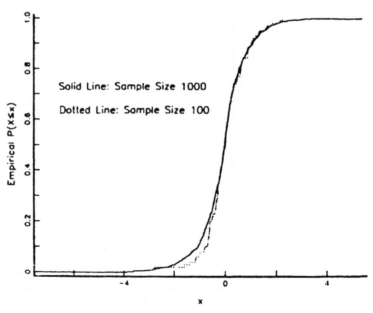

Figure 6.2.4 Empirical distribution for simulated Laplace(0, 0.654) random variables.

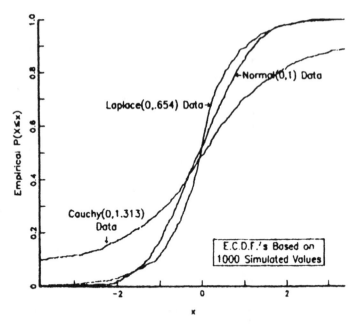

Figure 6.2.5 Comparison of the bodies of Normal(0, 1), Laplace(0, 0.654), and Cauchy(0, 1.313) ecdf's.

and it could be hard to identify the theoretical distribution associated with an arbitrary ecdf. Experience, will, however allow you to narrow the field of possibilities.

2. Once a limited number of possible theoretical distributions have been identified, estimates of the parameters in those distributions can be made, formal comparisons performed (see Section 6.2.4), and more sensitive graphical techniques (see Chapter 7) employed to decide on the parent population. Such methods are discussed throughout the remainder of this and the next chapters.

3. The number of observations determines the accuracy of our picture of the ecdf. Recall that $\text{var}\{\hat{F}_X(x)\} = F_X(x)\{1 - F_X(x)\}/m$, where m is the number of simulated data points. Thus the variance is decreasing as $1/m$ and in fact is less than or equal to $1/(4m)$. The effects of sample size will be considered again in the estimation of the density function in Section 7.1; it can be clearly seen in Figures 6.2.2, 6.2.3, and 6.2.4. Thus note the difference in these figures between the $m = 100$ and $m = 1000$ cases. Note too that while the discontinuous nature of the ecdf is clear for samples of size 100, for samples of size 1000 it is difficult for the eye or the computer graphics to resolve the jumps.

EXAMPLE   **6.2.2 The ecdf for the Waiting Times from the GI|GI|3 Queue**
We have generated 500 simulated waiting times of the 5000th customer in a three-server queue in order to consider what difference the order of service makes. The dotted line in Figure 6.2.6 shows the ecdf for the data on the first-come-first-served (or single-line discipline) case. The data has been truncated to show only waiting times less than 50 to allow examination of the body of the data.

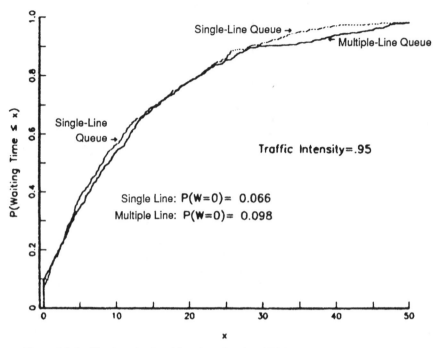

Figure 6.2.6 Five hundred waiting times for the 5000th customer in a three-server queue with Gamma(0.5, 0.351) arrivals and Gamma(0.5, 1) service.

Using the same input assumptions: Gamma(0.5, 0.351) arrivals and Gamma (0.5, 1) service and the "multiple line, go to the shortest line" discipline [(1) in Section 3.3], we can generate another independent sample of 500 waiting times for the 5000th customer. The ecdf for the data is plotted as a solid line in Figure 6.2.6.

Notice how similar overall the ecdf's for the two queueing disciplines are, indicating that the customer is usually unaffected by management's choice of queueing strategy. There is a slight domination by the single-line discipline ecdf over the shortest line ecdf, particularly at large values of the waiting time, which suggests that extreme waiting times are less likely if the former discipline is used. We will consider another way to look at such extreme values in Section 6.3, by tabulating quantiles for given probabilities. Other graphical methods for comparisons are given in Chapter 8.

Notice also that the distributions are not purely continuous but seem to have a discrete mass of probability at $x = 0$. This reflects all the people who arrive and find a server available immediately. The probability of nonwaiting (0.098) is higher for the multiple-line system [compare to $\hat{P}(W = 0) = 0.066$ for the single-line discipline]. The estimated standard deviations of these two probability estimates are, respectively, 0.013 and 0.011, and the estimates differ by 0.032. This difference probably will not be significant, since the standard deviation of the difference is $(0.013^2 + 0.011^2)^{1/2} = 0.017$ (see Conover, 1980, Chap. 4.1 for a formal 2×2 contingency test of this hypothesis).

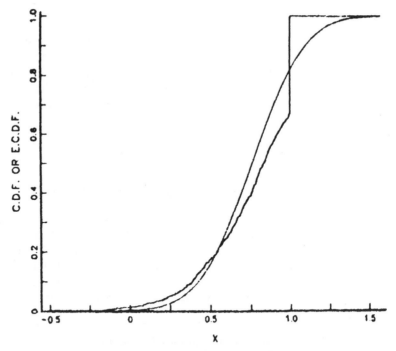

Figure 6.2.7 Comparison of empirical and theoretical cumulative distribution functions. The empirical curve is generated from 1000 mle estimates of δ from simulated samples of size 20 with $\delta = 0.8$. The theoretical curve is a Normal(0.762, 0.263) cumulative distribution function. The values 0.762 and 0.263 are the sample mean and standard deviation of the 1000 estimates of δ.

EXAMPLE   **6.2.3 The ecdf for the mle of the Polarization Parameter**

Using the 1000 estimates of the mle, $\hat{\delta}$, from Example 6.1.4 for the decay model with true polarization parameter value $\delta = 0.8$, an ecdf was computed and is displayed in Figure 6.2.7. Since maximum likelihood estimates are generally asymptotically Normal, a smooth Normal cdf is also superimposed on the figure, based on the mean and standard deviation of the data.

Notice the wide disparity between the cdf and the ecdf, indicating that using 20 points to estimate $\hat{\delta}$ does not ensure asymptotic Normality. In addition, the jump in the ecdf at $x = 1.0$ reveals that no values of $\hat{\delta}$ greater than 1 were recorded! This has occurred because the program to calculate $\hat{\delta}$ breaks down when $\hat{\delta} > 1$, and therefore, the mle is modified to set $\hat{\delta} = 1$, if the maximum is greater than 1. The default is statistically reasonable, since values of δ outside the range $[-1, 1]$ cannot physically occur.

6.2.4 Formal Goodness-of-Fit Tests

The estimated cumulative distribution function can be compared with any specified cumulative distribution function by means of statistical tests. We discuss formal tests here and informal graphical methods later in Chapter 7.

Usually we are interested in two types of testing situations.

1. We want to determine whether our data behave according to a prespecified distribution. This will be the case if we need to know the distribution of observed input data for a physical system to generate data from the same distribution for use in a simulation of that system. This also will be the case when we are validating a simulation program by comparing output for a special case to the known distribution for that case.

2. We want to determine whether our data have a Normal distribution. This can be useful when analyzing the output from a simulation. Tests of significance for summary statistics of that output distribution and comparisons to analogous values for other simulations will generally be sensitive to the Normality of the output.

Although many formal tests exist to handle these two situations, we consider only the Kolmogorov–Smirnov test for the first case and the Shapiro–Wilk test for the second. It is hoped that this restriction will reduce confusion due to the proliferation of nonparametric tests, while presenting clearly two formal tests with good general performance. Examples and references to alternative tests, comparisons between tests, and tables of significance points can be found in Conover (1980, Chap. 6).

i. The Kolmogorov–Smirnov Test Statistic

The Kolmogorov–Smirnov test statistic D_m for a sample of size m is based on the maximum difference between the observed ecdf $\hat{F}_X(x)$ and a hypothesized distribution $F_{H_0}(x)$ across all values of x: that is,

$$D_m = \sup_x |\hat{F}_X(x) - F_{H_0}(x)| \tag{6.2.18}$$

This standard Kolmogorov–Smirnov test assumes two things:

- That the sample used to estimate $\hat{F}_X(x)$ (i.e., $X_1, X_2, \ldots, X_m$) is iid with unknown *continuous* distribution function.
- That our choice of $F_{H_0}(x)$ does not depend directly on X_1 through X_m (i.e., we cannot use the sample to estimate parameters in the theoretical distribution if the null distribution of D_m is to hold).

The test statistic is used to test the hypothesis

$$H_0: F_X(x) = F_{H_0}(x)$$

against

$$H_1: F_X(x) \neq F_{H_0}(x)$$

although one-sided alternatives are possible (see Conover, 1980, Chap. 6).

Computationally, the difference $|\hat{F}_X(x) - F_{H_0}(x)|$ needs to be evaluated only at the $2m$ points immediately *before and after* the observed values, $X_1, X_2, \ldots, X_m$.

Table 6.2.2 Asymptotic quantiles $d_{1-\alpha}$ of $\sqrt{m}D_m$

$1 - \alpha$	0.80	0.85	0.90	0.95	0.99
$d_{1-\alpha}$	1.07	1.14	1.22	1.36	1.63

This is because $\hat{F}_X(x)$ changes only at those m observed values and $F_{H_0}(x)$ is monotonically increasing in x. Formally we have

$$D_m = \max(D_m^+, D_m^-) \tag{6.2.19}$$

where

$$D_m^+ = \sup_x \{\hat{F}_X(x) - F_{H_0}(x)\} = \max_{1 \leqslant i \leqslant m} \left\{ \frac{i}{m} - F_{H_0}(X_{(i)}) \right\} \tag{6.2.20}$$

and

$$D_m^- = \sup_x \{F_{H_0}(x) - \hat{F}_X(x)\} = \max_{1 \leqslant i \leqslant m} \left\{ F_{H_0}(X_{(i)}) - \frac{i-1}{m} \right\} \tag{6.2.21}$$

Note that the random variables appearing in expressions 6.2.20 and 6.2.21 for D_m^+ and D_m^- are $F_{H_0}(X_{(i)})$, $i = 1, \ldots, m$, which, by the probability integral transform, are the order statistics from an iid sample distributed uniformly on $(0, 1)$. Thus the distributions of D_m^+ and D_m^- will be distribution free under the null hypothesis: that is, these distributions will not depend on the distribution $F_{H_0}(x)$ when the null hypothesis is true.

Both the exact (small-sample) and asymptotic nonparametric distributions of the Kolmogorov–Smirnov statistic are known, so that once a value of D_m has been computed, it can be compared to the theoretical value and a decision to accept or reject the null hypothesis can be made. Clearly, the null hypothesis will be rejected if D_m, the distance between $F_{H_0}(x)$ and $\hat{F}_X(x)$, is too large.

Table 6.2.2 gives several quantile values for the asymptotic distribution of $\sqrt{m}D_m$, when $m \to \infty$. Thus the tabulated value $d_{1-\alpha}$ satisfies the equation $P\{\sqrt{m}D_m \leqslant d_{1-\alpha}\} = 1 - \alpha$. A more complete set of exact values appears in Conover (1980, Table A14), as well as quantiles for one-sided versions of the Kolmogorov–Smirnov test. The asymptotic values may be used for $m \geqslant 50$. Notice that like many other statistics we have encountered, D_m converges to zero at a rate proportional to $(1/m)^{1/2}$, and $\sqrt{m}D_m$ converges to a proper random variable.

EXAMPLE **6.2.4 Testing for the Normality of the Moment Estimator of the Polarization Parameter**

It is generally true that moment estimators are asymptotically Normally distributed (see Section 6.1). For the moment estimator $\tilde{\delta}$ of the polarization parameter δ, in Example 6.0.1, we may be interested in the following question: Is a sample of size

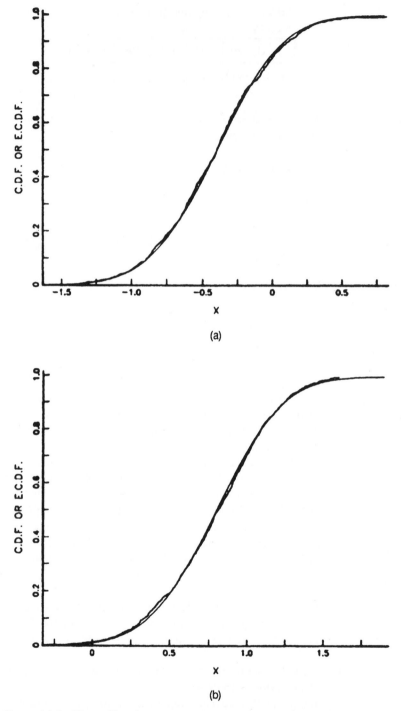

Figure 6.2.8   The ecdf's of $m = 1000$ simulated moment estimators of δ when $n = 20$. (a) $\delta = -0.4$. The solid curve is a Normal$(-0.4, 0.377)$ distribution function. (b) $\delta = 0.8$. The solid curve is a Normal$(0.8, 0.344)$ distribution function.

$n = 20$ large enough for $\tilde{\delta}$ to be approximately Normally distributed? If so, we can use Normal theory tests and confidence intervals for inference about $\tilde{\delta}$.

The moment estimator

$$\tilde{\delta} = 3\bar{X}$$

has a mean of δ and a variance of $(3 - \delta^2)/n$. As in Example 6.1.4, we consider only the models with $\delta = -0.4$ and $\delta = 0.8$ and use graphics and the Kolmogorov–Smirnov test to compare the empirical distribution of our moment estimates of δ to theoretical Normal$(-0.4, 0.377)$ and Normal$(0.8, 0.344)$ distributions, respectively.

For the case $\delta = -0.4$, the graphical match of the ecdf to a Normal$(-0.4, 0.377)$ distribution in Figure 6.2.8a is extremely close. In addition, the Kolmogorov–Smirnov statistic, $\sqrt{1000}D_{1000} = 0.614$, is smaller than 1.07, the 0.8 quantile of the null distribution of $\sqrt{1000}D_{1000}$ (Table 6.2.2), thus confirming the graphical fit.

Another interesting facet of the graph is that the estimator $\tilde{\delta}$ takes values less than -1.0, even though the true value δ is constrained to lie in $[-1, 1]$. Should these values be set to -1.0 even though this would introduce small-sample bias into the estimator? This is a possibility because of the constraint on δ.

Similar results are displayed for the moment estimator $\tilde{\delta}$ when $\delta = 0.8$ in Figure 6.2.8b. The empirical and theoretical distributions nearly coincide, and the Kolmogorov–Smirnov test statistic, $\sqrt{1000}D_{1000} = 0.946$, is again less than 1.07, the 0.8 quantile of the null distribution of $\sqrt{1000}D_{1000}$. Notice that once again about 30% of the values of $\tilde{\delta}$ are above the physically realizable limit of 1.0.

Some common continuous distributions you may want to use as testing alternatives, $F_{H_0}(x)$, have appeared in examples in the text and are detailed in Table 6.1.2. Two important cases are the Normal and exponential distributions. Adjustments to the distribution of D_m when the Normal distribution parameters μ and σ are estimated by $\bar{X}$ and S, and in the exponential case when the mean is estimated by $\bar{X}$, are known (Lilliefors, 1967 and 1969, respectively). Table 6.2.3 gives the asymptotic values; for small sample results see Conover (1980, Tables A15 and A16).

Adjustments to the distribution of D_m for discrete random variables are also known (Conover, 1980, Chap. 6) but are difficult to compute. The advantage over using the usual chi-square goodness-of-fit test (Conover, 1980) is that no grouping of data is required.

EXAMPLE **6.2.5 Testing the Nonzero Waiting Times for an Exponential Distribution**
There are many queueing models for which theory tells us that the limiting distribution of the waiting times will have a discrete component at $W = 0$ and will be exponentially distributed if they are greater than 0 (see, e.g., Ross, 1980, and Newell, 1982). Both sets of simulated data displayed in Figure 6.2.5 of Example 6.2.2 appear to have such a distribution. Hence we look only at the nonzero waiting times for each discipline and use Lilliefors's modification of the Kolmogorov–Smirnov test for which the asymptotic quantiles are given in Table 6.2.3 to see if those data sets have exponential distributions. This information will be important, for instance, in deciding if the queue has stabilized by the 5000th customer.

Table 6.2.3 Lilliefors-modified Kolmogorov–Smirnov statistic

Asymptotic quantiles of $\sqrt{m}D_m$ adjusted for parameter estimation $d_{1-\alpha}$ quantile		
$1 - \alpha$	Normal$(\bar{X}, S)$	Exponential$(1/\bar{X})$
0.95	0.886	1.0753
0.99	1.031	1.2743

For the single-line discipline queue, there are 467 *nonzero* waiting times with an average of 12.91, so we will be comparing our data to an exponential(1/12.91) distribution. In other words, we do not have a theoretical value for λ, but we estimate it as $1/\bar{X} = 1/12.91$. The graphical comparison in Figure 6.2.9a shows a close match. Also, since the formal Lilliefors test statistic

$$\sqrt{467}D_{467} = 0.6425$$

is well below the 0.95 cutoff point of 1.0753, we accept the hypothesis of an underlying exponential(1/12.91) distribution.

Likewise, for the 451 *nonzero* waiting times for the simulation with "shortest line" queueing discipline, we find an average of 14.46 and are led by the graphics in Figure 6.2.9b and the value of the Lilliefors test statistic

$$\sqrt{451}D_{451} = 0.6430$$

to conclude that the data follow an exponential(1/14.46) distribution.

One general problem with the Kolmogorov–Smirnov test should also be mentioned: it is not very powerful (i.e., we do not reject incorrect hypotheses as often as we would like) for small samples.

Because we are dealing with a simulation situation, we can always generate more observations and increase the power of the test. As with all hypothesis testing, though, we risk increasing the power so much that we reject alternatives that may, for practical purposes, be "close enough." For instance, for m large enough, we would be able to reject the hypothesis that we had a Normal(0, 1) population if our sample data actually came from a Normal(0.001, 1) distribution. Such a small shift in the mean, however, may have little or no significance to our purpose in identifying or approximating the underlying distribution.

ii. The Shapiro–Wilk Test for Normality

If you are interested in testing only whether sample data come from a Normal distribution, you can avoid having to specify the mean and standard deviation of that distribution (as we must for the Kolmogorov–Smirnov test) by using the Shapiro–Wilk test (Conover, 1980, Chap. 6). Thus it is a test of Normality, per se, with μ and σ as nuisance parameters, that has been shown to have high power against a broad range of alternatives.

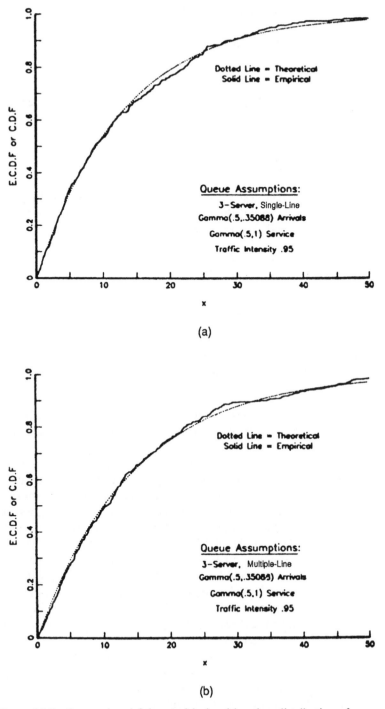

Figure 6.2.9 Comparison of the empirical waiting time distribution of nonzero waiting times (a) for a single-line queueing discipline with an exponential distribution (0.0744) and (b) for a multiple-line queueing discipline where the customer joins the "shortest" line with an exponential distribution (0.0691).

We again assume that

the X_i are iid $F_X(x)$

and we test

H_0: $F_X(x)$ is Normal(μ, σ) with μ and σ unspecified

against

H_1: $F_X(x)$ is not Normal

The test statistic $S-W$ is based on an analogue of the F statistic from the regression of the sample order statistics onto the theoretical order statistics of a Normal sample of size m:

$$S-W = \frac{\sum_{i=1}^{m} a_i(X_{(i)} - \bar{X})^2}{\sum_{i=1}^{m} (X_i - \bar{X})^2} \tag{6.2.22}$$

where

$$\bar{X} = \sum_{i=1}^{m} X_i/m$$

Formulas for the exact calculation of the coefficients a_i are known (Royston, 1982) but are impractical computationally. For $m \leqslant 50$, exact tabled values of the a_i are available (Conover, 1980, Table A17).

Further approximations of the a_i for m up to 2000 have been proposed and tested (Royston, 1982). These are accurate, very suitable to computer implementation, and an algorithm is available from the same source. (It is also implemented in the *Enhanced Simulation and Statistics Package* supplement to this book.)

Once a value for $S-W$ has been computed, it can be compared to the values of the quantiles of the distribution of the Shapiro–Wilk statistic. These are available for samples up to size 50, for example, in Table A18 of Conover (1980). The significance level must be interpolated from the first line of the table. Normality is rejected for values that are too high or too low. For values of m larger than 50, the significance level can be found by using a Normalizing transformation of $S-W$, as outlined in Royston (1982), and then comparing the Normalized value to standard Normal quantiles.

A program to compute the Shapiro–Wilk statistic, SWPC, is available on the floppy disk in the *Enhanced Simulation and Statistics Package* supplement to this book. When the sample size is greater than 6, a Normalized value of the Shapiro–Wilk statistic is also calculated.

For the Shapiro–Wilk test, the more sample points we generate, the more powerful the test. The only drawback is that the approximations for the a_i and the null distribution of $S-W$ become more difficult to compute beyond $m = 50$ and are unavailable beyond $m = 2000$.

EXAMPLE **6.2.6 Testing for the Normality of the Estimates of the Polarization Parameter**
Theoretically, the asymptotic distribution of an mle is known to be Normal, but the small-sample variance may be hard to compute analytically. This is the case for $\hat{\delta}$, the mle of the polarization parameter δ in Example 6.0.1. Concentrating on the model in Example 6.2.3 with $\delta = 0.8$, we may need to know if a sample of given size (e.g., $n = 20$) is large enough to use Normal theory tests and confidence intervals for $\hat{\delta}$. That is, we need to check whether the distribution of $\hat{\delta}$ is Normal(μ, σ). In this case, though, μ and σ will be treated as unknowns and estimated from the simulated values of $\hat{\delta}$ from Example 6.1.4 by $\bar{X} = 0.762$ and $S = 0.263$, respectively.

The graphical comparison in Figure 6.2.7 of the ecdf of the simulated values of $\hat{\delta}$ to a theoretical Normal$(\bar{X}, S)$ distribution shows distinct differences. The comparison can be made formal by computation of the Shapiro–Wilk statistic

$$S-W = 0.8399$$

Using the Normalizing transformation mentioned previously of Royston (1982), we obtain

$$\text{Normalized } (S-W) = 80.64$$

Since 80.64 falls far out in the tail of a standard Normal$(0, 1)$ distribution, we can reject the hypothesis that $\hat{\delta}$ has a Normal distribution.

To illustrate the sensitivity of the Shapiro–Wilk test, the moment estimates of δ for the model with $\delta = 0.8$ were also subjected to this test. The result,

$$S-W = 0.9806$$

when Normalized,

$$\text{Normalized}(S-W) = 2.80$$

leads us to reject the Normality of $\tilde{\delta}$ when sample size is 20. This conclusion will be seen to agree with a $Q-Q$ plot analysis in Section 7.2—the extreme tails diverge from those of a Normal distribution. Recall, however, that neither the ecdf plot nor the Kolmogorov–Smirnov test in Example 6.2.4 was sensitive enough to pick up the differences.

We now know how to estimate the percentiles and cumulative distribution function of a sample of iid observations, to inspect the ecdf graphically, and to test it to see if it resembles a given arbitrary or Normal distribution function. In Chapter 7, we will return to the idea of seeing the entire distribution at once when we look at another, sometimes more familiar, representation, the probability density function. We will also consider probability plotting methods.

6.3   Quantiles

The quantiles of a distribution (defined for continuous distributions in Equations 3.5.4 and 3.5.5) are related to the percentiles in the following way.

If $F_X(x)$ is a *continuous* cdf and α satisfies $0 < \alpha < 1$, then the equality

$$P(X \leqslant x_\alpha) = \alpha$$

means that x_α is, by definition, the α *quantile*. Intuitively, the random variable X takes values less than x_α with probability α. If the cdf is not continuous, the definition above does not give a quantile for all $\alpha \in (0, 1)$. Then x_α is defined as follows:

$$x_\alpha = \inf\{x \,|\, P(X \leqslant x) = \alpha\} \tag{6.3.1}$$

Notice that this definition is also valid for continuous $F_X(x)$, and corresponds to the original definition given in Equations 3.5.4 and 3.5.5.

Although knowing x_α for every value of α between 0 and 1 would be equivalent to knowing the distribution function, one generally looks at only a few quantiles or combinations of quantiles to get information about the location, shape, and dispersion of the distribution. We consider the estimation of quantiles in Sections 6.3.1 and 6.3.2 and go on to their interpretation in 6.3.3.

6.3.1   Estimation of Quantiles

Given the random sample $X_1, X_2, \ldots, X_m$ with order statistics $X_{(1)}, X_{(2)}, \ldots, X_{(m)}$ and α such that $0 < \alpha < 1$, the usual nonparametric estimator of the αth quantile is given by:

$$\hat{x}_\alpha = \begin{cases} X_{(m\alpha)} & \text{if } m\alpha \text{ is an integer} \\ X_{(\lfloor m\alpha \rfloor + 1)} & \text{if } m\alpha \text{ is not an integer} \end{cases} \tag{6.3.2}$$

where $\lfloor m\alpha \rfloor$ is the greatest integer less than or equal to $m\alpha$. The rationale for this estimator is that the ecdf is the nonparametric estimator of the cdf and, using the definition of Equation 6.3.1, $X_{(i)}$ is the α quantile of the ecdf of a continuous random variable when $(i - 1)/m < \alpha \leqslant i/m$, for $i = 1, 2, \ldots, m$. Thus the index of the order statistic that is the α quantile of the ecdf is $\lfloor m\alpha \rfloor + 1$.

EXAMPLE   6.3.1   Estimating Quantiles in Normal Samples

Generating 10 simulated Normal(0.8, 0.34) random variables produced the following ordered, observed values of $X_{(i)}$:

$$(0.08, 0.25, 0.47, 0.49, 0.73, 0.91, 0.9, 1.18, 1.26, 1.45)$$

The 0.05, 0.10, 0.5, 0.9, and 0.95 quantiles associated with these points are displayed in the first column of Table 6.3.1. If, instead, 1000 Normal(0.8, 0.34) points were generated, the quantiles in column 2 would have been produced. Both these columns

Table 6.3.1 Estimated quantiles of two Normal(0.8, 0.34) samples of different sizes and estimated quantiles of the 1000 simulated values of $\tilde{\delta}$

Quantile	Sample of 10 $N(0.8, 0.34)$ variates	Sample of 1000 $N(0.8, 0.34)$ variates	1000 Estimates of $\tilde{\delta}$	Theoretical $N(0.8, 0.34)$
$\hat{x}_{0.05}$	0.08	0.23	0.21	0.23
$\hat{x}_{0.10}$	0.08	0.36	0.33	0.36
$\hat{x}_{0.5}$	0.73	0.78	0.81	0.80
$\hat{x}_{0.90}$	1.26	1.22	1.24	1.24
$\hat{x}_{0.95}$	1.45	1.34	1.35	1.36

should be compared to each other and to the fourth column of the table, the exact theoretical quantiles of the Normal(0.8, 0.34) distribution. You should note how variable quantile estimates can be, especially in small samples.

Finally, the third column of Table 6.3.1 gives the estimated quantiles for the 1000 simulated values of $\tilde{\delta}$, the moment estimator of the polarization parameter δ, described in Example 6.0.1. In Example 6.2.3, we saw that $\tilde{\delta}$ has approximately a Normal(0.8, 0.34) distribution according to graphical views of the ecdf and the Kolmogorov–Smirnov test; by comparing columns three and four of the table, we also see close agreement of the quantiles considered in this example.

It can be seen from Example 6.3.1 that for small samples the estimator $\hat{x}_\alpha$ may not perform well, particularly for α close to 0 or 1 (i.e., for quantiles in the tails of the distribution where observed values are sparse). This will not cause extreme problems if the X_i are outputs from a simulation, since we can always generate more observations (at some cost in time). This poor performance may be an important issue, though, if the X_i are a limited set of real data used to model input or check output from a simulation. In such cases, the following alternative quantile estimate is useful (Mood, Graybill, and Boes, 1974, p. 512):

$$\text{let } K = [(m + 1)\alpha] + 1 \quad \text{and} \quad \beta = \{K - (m + 1)\alpha\}$$

Then set

$$\hat{x}_\alpha^* = \begin{cases} X_{(1)} & \text{if } K = 1 \\ \beta X_{(K-1)} + (1 - \beta)X_{(K)} & \text{if } 1 < K \leqslant m \\ X_{(m)} & \text{if } K = m + 1 \end{cases} \qquad (6.3.3)$$

Using this definition, we linearly interpolate between observed values when the quantile falls between these values. A similar scheme is given in Chambers et al. (1983, p. 12).

EXAMPLE **6.3.2 Estimating Quantiles by Interpolating Between Order Statistics**
When the 10 simulated Normal(0.8, 0.34) random variables in Example 6.3.1 are used to estimate the 0.05, 0.10, 0.5, 0.9, and 0.95 quantiles according to Equation 6.3.3, we get the following results:

α	K	β	Modified quantile estimate
0.05	1	—	$\hat{x}_{0.05}^{*} = X_{(1)} = 0.080$
0.10	2	0.9	$\hat{x}_{0.10}^{*} = 0.9\,X_{(1)} + 0.1\,X_{(2)} = 0.097$
0.50	6	0.5	$\hat{x}_{0.5}^{*} = 0.5\,X_{(5)} + 0.5\,X_{(6)} = 0.820$
0.90	10	0.1	$\hat{x}_{0.9}^{*} = 0.1\,X_{(9)} + 0.9\,X_{(10)} = 1.431$
0.95	11	—	$\hat{x}_{0.95}^{*} = X_{(10)} = 1.450$

Referring back to Table 6.3.1, we see that these estimates for $x_{0.10}$ and $x_{0.5}$ are closer to the true theoretical quantiles in column four than the estimates in column one, obtained through the original definition (Equation 6.3.2). The estimate of $x_{0.9}$ is now further away from the true value, but, in general, the definition of Equation 6.3.3 should produce results with better statistical properties than Equation 6.3.2, at the cost of some more computation.

Certain quantiles are of particular importance, specifically, the *median* $x_{0.5}$, the *lower quartile* $x_{0.25}$, and the *upper quartile* $x_{0.75}$, as discussed in Section 6.3.3, while their estimation usually follows the special definitions (see, for example, Mosteller and Tukey, 1977, Chap. 3, p. 47):

$$
\text{estimated median} = \begin{cases} X_{([m+1]/2)} & \text{if } m \text{ is odd} \\ \dfrac{X_{(m/2)} + X_{(m/2+1)}}{2} & \text{if } m \text{ is even} \end{cases} \tag{6.3.4}
$$

Once the estimated median has been computed there are

$$
l = \begin{cases} \dfrac{m-1}{2} & \text{if } m \text{ is odd} \\ \dfrac{m}{2} & \text{if } m \text{ is even} \end{cases}
$$

observed data points below the estimated median and the same number above the estimated median. The lower quartile is then estimated to be the median of the l points *below* the estimated median of the whole sample:

$$
\text{estimated lower quartile} = \begin{cases} X_{([l+1]/2)} & \text{if } l \text{ is odd} \\ \dfrac{X_{(l/2)} + X_{(l/2+1)}}{2} & \text{if } l \text{ is even} \end{cases} \tag{6.3.5}
$$

and, likewise, the upper quartile is estimated to be the median of the l points *above* the estimated median of the whole sample.

$$
\text{estimated upper quartile} = \begin{cases} X_{(m+1-[l+1]/2)} & \text{if } l \text{ is odd} \\ \dfrac{X_{(m+1-l/2)} + X_{(m-l/2)}}{2} & \text{if } l \text{ is even} \end{cases} \tag{6.3.6}
$$

EXAMPLE **6.3.3 Estimating the Median and Quartiles**

For a sample of size $m = 100$, the original definition of a quantile estimate given at Equation 6.3.2 tells us to estimate the median and quartiles by

$$\hat{x}_{0.5} = X_{(50)}$$
$$\hat{x}_{0.25} = X_{(25)}$$
$$\hat{x}_{0.75} = X_{(75)}$$

Using the interpolating scheme given at Equation 6.3.3, we get the estimates

$$\hat{x}^*_{0.5} = \frac{X_{(50)} + X_{(51)}}{2}$$

$$\hat{x}^*_{0.25} = \frac{3X_{(25)} + X_{(26)}}{4}$$

$$\hat{x}^*_{0.75} = \frac{X_{(75)} + 3X_{(76)}}{4}$$

The special definitions of Equations 6.3.4, 6.3.5, and 6.3.6 produce the estimates:

$$\text{estimated median} = \frac{X_{(50)} + X_{(51)}}{2}$$

$$\text{estimated lower quartile} = \frac{X_{(25)} + X_{(26)}}{2}$$

$$\text{estimated upper quartile} = \frac{X_{(75)} + X_{(76)}}{2}$$

When m is large enough, the order statistics are close to each other and all definitions produce numerically similar estimates.

6.3.2 Distributional Properties

All the following results pertain to the original definition (see Equation 6.3.2) *for the estimator $\hat{x}_\alpha$.* Since $\hat{x}_\alpha$ is simply an order statistic, its distribution, and therefore its mean and variance, are, in theory, known. However, in practice, the distribution depends heavily on the (unknown) distribution of the X_i, $F_X(x)$, and on the sample size m.

In particular, if $\hat{x}_\alpha = X_{(k)}$ for some k, it is known (Mood, Graybill, and Boes, 1974, Chap. 6) that

$$P(\hat{x}_\alpha \leqslant x) = F_{\hat{x}_\alpha}(x) = \sum_{j=k}^{m} \binom{m}{j} \{F_X(x)\}^j \{1 - F_X(x)\}^{m-j} \tag{6.3.7}$$

which may be awkward computationally.

One quantity that we may compute exactly for small m is a nonparametric $100(1 - \gamma)\%$ confidence interval for x_α. The interval is given (Conover, 1980, pp. 111–112) by

$$(X_{(j)}, X_{(k)}) \tag{6.3.8}$$

where j and k are chosen, not necessarily uniquely, to satisfy

$$\sum_{i=j}^{k-1} \binom{m}{i} \alpha^i (1 - \alpha)^{m-i} \geqslant 1 - \gamma \tag{6.3.9}$$

$$\sum_{i=j+1}^{k-1} \binom{m}{i} \alpha^i (1 - \alpha)^{m-i} < 1 - \gamma \tag{6.3.10}$$

$$\sum_{i=j}^{k-2} \binom{m}{i} \alpha^i (1 - \alpha)^{m-i} < 1 - \gamma \tag{6.3.11}$$

Equation 6.3.9 ensures a coverage probability of at least $1 - \gamma$, while Equations 6.3.10 and 6.3.11 check to see that there is no waste—that is, that the shorter intervals, $(X_{(j+1)}, X_{(k)})$ and $(X_{(j)}, X_{(k-1)})$ do not cover with the required probability, $1 - \gamma$. For large values of m, the confidence interval is still valid but the process of finding j and k may be prohibitively expensive computationally. In that case, the Normal approximation mentioned later (Equation 6.3.16) may be used.

We also have the result

$$E(\hat{x}_\alpha) = x_\alpha + \frac{a}{m} + O(m^{-2}) \tag{6.3.12}$$

which holds for many distributions of the X_i (Cramér, 1946). From Equation 6.3.12, we have assurance that the estimator will be asymptotically unbiased, but we should be wary of the bias in small samples.

Other asymptotic results that hold when $F_X(x)$ is strictly monotone with density $f_X(x) > 0$ include the following (Mood, Graybill, and Boes, 1974, Chap. 6):

$$E(\hat{x}_\alpha) = x_\alpha \qquad \text{as } m \to \infty \tag{6.3.13}$$

$$\text{var}(\hat{x}_\alpha) = \frac{\alpha(1 - \alpha)}{[m\{f_X(x_\alpha)\}^2]} \to 0 \qquad \text{as } m \to \infty \tag{6.3.14}$$

$$\hat{x}_\alpha \sim N\left(x_\alpha, \frac{\alpha^{1/2}(1 - \alpha)^{1/2}}{[m^{1/2}\{f_X(x_\alpha)\}]}\right) \qquad \text{as } m \to \infty \tag{6.3.15}$$

Unfortunately, m may have to be very large before the asymptotic Normality sets in, especially for extreme quantiles.

From these results we can construct a large sample, $100(1 - \gamma)\%$ confidence interval for x_α as:

$$\hat{x}_\alpha \pm z_{1-\gamma/2}\left[\frac{\alpha^{1/2}(1 - \alpha)^{1/2}}{\{m^{1/2}\hat{f}_x(\hat{x}_\alpha)\}}\right] \tag{6.3.16}$$

where $z_{1-\gamma/2}$ is the $100(1 - \gamma/2)\%$ quantile of the $N(0, 1)$ distribution and $\hat{f}_X(\hat{x}_\alpha)$ is an estimate of the density of the X_i, evaluated at the estimated αth quantile. Because of the instability of density estimates, particularly in the tails where $f_X(x_\alpha)$ can be very small and consequently $1/\{f_X(x_\alpha)\}$ can be very large, this interval should not be used unless you have a large sample and are very sure of your estimate $\hat{f}(\hat{x}_\alpha)$. Note that it requires estimating the density at a random point $\hat{x}_\alpha$, not at a fixed point x_α. Some experience with this kind of procedure is given in Heidelberger and Lewis (1984).

A nonparametric, *large-sample* approximation to Equations 6.3.9 to 6.3.11 for finding a confidence interval estimate for x_α, which does not require the density estimate used in Equation 6.3.15, is given in Conover (1980, p. 112). It will usually produce wider confidence intervals than the large-sample procedure (Equation (6.3.16).

6.3.3 Use and Interpretation of Quantiles

Once a representative sample of quantiles—for instance, $\hat{x}_{0.05}$, $\hat{x}_{0.10}$, $\hat{x}_{0.25}$, $\hat{x}_{0.5}$, $\hat{x}_{0.75}$, $\hat{x}_{0.90}$, $\hat{x}_{0.95}$—has been estimated from a sample $X_1, \ldots, X_m$, we can use the estimates to characterize the distribution $F_X(x)$ of the X_i's, in much the same way as sample moments were used in Section 6.1.

i. Location

The median (or 0.5 quantile) tells us about the center of the distribution, since 50% of the mass is on points less than $\hat{x}_{0.5}$ and 50% on points greater than $\hat{x}_{0.5}$. If the distribution is symmetric, mean and median will be the same and their estimates should be close. The distributional properties of the sample median, including tests and confidence intervals, follow from the general discussion in Section 6.3.2.

ii. Symmetry

If the distribution is symmetric, the pairs $(\hat{x}_{0.25}, \hat{x}_{0.75})$, $(\hat{x}_{0.10}, \hat{x}_{0.90})$, and $(\hat{x}_{0.05}, \hat{x}_{0.95})$ should be symmetric about $\hat{x}_{0.5}$. If they are not symmetric about $\hat{x}_{0.5}$, the direction of the asymmetry will tell us about the direction of the skew. For instance, if $(\hat{x}_{0.95} - \hat{x}_{0.5})$ exceeds $(\hat{x}_{0.5} - \hat{x}_{0.05})$, the right tail is longer than the left and we have a distribution with a positive skewness.

Comparison of the estimated mean and median will also give information about skewness. If the estimated mean is greater than the estimated median, the distribution has a positive skew; if the estimated mean is less than the estimated median, the distribution has a negative skew; and if the two are approximately equal, the distribution is symmetric. These measures are exploited in boxplots, which are presented in Section 7.1.

iii. Range and Spread

Combinations of the estimated quantiles, in particular the interquartile range (IQR),

$$IQR = \hat{x}_{0.75} - \hat{x}_{0.25} \tag{6.3.17}$$

which is used in boxplots, are often used to characterize the range or spread of the data. Such measures of range are robust alternatives to the sample variance. More information can be found in Mosteller and Tukey (1977, p. 207). The *IQR* values for several distributions are given in Table 6.1.2. Note, however, that while the sample *IQR* is straightforward to compute, the theoretical *IQR* is difficult to compute for many distributions.

iv. Heavy-Tailedness

How "heavy" the tails of a distribution are, or, equivalently, the relative likelihood of extreme values, can be examined by the magnitudes of the extreme quantiles, for instance,

$$\hat{x}_{0.05} \quad \text{and} \quad \hat{x}_{0.95}$$

Typically, heavy or light tails are measured relative to a Normal distribution, adjusted to have the same estimated mean and variance, as the distribution in question.

Now for any Normal(μ, σ) quantile $x_\alpha^{(\mu, \sigma)}$, we can take the corresponding Normal$(0, 1)$ quantile

$$x_\alpha^{(0, 1)}$$

from a standard table and adjust to obtain

$$x_\alpha^{(\mu, \sigma)} = \sigma x_\alpha^{(0, 1)} + \mu$$

When μ must be estimated by $\hat{\mu}$ and σ must be estimated by $\hat{\sigma}$, we make the adjustment

$$x_\alpha^{(\hat{\mu}, \hat{\sigma})} = \hat{\sigma} x_\alpha^{(0, 1)} + \hat{\mu}$$

This estimated adjusted Normal quantile can then be compared to the corresponding data quantile $\hat{x}_\alpha$.

If $\alpha > \frac{1}{2}$ and $\hat{x}_\alpha$ is greater than $x_\alpha^{(\hat{\mu}, \hat{\sigma})}$, the right tail of the observed data is heavier than a Normal tail. If $\hat{x}_\alpha < x_\alpha^{(\hat{\mu}, \hat{\sigma})}$, the tail is lighter. An analogous interpretation holds for the left tail if $\alpha < \frac{1}{2}$.

In addition to the pure descriptive uses of quantiles outlined above, we can compare estimated quantiles to theoretical quantiles for any distribution we suspect underlies our data. Usual practice is to estimate from the data any parameters in the suspected distribution (i.e., $\hat{\mu}$ and $\hat{\sigma}^2$ if we suspect the X_i are Normal); then the theoretical quantiles are computed from the suspected distribution with the estimated parameters substituted for the unknown real parameters. Finally, we compare the estimated quantiles to the theoretical ones. An example is given in Table 6.3.1. The idea of comparing estimated and theoretical quantiles can also be exploited graphically, as will be seen in Section 7.2.

EXAMPLE   **6.3.4 Interpretation of Quantiles**
From the comparison of the ecdf's of the single-line discipline and the multiple-line waiting times in Example 6.2.2, we saw that we may be able to distinguish between

Table 6.3.2   Estimated quantiles of the waiting time distributions

Discipline	Quantile level (α)								
	0.05	0.10	0.25	0.50	0.75	0.90	0.95	0.975	0.99
Single-line	0.00	0.58	2.94	8.11	18.30	28.18	35.10	43.830	54.00
Multiple-line	0.00	0.49	3.10	9.01	18.47	29.32	43.20	47.260	59.78

the two disciplines by considering the possibility of long waiting times. The extreme percentiles would give us some information, but they are "pinned" together from above by the condition $F_W(x) \leqslant 1$. Instead, we can turn to the extreme positive quantiles—for instance, 0.9, 0.95, 0.975 and 0.99.

Table 6.3.2 compares the estimated quantiles for these four values of α (based on 500 independent replications of the 5000th waiting times) and presents other commonly used quantiles. Note that the waiting time distributions are a combination of a discrete random variable at $W = 0$ and a continuous random variable for $W > 0$. This accounts for the zero values at $\alpha = 0.05$. (The mean for the single-line discipline is 12.1; for the multiple-line it is 13.0.)

In general the table indicates that waiting times are longer for the multiple-line discipline. This result must be tempered by the possibility that, because m is only 500, the difference is not statistically significant (see Exercise 6.10). Both disciplines produce extremely right-skewed waiting times with medians that are much lower than the respective means (single-line discipline median = 8.11; multiple-line median = 9.01).

The tabular quantities in Table 6.3.2 should be compared with Figure 6.2.6, where the quantiles may be read off by starting from a fixed value (α) on the vertical axis and finding the corresponding inverse function value on the horizontal axis. The table is given because of the difficulty of reading exact values from the graph.

6.3.4   Precision and Computational Considerations

Examination of Equation 6.3.14 for the variance of an order statistic estimate $\hat{x}_\alpha$ of a quantile x_α shows that the precision (standard deviation) of the estimate decreases as $1/m^{1/2}$. This is the same rate of convergence as for the estimate $\bar{X}$ of the mean, and it implies that to halve the precision of the estimate, the sample size must be quadrupled.

It is for this reason that attaining a given precision in a simulation may require very large sample sizes. This problem is aggravated in the tails of the distribution when $f(x_\alpha)$ is very small, making the constant in Equation 6.3.14 very large. Now the time to sort m objects is generally of the order of $m\log(m)$ and memory proportional to m is required to store sorted values in order to find a given order statistic. This can be a real problem in small computers.

Schemes for estimating quantiles that are linear in time and use little memory have been proposed but are beyond the scope of the present volume. A summary of some of the schemes is given in Heidelberger and Lewis (1984), and a very sophisticated scheme is given in Robinson (1975).

Exercises ^{C}**6.1.** Compute the coefficient of variation, the coefficient of skewness γ_1, and the coefficient of kurtosis γ_2, for the moment estimator $\tilde{\delta}$, given at Equation 6.0.3. Plot these, for $n = 20$, as a function of δ. What information can be obtained about the distribution of $\tilde{\delta}$ for $n = 20$ from these results?

What are the asymptotic results ($n \to \infty$) for γ_1 and γ_2 for $\tilde{\delta}$, and what does this say about the distribution of $\tilde{\delta}$ as $n \to \infty$?

6.2. Derive the result of Equation 6.0.6 and show that it has a unique solution $\tilde{\delta}$ for $-1 \leqslant \delta \leqslant 1$.

^{C}**6.3.** Simulate the distribution of the sample coefficient of variation $\tilde{C}(x) = S/\bar{X}$ from an exponential(λ) and tabulate its distribution for sample sizes $n = 5, 10, 20, 40, 50, 60, 70, 80$. To tabulate the distribution, estimate the quantiles x_α of the distribution for $\alpha = 0.05, 0.10, 0.25, 0.5, 0.75, 0.90, 0.95$ using $m = 100$ replications and the estimator of Equation 6.3.2. Tabulate also the estimated mean and standard deviation of $\tilde{C}(x)$.

Is there any reason for the distribution of $\tilde{C}(x)$ to be independent of the exponential parameter λ?

^{T}**6.4.** For the exponential(λ) distribution with $\mu = 1/\lambda$ the x_α quantiles, $0 < \alpha < 1$ are given by

$$x_\alpha = -\mu \ln(1 - \alpha)$$

This suggests a *parametric* estimator of x_α from a sample $X_1, \ldots, X_m$, namely,

$$\tilde{x}_\alpha = -\bar{X} \ln(1 - \alpha)$$

which will be a maximum likelihood estimate of x_α. (Why?)
   a. What is the distribution of $\tilde{x}_\alpha$?
   b. Give the mean, variance, coefficient of skewness, coefficient of kurtosis, and coefficient of variation of $\tilde{x}_\alpha$.
   c. Plot $\sqrt{m}\sigma(\tilde{x}_\alpha)$ for $0 < \alpha < 1$. What conclusions can you draw from this plot?
   d. Compute the asymptotic standard deviation of the usual nonparametric estimator x_α given in Equation 6.3.2, using Equation 6.3.14. The plot should actually be of

$$\sqrt{m}\sigma(\hat{x}_\alpha)$$

Compare the result to that obtained in part c to see what the parametric assumption has bought you in extra precision.

^{C}**6.5.** Generate a sample of size 20 from an exponential(1) distribution and compute and compare estimates of the three quartiles using (a) the maximum likelihood estimate in Exercise 6.4, (b) the definitions of Equations 6.3.5 and 6.3.6, and (c) the estimators $\hat{x}_\alpha$ and $\hat{x}_\alpha^*$ in Section 6.3.1.

6.6. Derive the result in Equation 6.3.7 and verify that it is correct for the maximum of the sample, which is $X_{(n)}$ and the minimum of the sample, which is $X_{(1)}$.

6.7. Prove that if x_α is the α quantile of the continuous random variable X, then the α quantile y_α of $y = \sigma X + \mu$ is just $y_\alpha = \sigma x_\alpha + \mu$.

^{T}**6.8.** Derive the result in Equation 6.2.17 for the covariance of the ecdf at two points x_1 and x_2.

^{T}**6.9.** Construct a formal 2×2 contingency table test of the hypothesis that the two probabilities of zero waiting time (0.098 and 0.066) in Example 6.2.2 are equal.

6.10. Use theoretical results on the distributions of sample quantiles to see, approximately, whether the sample quantiles in Table 6.3.2 are different by a statistically significant amount. You might want to assume, for convenience, that the underlying distribution, given that the waiting time is greater than zero, is exponential, though the data suggest a more skewed distribution.

References

Chambers, J. M., Cleveland, W. S., Kleiner, B., and Tukey, P. A. (1983). *Graphical Methods for Data Analysis*. Wadsworth & Brooks/Cole: Pacific Grove, CA.

Conover, W. J. (1980). *Practical Nonparametric Statistics*, 2nd ed. Wiley: New York.

Cramér, H. (1946). *Mathematical Methods of Statistics*. Princeton University Press: Princeton, NJ.

Heidelberger, P., and Lewis, P. A. W. (1984). Quantile Estimation in Dependent Sequences. *Operations Research, 32*(1), 185–209.

Hoaglin, D. C., Mosteller, F., and Tukey, J. W. (1983). *Understanding Robust and Exploratory Data Analysis*. Wiley: New York.

Johnson, N., and Kotz, S. (1970). *Distributions in Statistics*, Vols 1, *Discrete Distributions*, and 2, *Continuous Univariate Distributions*. Wiley: New York.

Lehmann, E. (1959). *Testing Statistical Hypotheses*. Wiley: New York.

Lilliefors, H. W. (1967). On the Kolmogorov–Smirnov Test for Normality with Mean and Variance Unknown. *Journal of the American Statistical Association, 62*, 399–402.

Lilliefors, H. W. (1969). On the Kolmogorov–Smirnov Test for the Exponential Distribution with Mean Unknown. *Journal of the American Statistical Association, 64*, 387–389.

Mood, A. M., Graybill, F. A., and Boes, D. C. (1974). *Introduction to the Theory of Statistics*. McGraw-Hill: New York.

Mosteller, F., and Tukey, J. W. (1977). *Data Analysis and Regression*. Addison-Wesley: Reading, MA.

Newell, G. F. (1982). *Applications of Queueing Theory*, 2nd ed. Chapman & Hall: London.

Robinson, D. W. (1975). Nonparametric Quantile Estimation Through Stochastic Approximation. Ph.D. thesis, Naval Postgraduate School, Monterey, CA.

Ross, S. M. (1980). *Introduction to Probability Models*, 2nd ed. Academic Press: New York.

Royston, J. P. (1982). An Extension of Shapiro and Wilk's *W*-Test for Normality to Large Samples. *Applied Statistics, 31*(2), 115–124.

Descriptions and Quantifications of Univariate Samples: Graphical Summaries

7.0 Introduction

This chapter continues the analysis started in Chapter 6 of samples of independent, univariate random variables. In Chapter 6 the focus was on summary statistics such as moments and quantiles, which provided numerical estimates to describe specific aspects of a distribution. Even the graphical technique illustrated in Section 6.2—plotting the ecdf—was associated with a test statistic (e.g., the Kolmogorov-Smirnov or Shapiro–Wilk statistic), which objectively "told" us whether our data came from a particular distribution.

In this chapter the goal remains the same: to describe the typical behavior of a data sample. The emphasis, however, is on exploratory graphical methods that require the analyst to draw *subjective* conclusions about the data. For example, Figure 7.0.1 shows the histogram (Section 7.1.4) of a waiting time distribution that is centered *around* 10 minutes and has a *long* right tail with *many* waiting times greater than 30 minutes. No parametric fit to any particular theoretical distribution is attempted or even desired.

This chapter starts with graphical representations of the probability density function (pdf), ranging from common boxplots and histograms to sophisticated kernel density estimates. Such qualitative plots can show many of the quantitative properties measured by the moment summaries and quantile estimates covered in Chapter 6: distributional center, spread, symmetry, and heaviness of tails. However, they do not directly give any probability information. Instead, the plots are often accompanied by the corresponding numerical summaries (see, e.g., the output from the program HISTPC, which is displayed in Figure 7.1.6) and by hypothesis test results for fitting a parametric family of distributions—for instance, if the Normal distribution is chosen from Tables 6.1.2 and 6.1.3, then the Shapiro–Wilk test statistic could be calculated, as illustrated later (bottom left-hand corner of Figure 7.1.6).

If we are interested in graphically comparing the distribution of our data with a specific theoretical distribution, the quantile (Q–Q) and percentile (P–P) plots discussed in Section 7.2 are very useful. A P–P plot is, essentially, an easier way to

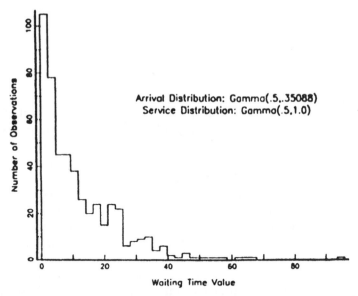

Figure 7.0.1 Histogram of 500 simulated waiting times for a three-server single-line discipline queue.

compare an ecdf with a theoretical cdf than the plots presented in Section 6.2. A $Q-Q$ plot contains exactly the same information as a $P-P$ plot, but emphasizes differences in the tails of the distributions. $P-P$ or $Q-Q$ plot analysis is based on the knowledge that the $P-P$ or $Q-Q$ plot will be linear if we guess the correct family of parametric distributions for our data. However, $Q-Q$ plots are broader in that their linearity is preserved if the distributional assumption is correct but the data are scaled and shifted (linearly transformed).

Looking ahead, in Chapter 8 we will bring together ideas from this chapter and from Chapter 6 as we consider the modeling, implementation, and analysis of statistical simulations. The emphasis will be on the analysis of the simulation output, as performed by the package SMTBPC. The output display in SMTBPC combines the ideas of boxplots (Section 7.1), numerical summaries (Section 6.1), and asymptotic expansions of an estimator's mean and variance (Section 6.1), to see changes in the distribution of a statistical estimator as sample size increases.

7.1 Numerical and Graphical Representations of the Probability Density Function

For many people who have been exposed only briefly to statistics, the bell-shaped Normal density curve rather than the Normal cumulative distribution function is the most memorable figure. Even for experienced statisticians, the probability density function (pdf) may be a more familiar clue to the distribution of a data set than the cumulative distribution function.

In areas such as reliability and survival analysis, both the pdf and the cdf are necessary to define the commonly used hazard rate function (failure rate, force of mortality) of a random variable,

$$h_X(x) = \frac{f_X(x)}{1 - F_X(x)} \tag{7.1.1}$$

The hazard function can act as a "signature" of the underlying data distribution and give information about system performance (Barlow and Proschan, 1975; Lawless, 1982; Cox and Oakes, 1984). As examples, the exponential(λ) distribution corresponds to a constant hazard rate λ, while the Weibull(k, β) distribution (Table 6.1.2) has a hazard rate that increases or decreases as a power of x (see Exercise 7.1).

In other contexts the value of the pdf may be required at a fixed value. This occurs in some rather sophisticated quantile estimation methods (see Heidelberger and Lewis, 1984) and in quantile estimation methods that use stochastic approximations (Lewis, 1971; Robinson, 1975). Also recall that in Section 6.3, an estimate of the density was needed for large-sample confidence intervals for quantiles.

For this variety of reasons, we concentrate in this section on how to estimate the pdf (or the probability function for discrete variables) and represent it graphically.

7.1.1 "Density" Estimation for Discrete Random Variables; The Probability Function

When the random variable X with cdf $F_X(x)$ can take only the discrete values $\{x_0 < x_1 < x_2 \cdots\}$, then the pdf of X can be given as the sequence

$$\{p_0, p_1, p_2, \ldots\} \tag{7.1.2}$$

where

$$p_0 = P(X = x_0) = F(x_0) \tag{7.1.3}$$

and

$$p_k = P(X = x_k) = F(x_k) - F(x_{k-1}) \qquad \text{for } k = 1, 2, \ldots \tag{7.1.4}$$

Notice that we have directed the notation toward a positive-valued random variable X, as will usually be the case. We could easily, however, extend the definition for X discrete valued on $\{\cdots x_{-1} < x_0 < x_1 \cdots\}$ with probability function $\{\ldots, p_{-1}, p_0, p_1, \ldots\}$. Most discrete data arise from counting processes, and the data are therefore inherently nonnegative.

EXAMPLE   7.1.1 A Poisson Random Variable
If X is a Poisson random variable with parameter λ: that is,

$$X \sim \text{Poisson}(\lambda)$$

then X can take values in the set $\{0, 1, 2, \ldots\}$ so that

$$x_k = k \qquad \text{for } k = 0, 1, 2, \ldots$$

By definition of the Poisson(λ) distribution (Exercise 4.4)

$$p_k = P(X = k) = \frac{\lambda^k e^{-\lambda}}{(k!)} \qquad k = 0, 1, 2, \ldots \tag{7.1.5}$$

Nonparametric estimation of the probability function $\{p_k\}$ of a discrete-valued random variable X is straightforward. If $I(\cdot)$ is the indicator function defined in Section 6.2.1 and $X_1, X_2, \ldots, X_m$ is an iid sample with $X_j, j = 1, 2, \ldots, m$, having probability function $\{p_k\}$, then we estimate p_k by

$$\hat{p}_k = \sum_{j=1}^{m} \frac{I(X_j; x_k) - I(X_j; x_{k-1})}{m} \qquad k = 0, 1, 2, \ldots \tag{7.1.6}$$

The mean,

$$E(\hat{p}_k) = p_k \tag{7.1.7}$$

variance,

$$\text{var}(\hat{p}_k) = \frac{p_k(1 - p_k)}{n} \tag{7.1.8}$$

and asymptotic Normality of the estimate $\hat{p}_k$ follow from the theory for iid Bernoulli random variables exactly as in Section 6.2.1.

Even though we are working with indicator functions in both cases, be careful not to confuse the notation in this chapter: that is,

$$p_k = P(X = x_k) \tag{7.1.9}$$

with that in Sections 6.2.1 to 6.2.3:

$$p = p(x) = P(X \leqslant x) \tag{7.1.10}$$

Note however the equivalences

$$p_k = p(x_k) - p(x_{k-1}) \tag{7.1.11}$$

and

$$\hat{p}_k = \hat{p}(x_k) - \hat{p}(x_{k-1}) \tag{7.1.12}$$

Of course, if X was known to be Poisson(λ), then the alternative parametric mle estimates $\tilde{\lambda} = \bar{X}$ and $\tilde{p}_k = \tilde{\lambda}^k \exp(-\tilde{\lambda})/(k!)$ could be used! However, the exact distribution of discrete random variables occurring in simulation studies (either in the input modeling phase or in the output analysis) is seldom known, so that nonparametric estimation is usually appropriate.

7.1.2   Density Estimation for Continuous Random Variables

A continuous random variable X can take on a continuum of possible values; hence the probability of any particular value is zero:

$$P(X = x) = 0 \qquad \text{for all } x$$

Therefore, the pdf, $f_X(x)$, of an (absolutely) continuous random variable is defined as the derivative of its cdf, $F_X(x)$, where we are assuming that the derivative exists for all x:

$$f_X(x) = \frac{dF_X(x)}{dx} = \frac{dP(X \leq x)}{dx}$$

Equivalently, we can start with any continuous function $f_X(x)$ where, to be a density, $f_X(x)$ must satisfy Equations 7.1.13 a and b.

$$f_X(x) \geq 0 \qquad \text{for all } x \tag{7.1.13a}$$

$$\int_{-\infty}^{\infty} f_X(x)\,dx = 1 \tag{7.1.13b}$$

We then integrate to get the corresponding cdf.

$$F_X(x) = \int_{-\infty}^{x} f_X(y)\,dy$$

EXAMPLE   **7.1.2 The Exponential Density**
We have already seen (Equation 3.2.3) that if X is an exponential(λ) random variable, then

$$F_X(x) = \begin{cases} 1 - e^{-\lambda x} & x \geq 0 \\ 0 & x < 0 \end{cases}$$

The pdf of X is given by

$$f_X(x) = \frac{dF_X(x)}{dx} = \begin{cases} \lambda e^{-\lambda x} & x \geq 0 \\ 0 & x < 0 \end{cases}$$

EXAMPLE   **7.1.3 The Normal Density**
If X is a Normal(0, 1) random variable, then the density is defined to be (Equation 4.2.5)

$$f_X(x) = (2\pi)^{-1/2} \exp\left(-\frac{x^2}{2}\right)$$

The cdf is expressed in terms of the density as

$$F_X(x) = \int_{-\infty}^{x} (2\pi)^{-1/2} \exp\left(-\frac{y^2}{2}\right) dy$$

This cumulative distribution function $F_X(x)$ does not have a closed-form solution but has been evaluated numerically and tabulated for various values of x (see, e.g., Conover, 1980, Table A1, p. 428). This function is also available in many computer systems in a library of standard functions, and is usually denoted $\Phi(x)$, as in Equation 3.4.5.

   Estimation of the density function of a continuous random variable is difficult and, unless the sample size is very large, the results can be very disappointing. For the methods that follow, we assume that we have an iid sample $X_1, X_2, \ldots, X_m$, where each X has pdf $f_X(x)$ and cdf $F_X(x)$. The ordered observed values of $X_1, X_2, \ldots, X_m$ will be denoted by $X_{(1)}, X_{(2)}, \ldots, X_{(m)}$.

   One simple and intuitive approach to density estimation is to try to approximate the derivative of the (continuous) cdf by using the (discontinuous) ecdf where we have actual data information. Such estimates are called nearest-neighbor estimates, and in what follows they are denoted by $\hat{f}_i(x)$, $i = 1, 2$, where the index i designates nearest-neighbor estimates of different types.

i. Nearest-Neighbor Estimates

For the first estimator, $\hat{f}_1(x)$, if x is a point that lies between $X_{(i-1)}$ and $X_{(i)}$, we set

$$\hat{f}_1(x) = \frac{\hat{F}_X(X_{(i)}) - \hat{F}_X(X_{(i-1)})}{\{X_{(i)} - X_{(i-1)}\}} = \frac{1/m}{\{X_{(i)} - X_{(i-1)}\}} \qquad (7.1.14)$$

where $\hat{F}_X(\cdot)$ is the ecdf defined in Section 6.2.3. Recall that $\hat{F}_X(\cdot)$ has jumps of height $1/m$ at each of the points $X_{(i)}$, so that $\hat{F}_X(X_{(i)}) - \hat{F}_X(X_{(i-1)}) = 1/m$. If $x < X_{(1)}$ or $x > X_{(m)}$, then we have no information about the density and set $\hat{f}_1(x) = 0$.

   When the sample size m is large or when the cdf can be safely assumed to be smooth, the variability of the estimated pdf can be reduced (at the cost of increased bias), by increasing the number of observed points that go into the estimate of the derivative. In general, we can define, for fixed k,

$$\hat{f}_2(x) = \frac{\hat{F}_X(X_{(i+k)}) - \hat{F}_X(X_{(i-k-1)})}{X_{(i+k)} - X_{(i-k-1)}} \qquad \text{if } X_{(i-1)} < x < X_{(i)}; i = (k+2), \ldots, (n-k)$$

$$= \frac{2k+1}{m\{X_{(i+k)} - X_{(i-k-1)}\}} \qquad (7.1.15)$$

where $\hat{f}_2(x)$ is defined to be 0 if $x < X_{(k+1)}$ or $x > X_{(m-k)}$. Notice that for $k = 0$, $\hat{f}_2(x) \equiv \hat{f}_1(x)$. The distributional properties of $\hat{f}_1(x)$ and $\hat{f}_2(x)$ depend on the underlying $f(x)$, on the sample size m, and on k. Unfortunately there is little theory to guide us on an optimal choice of k other than the general principle that larger values of k will reduce variance at the cost of increased bias. Note also that unless Equation 7.1.15 is modified for extreme values of x, large values of k will force us to estimate the density to be zero for a broad range of x: that is, for $x < X_{(k+1)}$ and $x > X_{(m-k)}$. Hence, nearest-neighbor estimates are recommended only if simplicity is an important factor—for instance, if preliminary hand calculations are being done.

EXAMPLE   **7.1.4 Comparison of Nearest-Neighbor Density Estimates**
We now return to the problem introduced in Example 6.0.1: estimating the polarization parameter δ in a muon decay distribution. Through various examples in Chapter 6 (6.1.4, 6.2.4, and 6.2.6), we saw that when the true value of δ is 0.8, the moment estimator $\tilde{\delta}$ has a distribution that is close to (but not exactly) Normal(0.797, 0.343).

Figure 7.1.1 shows the nearest-neighbor density estimate $\hat{f}(x)$ for the 1000 moment estimates of $\tilde{\delta}$ generated in Example 6.1.4 under the model with $\delta = 0.8$. The graph clearly shows the high variability of $\hat{f}_1(x)$ and bears little resemblance to a Normal(0.797, 0.343) density.

If we try to reduce the variability of the density estimate by using a wider span of data, say $\hat{f}_2(x)$ with $k = 49$, the graph in Figure 7.1.2 results. Although the curve is much smoother, it is not smooth, and a great deal of the density is estimated to be zero.

ii. Kernel Density Estimates

When a more sophisticated density estimate is desired, we can choose from a group called *kernel density estimates*, whose statistical properties are better understood than the properties of the estimates in the previous subsection.

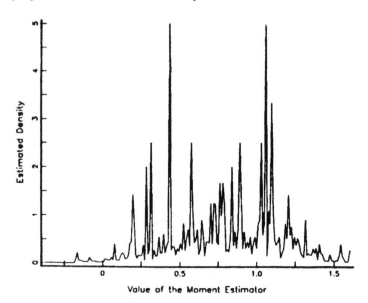

Figure 7.1.1 Estimated density of the moment estimator of the polarization parameter (true value: $\delta = 0.8$). First nearest neighbor ($k = 0$). Sample size is 1000.

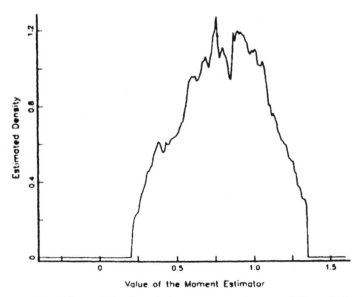

Figure 7.1.2 Estimated density of the moment estimator of the polarization parameter (true value: $\delta = 0.8$). 50th nearest neighbor. Sample size is 1000.

Kernel density estimates (Rosenblatt, 1956, 1971; Parzen, 1962) started with the following analogue of $\hat{f}_1(x)$:

$$\hat{f}_3(x) = \frac{\hat{F}_X(x + b) - \hat{F}_X(x - b)}{2b} \tag{7.1.16}$$

where $\hat{F}_X$ is the ecdf and b is, for given sample size, a fixed smoothing parameter (sometimes called bandwidth) that must be chosen by the user. Since we can represent $\hat{F}_X(x + b) - \hat{F}_X(x - b)$ as

$$\hat{F}_X(x + b) - \hat{F}_X(x - b) = \sum_{i=1}^{m} \frac{I(|X_i - x|; b)}{m}$$

we can use the properties of indicator functions to calculate

$$E(\hat{f}_3(x)) = \frac{F_X(x + b) - F_X(x - b)}{2b} \tag{7.1.17}$$

and

$$\mathrm{var}(\hat{f}_3(x)) = \frac{[F_X(x + b) - F_X(x - b)]\{1 - [F_X(x + b) - F_X(x - b)]\}}{4mb^2} \tag{7.1.18}$$

In this case the tradeoff between bias and variance is explicit:

If we want the estimate to be unbiased, then, in general, we must let b get small, since

$$\lim_{b \to 0} E(\hat{f}_3(x)) = \lim_{b \to 0} \frac{F_X(x + b) - F_X(x - b)}{2b} = \frac{dF_X(x)}{dx} = f_X(x) \qquad (7.1.19)$$

However, $\text{var}(\hat{f}_3(x))$ increases at rate $1/b$ as b goes to zero. Hence, before we can let b get too small, we must increase the sample size m so that $mb \to \infty$ as $m \to \infty$ and $b \to 0$. This will ensure that $E(\hat{f}_3(x)) \to f_X(x)$ and $\text{var}(\hat{f}_3(x)) \to 0$. We could, for instance, let $b = Cm^{-2/3}$, where C is a finite, positive constant. For a derivation of these results, see Rosenblatt (1971).

If we define

$$K_3(y) = \begin{cases} \frac{1}{2} & \text{if } |y| \leq 1 \\ 0 & \text{if } |y| > 1 \end{cases} \qquad (7.1.20)$$

the estimate $\hat{f}_3(x)$ can be written in an equivalent form:

$$\hat{f}_3(x) = \frac{1}{mb} \sum_{i=1}^{m} K_3\left(\frac{X_i - x}{b}\right) \qquad (7.1.21)$$

The function $K_3(y)$ is called a *kernel*, in particular the *boxcar kernel*, and $\hat{f}_3(x)$ can be thought of as a weighted average of the observed data points around x, where the weighting is determined by K_3. This idea can be generalized by allowing

$$\hat{f}(x) = \frac{1}{mb} \sum_{i=1}^{m} K\left(\frac{X_i - x}{b}\right) \qquad (7.1.22)$$

where $K(y)$ is any function that satisfies $K(y) \geq 0$ for all y and $\int_{-\infty}^{\infty} K(y)\,dy = 1$. Along with $K_3(y)$ above, the following kernels are generally accepted as useful:

$$K_4(y) = (2\pi)^{-1/2} \exp\left(-\frac{y^2}{2}\right) \qquad \text{for } |y| < \infty \qquad (7.1.23)$$

This is called a *Normal kernel* and is widely used when computation is not a problem.

$$K_5(y) = \tfrac{15}{16}(1 - y^2)^2 \qquad \text{for } |y| \leq 1 \qquad (7.1.24)$$

This is called a *quartic kernel*. It is smooth like the Normal kernel but easier to compute because of the restriction $|y| \leq 1$.

$$K_6(y) = \tfrac{3}{4}(1 - y^2) \qquad \text{for } |y| \leq 1 \qquad (7.1.25)$$

This is called the *Epanechnikov kernel* and is optimal for a certain measurement of error (Epanechnikov, 1969). The kernel $K_5(y)$, however, is very close to optimal, and its smoothness at $y = 1$ and -1 may be useful when it comes to some methods for choosing b (Tapia and Thompson, 1978).

Under certain conditions on b, $K(y)$, and $f_X(x)$—in particular that $b \to 0$ as $m \to \infty$ and that $mb \to \infty$ as $m \to \infty$—it has been shown that kernel estimates are asymptotically unbiased and consistent (Parzen, 1962). In particular, if $f_X(x)$ is continuous, then $K_3(y)$, $K_4(y)$, $K_5(y)$, and $K_6(y)$ all provide asymptotically unbiased and consistent estimates of $f_X(x)$.

Choice of the smoothing parameter b, usually called the bandwidth parameter, depends on the sample size m, on the underlying density $f_X(x)$, and on the kernel $K(y)$. A gross starting value for b can be given by

$$b = 2Sm^{-1/5} \tag{7.1.26}$$

where S is the usual standard deviation of the data sample (other robust measures of standard deviation can be substituted). See Section 7.1.3, though, for refinements of this formula and more accurate methods for finding an optimal value of b.

EXAMPLE   **7.1.5 Graphical Comparison of Density Estimates**
In Example 7.1.4, the nearest-neighbor density estimates gave us a quick but unsatisfactory look at the density of 1000 moment estimates of the polarization parameter δ. Refined estimates of the density of that same data are displayed in Figures 7.1.3 and 7.1.4.

In those figures, the four kernel estimators—boxcar (K_3), Normal (K_4), quartic (K_5), and Epanechnikov (K_6)—can be compared to each other and to what we

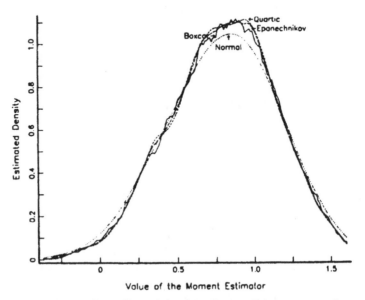

Figure 7.1.3 Comparison of kernel density estimates of the moment estimator of the polarization parameter (true value: $\delta = 0.8$). Here, $b = 0.15$, and sample size is 1000.

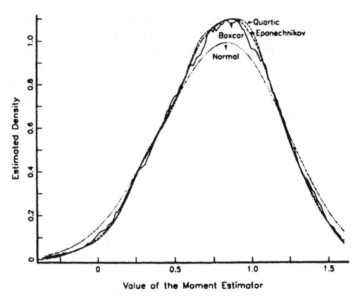

Figure 7.1.4 Comparison of kernel density estimates of the moment estimator of the polarization parameter (true value: $\delta = 0.8$). Here $b = 0.20$ and sample size is 1000.

expect to see [i.e., an approximately Normal $(0.797, 0.343)$ distribution for the $\tilde{\delta}$]. The bandwidth used to produce Figure 7.1.3, $b = 0.15$, is close to optimal for the boxcar and Normal kernels, while the bandwidth for Figure 7.1.4, $b = 0.20$, is close to optimal for the quartic and Epanechnikov kernels (both are close to the suggested $2Sm^{-1/5} = 0.172$).

Although the boxcar kernel estimate is more variable than the other density estimates and exhibits a ripple effect, all four estimates are very similar in shape in Figures 7.1.3 and 7.1.4. The curves are reminiscent of a Normal density but show a slightly longer left tail than the right (reflecting the measured sample skewness of -0.207 in Example 6.1.4).

There are many other competing methods to estimate densities (Tapia and Thompson, 1978), including

- Variable window size estimates (Wegman, 1970)
- Generalized kernellike estimates (Watson and Leadbetter, 1963)
- Smoothed functional or spline estimates (Wahba, 1975)

Theoretical comparisons among all methods are not possible, and exhaustive simulation results are not available. One simulation study that was published (Scott and Factor, 1981) shows that the Normal kernel estimate, with b intelligently chosen, performs comparatively well. It would be expected that other kernels, particularly K_5 and K_6, would perform similarly well.

Hence, we will restrict our attention to the density estimates already presented and proceed to the question of choosing the smoothing parameter b.

7.1.3   Choosing the Smoothing Parameter in a Density Estimate

We consider here two ways to choose the value of the smoothing parameter b for any general density estimate, $\hat{f}(x; b)$. Notice that the notation has changed to emphasize the dependence of the estimate on the smoothing parameter b. Again, other methods to choose b exist (Tapia and Thompson, 1978) but are left as outside reading to avoid confusion.

i. Interactive Selection

The most direct way to choose b is to produce plots of $\hat{f}(x; b)$ for various values of b and pick, by eye, the plot that follows the data (i.e., has low bias) but still remains smooth (i.e., has low variance). This rule can be implemented easily by anyone, especially if interactive graphics are available, but the final decision is subjective and can be guided only by experience.

As an aid to the search, the starting value $b = 2Sm^{-1/5}$ was suggested in the preceding section. This choice was based on the fact that the integrated mean square error of the density estimate can be minimized by choosing

$$b = \alpha(K)\beta(f)m^{-1/5} \tag{7.1.27}$$

where $\alpha(K)$ depends on the kernel and $\beta(f)$ depends on the underlying density of the data (see Tapia and Thompson, 1978, for definitions and full account of the result). For a given set of data, the density, $f_X(x)$, and therefore $\beta(f)$ are unknown, but we can solve for $\beta(f)$ for "typical" distributions and use these values as guides. In particular, for:

$$X \sim \text{Normal}(\mu, \sigma) \qquad \beta(f) = 1.364\sigma$$
$$X \sim \text{exponential}(\lambda) \qquad \beta(f) = 1.149/\lambda$$

Similarly, for the kernels considered here, we can solve for $\alpha(K)$:

$$\alpha(K_3) = 1.3510$$
$$\alpha(K_4) = 0.7764$$
$$\alpha(K_5) = 2.0362$$
$$\alpha(K_6) = 1.7188$$

This leads us to Table 7.1.1, which gives "optimal" bandwidths. Taking 2 to be a

Table 7.1.1   Optimal kernel density estimate bandwidths, b (to be multiplied by $m^{-1/5}$)

	Kernel			
Data distribution	K_3 (boxcar)	K_4 (Normal)	K_5 (quartic)	K_6 (Epanechnikov)
Normal (μ, σ)	$1.84\,\sigma$	$1.06\,\sigma$	$2.78\,\sigma$	$2.34\,\sigma$
Exponential (λ)	$1.55/\lambda$	$0.89/\lambda$	$2.34/\lambda$	$1.97/\lambda$

"typical" coefficient and noting that S estimates σ in the Normal case and $(1/\lambda)$ in the exponential case, we can reduce to the rule $b = 2Sm^{-1/5}$. When working with a specific kernel, however, use the correct value of $\sigma(K)$ for that kernel.

EXAMPLE   **7.1.6   Effects of Bandwidth on the Epanechnikov Kernel**

To see what effect changes in the bandwidth b have on the resulting density estimate, the Epanechnikov kernel was used to estimate the density of the moment estimator of the polarization parameter δ. The same 1000 estimates $\tilde{\delta}$ from the model with $\delta = 0.8$ that were used in Examples 7.1.4 and 7.1.5 went into the density estimates displayed in Figure 7.1.5, which shows high variability when the bandwidth is small ($b = 0.05$), reasonable smoothness at $b = 0.10$, and very similar, smooth estimates at $b = 0.2$ and 0.3. Notice also the "node" on the $b = 0.05$ and 0.10 densities before the point $\tilde{\delta} = 0.5$. This is not consistent with a Normal density and may signify too small a bandwidth; it may also be a real effect that warrants further investigation (see Example 7.1.7).

As has been noted before, the data are approximately Normal with $S = 0.343$, so that an optimal bandwidth should be

$$b = [(1.3643)(0.343)][1.7188][1000]^{-1/5}$$
$$= 0.202$$

Keep in mind, though, that this "optimality" is asymptotic and applies to one particular measure of error; it does not replace a visual search for hidden structure.

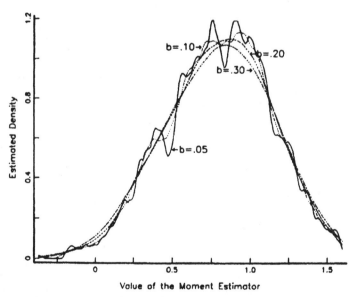

Figure 7.1.5   Effects of bandwidth (b) on the density estimate of the moment estimator of the polarization parameter δ, when using an Epanechnikov kernel (true value: $\delta = 0.8$). Sample size is 1000.

ii. Modified Maximum Likelihood

In this procedure we choose a value of b that maximizes the expression

$$L'(b) = \prod_{i=1}^{m} \hat{f}_i(X_i; b) \tag{7.1.28}$$

where we define

$$\hat{f}_i(X_i; b) = \frac{1}{(m-1)b} \sum_{\substack{j=1 \\ j \neq i}}^{m} K\left(\frac{X_i - X_j}{b}\right) \tag{7.1.29}$$

for our prechosen kernel $K(y)$.

The expression $\hat{f}_i(X_i; b)$ is just the estimated likelihood of the point X_i, if $\hat{f}$ had been estimated without X_i in the data sample. Intuitively, if b is properly chosen, the estimated density $\hat{f}_i$ should be able to accommodate the existence of X_i, and this accommodation would be reflected in a high value of $\hat{f}_i(X_i; b)$.

This procedure, due to Robertson (1967) and Van Ryzin (1973), requires a great deal of computation time as compared to interactive selection and more computation time than other competitors (Scott and Factor, 1981). It does, however, give a fixed result that is usually superior to that of other methods when no outliers are present in the data (Scott and Factor, 1981). Remember, however, that in simulated output there may be a long-tailed distribution but there are never outliers in the sense of false data values. Thus this procedure could be an attractive one for density estimation.

7.1.4 Graphical Representations of the Density of a Continuous Random Variable: Graphing the Smoothed Density Estimate, Histograms, Boxplots, and Stem-Leaf Plots

If values of $\hat{f}(x)$ for one of the density estimators suggested in Section 7.1.2 (or a good competitor) are available for all values of x, or at least a fine grid of values of x, then a plot of $\hat{f}(x)$ versus x is the natural graphical representation of the probability density function of a continuous random variable.

Frequently, however, we will want a quicker, simpler method—perhaps one that can be done by hand. In such cases, we can consider a histogram, a stem-leaf plot, or a boxplot of our data sample. As before, our random sample will be denoted $\{X_1, X_2, \ldots, X_m\}$ with ordered values $\{X_{(1)}, X_{(2)}, \ldots, X_{(m)}\}$.

i. The Histogram

The histogram is the most familiar density representation. It requires setting finite boundaries $a_1 \leqslant X_{(1)}$ and $a_2 \geqslant X_{(m)}$ and then dividing that range, $[a_1, a_2]$, into k equal-sized bins: $[a_1, a_1 + b)$, $[a_1 + b, a_1 + 2b)$, $\ldots$, $[a_2 - b, a_2]$, where $b = (a_2 - a_1)/k$ and k is an integer chosen by the user.

The choice of b (or equivalently k) is analogous to the choice of the smoothing parameter (bandwidth parameter) b in the continuous density estimators. For histograms, the choice is usually made through interactive inspection, although having more than five observations in a bin is a common rule of thumb. There is also recent work on this problem by Scott (1979) and Freedman and Diaconis (1981a,b) as summarized in Hoaglin, Mosteller, and Tukey (1983, pp. 22–29).

If the same approach is taken for histograms as for density estimates (i.e., attempting to minimize integrated mean square error), the optimal bandwidth is given by

$$b = \left[6 \bigg/ \int_{\infty}^{\infty} f_X'(x)^2 dx \right]^{1/3} m^{-1/3} \tag{7.1.30}$$

where $f_X'(x)$ is the derivative of the underlying density of the data (Scott, 1979). In specific cases, we have:

- If $X \sim \text{Normal}(\mu, \sigma)$ then $b = 3.49 \sigma m^{-1/3}$.
- If $X \sim \text{exponential}(\lambda)$ then $b = (2.28/\lambda) m^{-1/3}$.

The terms σ and $(1/\lambda)$ can be estimated as the standard deviation of the data (commonly by the sample standard deviation S).

Alternative measures of error lead to the rules

$$b = 1.66 \sigma (\ln m)^{1/3} m^{-1/3}$$

or

$$b = 2(IQR) m^{-1/3}$$

for Normal(μ, σ) data (Freedman and Diaconis, 1981a,b), where the interquartile range is defined at Equation 6.3.17.

Once the range has been divided, we simply count the number of observed data points that fall into each bin and plot that number as the height of a bar centered in that bin. Mathematically,

histogram(x)

$$= \begin{cases} 0 & \text{if } x < a_1 \text{ or } x > a_2 \\ \sum_{j=1}^{m} I[X_j; a_1 + (i+1)b] - I[X_j; a_1 + ib] & \text{if } x \in [a_1 + ib, a_1 + (i+1)b); \\ & \text{for } i = 0, 1, 2, \ldots, k-1 \end{cases} \tag{7.1.31}$$

Scaling of the histogram is often done by dividing the height of each bar by the sample size m. In this case, the height of each scaled bar will correspond to the empirical probability of the random variable falling within the given bin range. An alternative scaling scheme requires the height of each bar to be divided by bm. In this way the area under the histogram integrates to one, and the histogram actually estimates the density $f_X(x)$.

An example of a very complete histogram program is given in the software supplement to this book (*Enhanced Simulation and Statistics Package*) as the program HISTPC. Its features include:

• Numerous summary statistics (see Section 6.1) to quantify the data in the histogram.

• Printed counts of the number of observations in each bin. This will be important when the counts are disparate and some of the bars (scaled by *m*) become very small. Also, because of the limited resolution of the printer output, one symbol on the printer may represent a wide range of possible numbers of observations if the sample size is large.

• Special bars of symbols overlayed in the histogram to mark the locations of the median and the upper and lower quartiles.

• Portability. The histogram is printed on a standard printer and requires no special graphics output hardware.

• An optional overlayed "smooth" of the density. This "smooth" density is just a kernel estimate using a triangular kernel:

$$K_7(y) = \begin{cases} 1 - |y| & \text{if } |y| < 1 \\ 0 & \text{otherwise} \end{cases} \tag{7.1.32}$$

Because of the printer resolution, the actual kernel used has little effect on the picture seen, so a computationally simple kernel was chosen.

EXAMPLE **7.1.7 Comparison of the Histogram Densities of the Moment and mle Estimates of the Polarization Parameter**

The program HISTPC was used to produce the histograms in Figures 7.1.6 and 7.1.7, comparing the densities of 1000 moment and mle estimates of the polarization parameter δ from a muon decay distribution (see Examples 6.0.1 and 6.1.4). Each (moment and mle) estimate was computed from $n = 20$ simulated points from a decay distribution with true parameter value $\delta = 0.8$.

The graphics are supported by extensive numerical summaries (you can match some of these to the results in Example 6.1.4), by boxplots, to be discussed shortly, and by the Shapiro–Wilk test for Normality, described in Chapter 6.

Figure 7.1.6 can be compared to the smooth density estimates in Examples 7.1.4, 7.1.5, and 7.1.6. Notice the generally Normal shape, slight left skew, and "node" at about $\tilde{\delta} = 0.4$. Since we would not expect the estimation method to produce a bimodal distribution, we will assume that although the node is supported by real data, it represents variability in our histogram estimate and is not a secondary node in the distribution (recall that this point was also ambiguous with the Epanechnikov estimates, which changed with bandwidth).

Figure 7.1.7 is much more striking in that it dramatizes the truncation rule in our mle estimation routine (i.e., if $\hat{\delta} > 1$, set $\hat{\delta} = 1$). There is certainly no asymptotic Normality, and it would probably be better not to even try to identify any particular

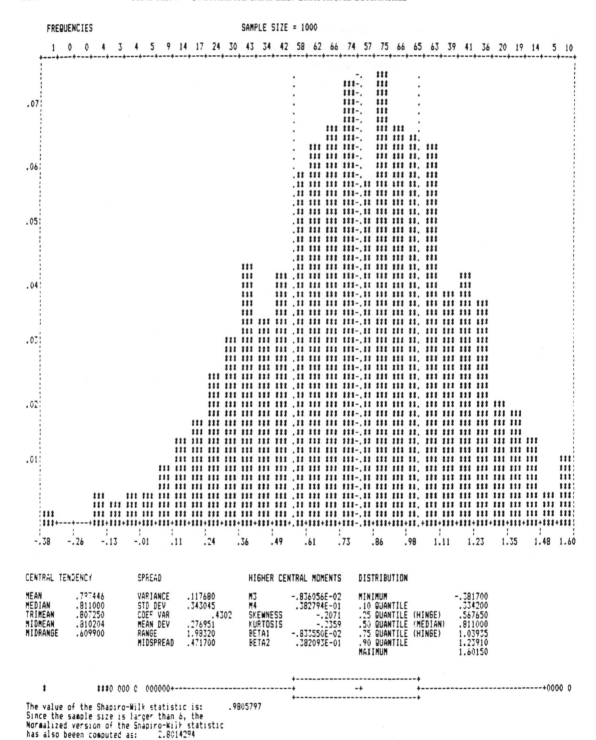

***Figure* 7.1.6**   Histogram of 1000 moment estimates $\tilde{\delta}$ of the polarization parameter from a muon decay distribution (true value: $\delta = 0.8$).

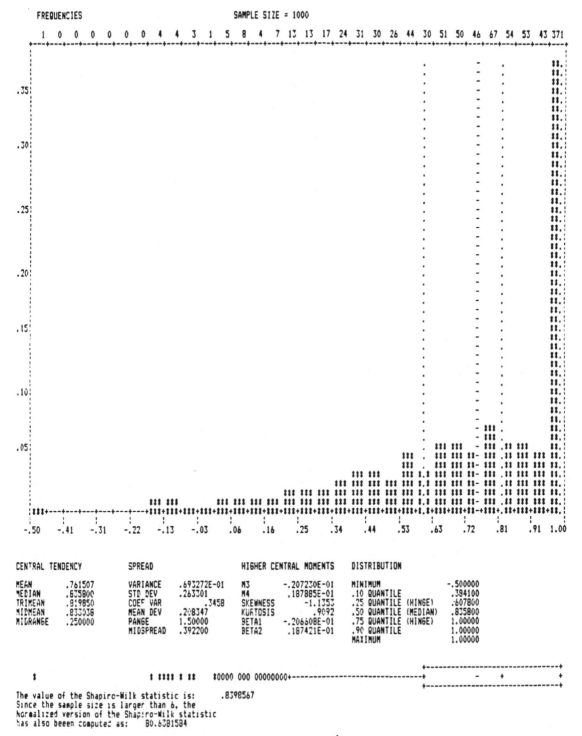

***Figure* 7.1.7** Histogram of 1000 mle estimates $\hat{\delta}$ of the polarization parameter from a muon decay distribution (true value: $\delta = 0.8$).

distribution. Rather, one would use the histogram to see the discrete component at $\hat{\delta} = 1$, the outlying point $\hat{\delta} = -0.50$, and the general shape of the remaining data.

ii. The Stem-Leaf Plot

The *stem-leaf plot* is a more detailed version of the basic histogram, usually used to examine small data sets by hand. As in the histogram, the range of the data is divided into bins of equal width. Instead of counting the observations in a given bin, though, we write, in the vertical bar, the actual data values (or at least the next significant digit of each data value). The outline of the bar itself can be omitted, since it is filled with numbers. The final stem-leaf plot is then usually presented with the "bars" running horizontally. Examples and details can be found in Hoaglin, Mosteller, and Tukey (1983, Chap. 1).

iii. The Boxplot

The *boxplot* is an (almost) one-dimensional picture that concentrates on quantiles to describe the distribution. The one-dimensional property masks some details of the distribution, but makes it very convenient to compare boxplots from many different samples—we simply plot all the boxplots one next to the other, using the same scale.

For illustration, assume that the sample size m is such that $(0.75m)$ and $(0.25m)$ are integers. One version of the boxplot is then constructed as follows:

1. Draw a rectangle with end points at the order statistics $X_{(0.75m)}$ and $X_{(0.25m)}$. The length of the rectangle

$$X_{(0.75m)} - X_{(0.25m)} = IQR$$

is usually called the *interquartile range* (Equation 6.3.17) (or interquartile distance).

2. Plot $-$ at the *mean* of the data sample and $+$ at the *median*. Notice that the median will always fall within the rectangle but the mean, for very skewed data, may not.

3. From each end of the boxplot, draw a line that terminates with the symbol $+$ at the largest magnitude data point within distance IQR of the end of the boxplot. This line is sometimes called a *whisker*.

4. If there are any data points greater than $1\,IQR$ away from either end of the boxplot but less than $1.5IQR$, plot them with the symbol O. These are called *outliers*—for simulations, this means that they are out in the tails of the distribution, not that they are false values.

5. If there are any data points greater than $1.5IQR$ away from the rectangle end points, plot them with the symbol *. These are called *far outliers*.

Boxplots show very quickly if the data are skewed and also if the tails are long. As a point of reference, we would expect a total of 5 outliers plus far outliers for every 100 Normal(μ, σ) points plotted in a boxplot. More details can be found in Tukey (1977). The examples of boxplots shown beneath the HISTPC outputs in

Figures 7.1.6 and 7.1.7 reveal the general shapes of the two densities, but be careful about interpreting these (or any automated boxplot printouts) in detail. In particular, if you try to count the number of outliers and far outliers to check for Normality in Figure 7.1.6, you will miss the fact that many of the symbols represent multiple values, overlaid because of printer precision (check the bin counts at the top of the histogram!). A more imaginative use of boxplots is presented in Section 8.2.1.

At least one of the four methods presented in this section—the smoothed density $\hat{f}(x)$, the histogram, the stem-leaf plot, or the boxplot—should be used to look at the probability density function of any data set. (Note that HISTPC gives a histogram, a smoothed density estimate, and a boxplot.) The effectiveness of these methods, though, depends highly on the size of the sample we have to work with. In simulation, we normally have some control of the sample size, unless we are checking for compatibility with input or output from a physical system. Hence, we should be aware of the problem and adjust for it. In Section 7.1.6, we will look at some examples of the effects of sample size on these methods for looking at densities.

7.1.5   Graphical Representations of the Probability Function of a Discrete Random Variable

When displaying the density of a discrete random variable, any of the methods mentioned in Section 7.1.4 can be used. It is possible, however, that all those methods can hide the discretization of the data. In addition, if the bins are centered at the points $\{x_0, x_1, x_2, \dots\}$, then the plot of $\{\hat{p}_0, \hat{p}_1, \hat{p}_2, \dots\}$, the histogram, and the stem-leaf plot are essentially identical.

An alternative approach is provided by a method such as that implemented by the program LISTPC in the software supplement. Through LISTPC we get a listing of all the values taken by the data, along with a count of the number of times that each given value appears. This program can also be useful in looking at allegedly continuous data that may have been artificially discretized—for instance, by rounding. Again, this problem will not usually come up with simulation output but may occur when inspecting input or output from a physical system. Problems with small sample sizes parallel those for estimating the pdf of a continuous random variable.

EXAMPLE   **7.1.8. Probability Function Estimates for Discrete Random Variables**
To validate the simulation model that produced the histogram of waiting times in Figure 7.0.1, we are given 100 real, observed waiting times. We expect the distribution to be approximately exponential, and calculations show that the sample standard deviation is $S = 11.6$. Hence a reasonable bin width is

$$b = \left(\frac{2.28}{\lambda}\right) m^{-1/3} = 2.28(11.6)(100^{-1/3})$$
$$= 5.698$$

which leads to the histogram in Figure 7.1.8. Allowing for differences in sample size and bandwidth, this is a reasonable match to the histogram in Figure 7.0.1, and we may wish to use formal comparison methods.

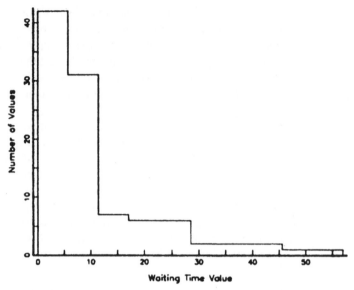

Figure 7.1.8 Histogram for 100 real, observed waiting times. The bin width is $5.698 = b$.

SERIAL NUMBER	ORDERED DATA	FREQUENCIES	PERCENT	PROBABILITY GRAPH
1	.00000000E+00	17	.17000	‡‡‡
18	.10000000E+01	5	.05000	‡‡‡‡‡‡‡‡‡‡‡‡
23	.20000000E+01	7	.07000	‡‡‡‡‡‡‡‡‡‡‡‡‡‡‡‡‡
30	.30000000E+01	9	.09000	‡‡‡‡‡‡‡‡‡‡‡‡‡‡‡‡‡‡‡‡‡‡
39	.40000000E+01	3	.03000	‡‡‡‡‡‡‡
42	.50000000E+01	1	.01000	‡‡
43	.60000000E+01	4	.04000	‡‡‡‡‡‡‡‡‡
47	.70000000E+01	9	.09000	‡‡‡‡‡‡‡‡‡‡‡‡‡‡‡‡‡‡‡‡‡‡
56	.80000000E+01	5	.05000	‡‡‡‡‡‡‡‡‡‡‡‡
61	.90000000E+01	3	.03000	‡‡‡‡‡‡‡
64	.10000000E+02	8	.08000	‡‡‡‡‡‡‡‡‡‡‡‡‡‡‡‡‡‡‡
72	.11000000E+02	2	.02000	‡‡‡‡‡
74	.13000000E+02	1	.01000	‡‡
75	.14000000E+02	2	.02000	‡‡‡‡‡
77	.16000000E+02	2	.02000	‡‡‡‡‡
79	.17000000E+02	2	.02000	‡‡‡‡‡
81	.18000000E+02	1	.01000	‡‡
82	.19000000E+02	2	.02000	‡‡‡‡‡
84	.21000000E+02	2	.02000	‡‡‡‡‡
86	.22000000E+02	1	.01000	‡‡
87	.23000000E+02	1	.01000	‡‡
88	.24000000E+02	1	.01000	‡‡
89	.25000000E+02	1	.01000	‡‡
90	.27000000E+02	2	.02000	‡‡‡‡‡
92	.28000000E+02	1	.01000	‡‡
93	.29000000E+02	1	.01000	‡‡
94	.31000000E+02	1	.01000	‡‡
95	.36000000E+02	1	.01000	‡‡
96	.38000000E+02	1	.01000	‡‡
97	.43000000E+02	1	.01000	‡‡
98	.44000000E+02	1	.01000	‡‡
99	.47000000E+02	1	.01000	‡‡
100	.54000000E+02	1	.01000	‡‡

Figure 7.1.9 LISTPC output for real, observed waiting times.

Unfortunately, as the LISTPC output in Figure 7.1.9 shows, the observed data were preprocessed or truncated to the nearest integer before being given to us. The distribution, if it was exponential before, is now geometric, and formal tests comparing the exponential distribution to the simulation results might produce misleading results.

7.1.6 Effects of Sample Size

It has been shown (Tapia and Thompson, 1978) that if we let $\hat{f}(x)$ represent any of the kernel estimators mentioned, the integrated mean square error

$$\int_{-\infty}^{\infty} E[\hat{f}(x) - f_X(x)]^2 dx$$

decreases as $m^{-4/5}$ when b is optimally chosen via Equation 7.1.27. Similarly, the histogram estimate of the density (histogram$(x)/bm$) has an integrated mean square error that decreases as $m^{-2/3}$ when b is optimally chosen via Equation 7.1.30. Although these results ensure that increasing the sample size will eventually eliminate the estimation error, graphics are more useful for judging performance in small samples.

Now let us compare the histogram of 50 iid simulated Normal(0, 1) random variables in Figure 7.1.10a to the histogram of 50 iid simulated Laplace(0, (0.5)^{1/2}) random variables in Figure 7.1.10b. (Notice that both distributions have mean 0 and variance 1.) Even with the triangular kernel density estimate superimposed, there is little in the pictures to help us distinguish between the two distributions.

Figures 7.1.11 and 7.1.12 are analogous to Figure 7.1.10 except that the sample sizes have been increased to 500 and 15,000, respectively. With a sample of size 15,000 we can clearly distinguish Normal data from Laplace data (when the scales are the same and the histograms are side by side). At sample size 500, the long tails and peaked body of the Laplace density are already becoming distinctive, and there is a marked difference from the Normal density. However, without numerical summaries and comparative figures, would you have been able to decide that Figure 7.1.11b represented Laplace data? Experience will make such decisions easier.

7.2 Alternative Graphical Methods for Exploring Distributions

The probability density function plot, the histogram, the stem-leaf plot, and the boxplot of Section 7.1 are all useful exploratory, graphical tools for looking at and understanding the distribution of a data sample. The Kolmogorov–Smirnov and Shapiro–Wilk tests of Section 6.2, on the other hand, are common confirmatory techniques (see Tukey, 1977, for a discussion of exploratory versus confirmatory statistics) to test whether the data follow a prespecified distribution. In this section we look at two graphical procedures for determining whether the data follow a prespecified distribution, namely quantile $(Q-Q)$ plots and percentile $(P-P)$ plots. These techniques fall somewhere between the exploratory and confirmatory approaches.

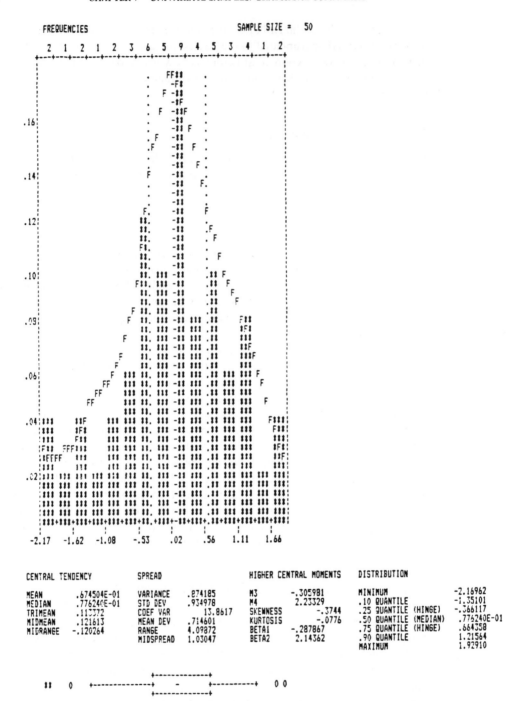

Figure 7.1.10a Effect of sample size: histogram of 50 iid simulated Normal(0, 1) random variables, produced by HISTPC with automatic bin width determination. Overlaid kernel density estimate uses triangular kernel.

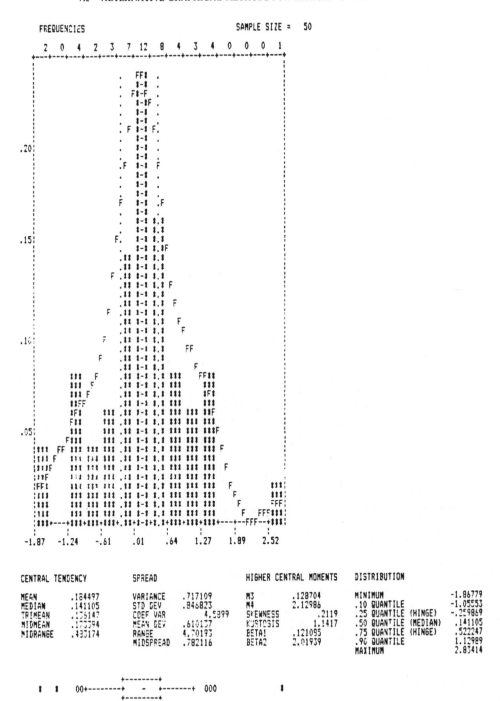

Figure 7.1.10b   Effect of sample size: histogram of 50 iid simulated Laplace$(0, (0.5)^{1/2})$ random variables, produced by HISTPC with automatic bin width determination. Overlaid kernel density estimate uses triangular kernel.

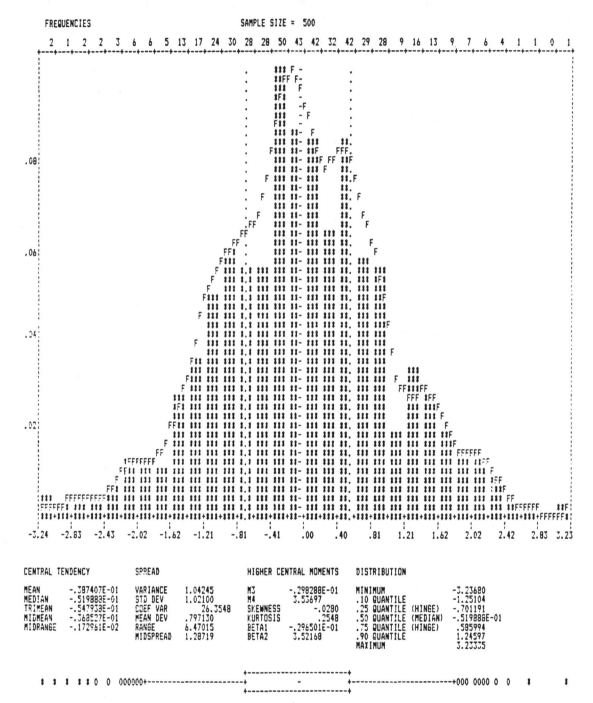

Figure 7.1.11a   Effect of sample size: histogram of 500 iid simulated Normal(0, 1) random variables, produced by HISTPC with automatic bin width determination. Overlaid kernel density estimate uses triangular kernel.

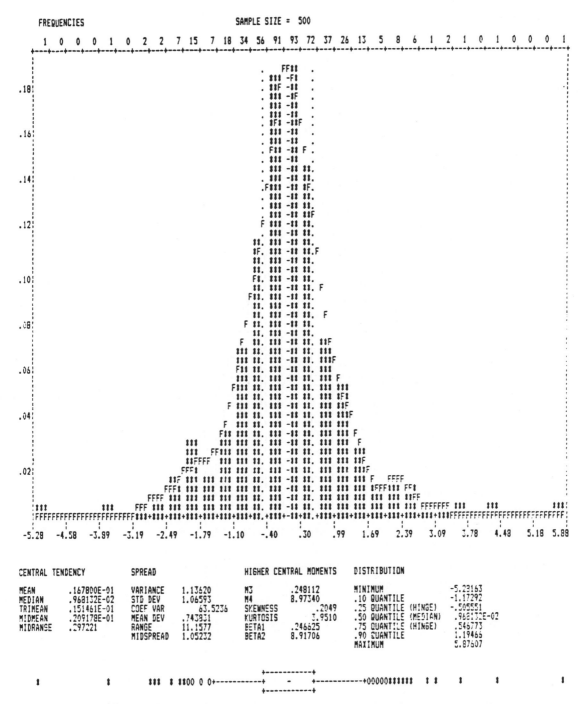

Figure 7.1.11b Effect of sample size: histogram of 500 iid simulated Laplace(0, (0.5)^(1/2)) random variables, produced by HISTPC with automatic bin width determination. Overlaid kernel density estimate uses triangular kernel.

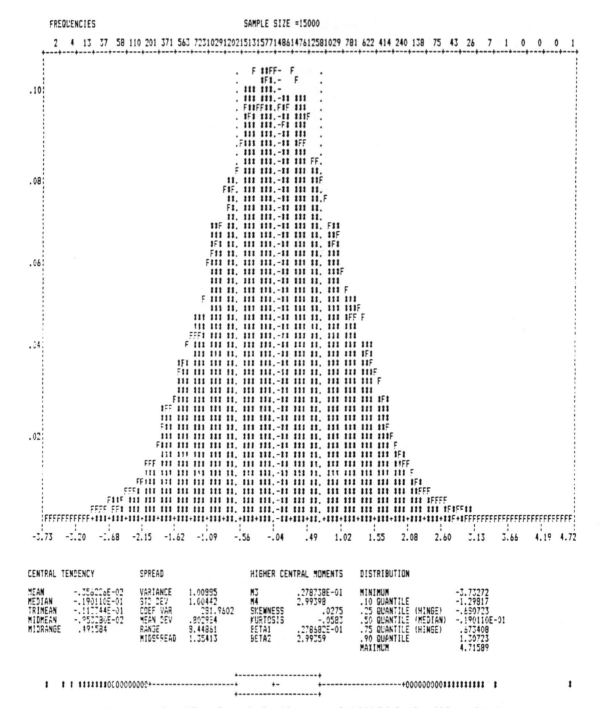

Figure 7.1.12a　Effect of sample size: histogram of 15,000 iid simulated Normal(0, 1) random variables, produced by HISTPC with automatic bin width determination. Overlaid kernel density estimate uses triangular kernel.

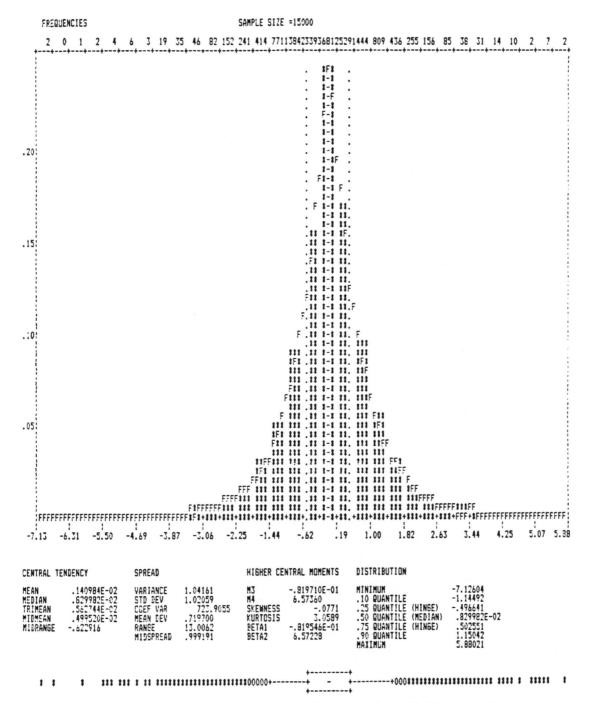

Figure 7.1.12b Effect of sample size: histogram of 15,000 iid simulated Laplace(0, (0.5)$^{1/2}$) random variables, produced by HISTPC with automatic bin width determination. Overlaid kernel density estimate uses triangular kernel.

We will plot either the quantiles or the percentiles of our data against those of a prespecified theoretical distribution, but our conclusions will be subjective ones based on the linearity of the graphs, not on test statistics and confidence levels. We will see, too, that departures from linearity of certain types suggest alternative hypotheses for the distribution of the data.

7.2.1   Quantile Plots (Q–Q Plots)

Producing a quantile or Q–Q plot is in principle relatively simple. If $X_{(1)}$, $X_{(2)}$, ..., $X_{(m)}$ represent the ordered random sample and $F(x)$ is believed to be the distribution of X, then we plot

$$X_{(k)} \text{ versus } F^{-1}\left(\frac{k}{m+1}\right) \qquad k = 1, 2, \ldots, m$$

where $F^{-1}(\alpha)$ is the inverse cdf defined in Section 3.5. Thus we will be plotting $X_{(k)}$, the $\{k/m + 1\}$ quantile of the ecdf, against $F^{-1}\{k/m + 1\}$, the $\{k/m + 1\}$ quantile, of the hypothesized distribution $F(x)$. Note that the order statistics *do not have to be distinct*, although they will in principle be so if the data are continuous.

Interpretation of a Q–Q plot depends heavily on experience and can be difficult, especially for small samples. There are, however, some general principles and observations that can make the task easier.

i. First Principle

If $F(x)$ is the correct distribution for X, then, according to Section 6.3.2, $X_{(k)}$ converges to the quantile $x_{(k/(m+1))}$ and we have the relationship

$$X_{(k)} \rightarrow x_{(k/(m+1))} = F^{-1}\left(\frac{k}{m+1}\right) \tag{7.2.1}$$

Hence, since for each k the points within the plotted pair are approximately equal under the null hypothesis, the Q–Q plot will be approximately a straight line of slope one, passing through the origin.

A trick that makes it easier to gauge whether the line is straight is to hold the plot at eye level and look at the points from the $(0, 0)$ corner. Since your eyes are looking forward in a straight line, it easy to see if the points curve away.

The question of how much the points can "reasonably" deviate from a straight line depends on the actual distribution of the X_i and on the sample size m, and brings us back to confirmatory statistics. Procedures to produce asymptotic, pointwise bounds are indicated in the exercises and in Chambers et al. (1983, Chap. 6), but they are not in the spirit of the current discussion on exploratory techniques, nor are they commonly used.

EXAMPLE   **7.2.1 Quantile Plots with Correctly Specified Theoretical Distributions**
Figures 7.2.1a and 7.2.1b show Q–Q plots for 100 simulated Normal(0, 1) random variables and 100 simulated exponential(1) random variables, respectively. The

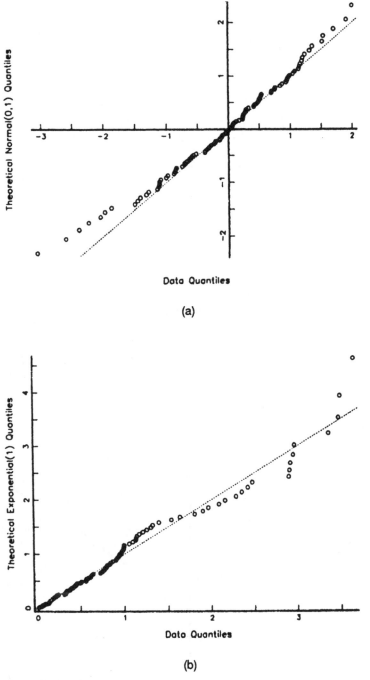

Figure 7.2.1   Two Q–Q plots, each for 100 simulated random variables versus a theoretical distribution. The dotted line through the origin has slope 1. (a) Normal$(0,1)$ random variables and Normal $(0,1)$ theoretical distribution. (b) exponential (1) random variables and theoretical exponential (1) distribution.

ordered data values, which are the $k/(m + 1)$ quantiles of the ecdf, are in each case plotted against the correct theoretical quantiles. Notice the variability in the Q–Q plots of only 100 points, as evidenced by departures from the superimposed line $y = x$ in each graph. The variability is greatest in the tails, where the quantile estimates are themselves most highly variable.

ii. Second Principle

If $F(x)$ is not the correct distribution for X, there may still be information in the Q–Q plot to help us identify the correct distribution easily.

a. The Q–Q plot may show a straight line of slope 1 that does not pass through the origin. This plot is represented by the equation

$$F^{-1}\left(\frac{k}{m + 1}\right) \cong X_{(k)} + a \tag{7.2.2}$$

and indicates that we have incorrectly specified the (additive) mean in the theoretical distribution. Averaging the data or using the x-intercept from the Q–Q plot would give a better idea (estimate) of the true mean.

Figure 7.2.2 shows a Q–Q plot of 100 simulated Normal(2, 1) random variables versus a theoretical Normal(0, 1) distribution. The fact that the scatter crosses the x-axis at about 2, but not exactly at 2, is not surprising because the data median, which determines the point of crossing, has a standard deviation of 0.125.

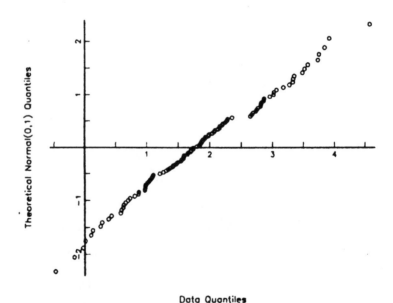

Data Quantiles

Figure 7.2.2   A Q–Q plot for 100 simulated Normal(2, 1) random variables versus a theoretical Normal(0, 1) distribution. Note the shift in the curve with the scatter crossing the x-axis at about 2.

b. The Q–Q plot may show a straight line, passing through the origin, that has slope $\lambda \neq 1$. This plot is represented by the equation

$$F^{-1}\left(\frac{k}{m+1}\right) \cong \lambda X_{(k)} \tag{7.2.3}$$

and indicates that we have incorrectly specified the (multiplicative) scale parameter. If $F(x)$ is the distribution of a random variable we call Y, then what we really want is the distribution of Y/λ. Often the distribution of Y/λ is easy to calculate.

Notice that the scale parameter λ can be related to the usual variance parameter in two-sided distributions such as the Normal and logistic, but, for positive-valued distributions such as the exponential and Gamma, λ appears in both the mean and variance.

EXAMPLE **7.2.2 Q–Q Plots with Incorrectly Specified Scale Parameters**
Figure 7.2.3 shows a Q–Q plot for 100 simulated Normal$(0, 10)$ random variables against theoretical Normal$(0, 1)$ quantiles. The slope of the points given the different scales of the x-axis and the y-axis is approximately $(1/10)$, indicating that the actual distribution of the simulated data is that of a Normal$(0, 1)$ random variable divided by $(1/10)$: that is, a Normal$(0, 10)$.

Figure 7.2.4 shows a Q–Q plot for 100 simulated exponential(2) random variables against theoretical exponential(1) quantiles. The slope of the points is approximately 2, indicating that the actual distribution of the simulated data is that of an exponential(1) random variable divided by 2: that is, an exponential(2).

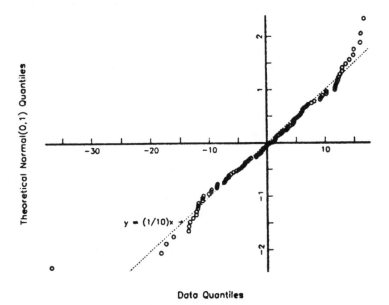

Figure 7.2.3 A Q–Q plot for 100 simulated Normal$(0, 10)$ points versus a theoretical Normal$(0, 1)$ distribution. Note that the slope of the dotted line is 1/10.

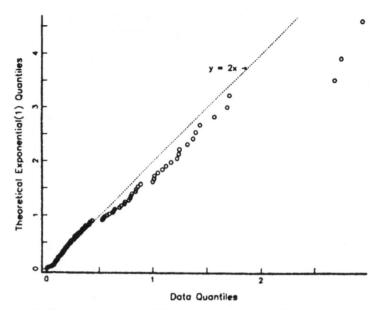

Figure 7.2.4   A Q–Q plot for 100 simulated exponential(2) random variables versus a theoretical exponential(1) distribution. Note that the slope of the dotted line is 2.

Notice that in both examples it was sufficient to guess the general underlying distribution, not all the *parameters associated with it, to produce an effective Q–Q plot. Thus the Q–Q plot can be used to test for a distribution, per se, up to scale and shift parameters.*

c. The Q–Q plot may show a combination of forms a and b, above, or some other simple relationship between the theoretical $x_{(k/(m+1))}$ quantiles and the $k/(m + 1)$ quantiles of the ecdf:

$$F^{-1}\left(\frac{k}{m + 1}\right) \cong g(X_{(k)}) \tag{7.2.4}$$

Hence, if Y has distribution $F(x)$, then $X_{(k)}$ has the same distribution as $g^{-1}(Y)$. This presumes that g^{-1} exists and, to be useful, that $g^{-1}(Y)$ has some standard distribution. In addition, note that Equation 7.2.4 says that the quantile $X_{(k)}$ of the ecdf equals the quantile of X transformed by $g^{-1}(\cdot)$. Now, this is the same as the quantile of $g^{-1}(X) = Y$ if $g^{-1}(\cdot)$ is monotone, since the α quantile of Y will equal $g^{-1}(x_\alpha)$.

The most common set of transformations that are considered consists of the power transformations:

$$g(x) = x^p$$

and, many times, the log or exponential transformations

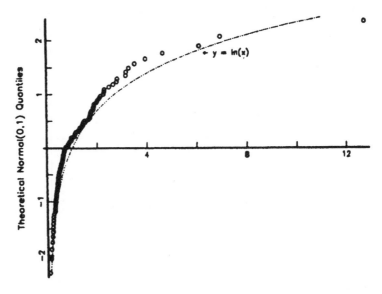

Figure 7.2.5 A Q–Q plot for 100 simulated Lognormal$(0, 1)$ random variables versus a theoretical Normal$(0, 1)$ distribution. The dotted line plots $y = \ln(x)$.

$$g(x) = \ln(x) \quad \text{or} \quad g(x) = \exp(x)$$

The log transformation is of course only useful for positive-valued random variables.

EXAMPLE   **7.2.3 Q–Q Plots with a Nonlinear but Identifiable Form**
Figure 7.2.5 shows a Q–Q plot that assumes a theoretical Normal$(0, 1)$ distribution but actually has 100 simulated Lognormal$(0, 1)$ random variables. The scatter suggests the relationship

$$F^{-1}\left(\frac{k}{m+1}\right) \cong \ln(X_{(k)}) \tag{7.2.5}$$

and such a curve has been superimposed on the plot. If we accept this transformation then

$$g^{-1}(\text{Normal}(0, 1)) = \exp(\text{Normal}(0, 1))$$
$$\sim \text{Lognormal}(0, 1)$$

and we come to the right conclusion: the X's are distributed Lognormal$(0, 1)$.

iii. Third Principle

If $F(x)$ is not the correct distribution for X and there is no simple, apparent transformation of the types mentioned in subsection ii, we may still learn some (comparative) facts about the distribution of X. In particular, the Q–Q plot will show some of the ways in which the distribution of the data differs from $F(x)$.

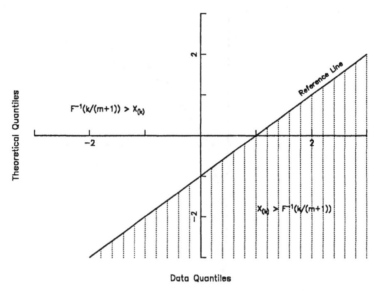

Figure 7.2.6   Figure for interpreting the locations of data points in a Q–Q plot.

We have already seen that the location and angle of the scatter are affected by correct or incorrect specification of the theoretical mean and variance. To look beyond these factors, we can take one of two options:

- Force $F(x)$ to have the same mean and variance as the data and thereby retain the 45° line as the reference
- Ignore the data mean and variance and use a straight line through the scatter, not necessarily with slope 1 nor passing through the origin, as the reference

With either approach, points in the Q–Q plot below the reference line indicate data values that are larger than comparable quantiles in the theoretical distribution. Points in the plot above the reference line indicate data values that are smaller than theoretical quantiles. Figure 7.2.6 should help reinforce these facts.

Often pictures of the following types occur:

- The Q–Q plot forms an S-shaped curve with points above the reference line at extreme low quantiles, below the line at the lower middle quantiles, above the line at the upper middle quantiles, and below the line at high quantiles. Thus, both tails of the data distribution are symmetrically heavier (longer) than those assumed in the theoretical distribution while the body is more peaked. In terms of moments, the kurtosis is greater in the data distribution than in $F(x)$, but the skewness is the same.

Figure 7.2.7 shows such a situation, where the quantiles of 100 simulated Laplace $(0, 1/\sqrt{2})$ random variables are plotted against theoretical Normal$(0, 1)$ quantiles. Notice that the distributions have been chosen to have equal means and variances so that a 45° line through the origin is the correct reference line.

- Alternatively, the scatter may show an S-shaped curve with points below the reference line at extreme low quantiles and above the line at extreme upper quan-

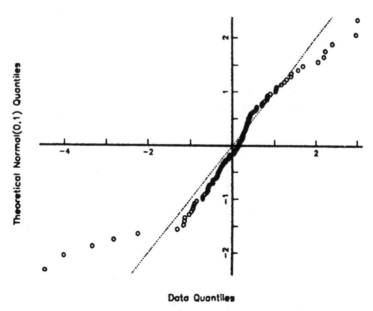

Figure 7.2.7   A $Q-Q$ plot for 100 simulated Laplace $(0, 0.707)$ points versus a theoretical Normal$(0, 1)$ distribution.

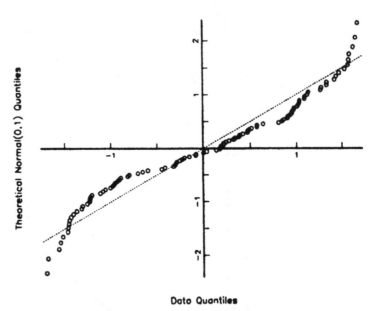

Figure 7.2.8   A $Q-Q$ plot for 100 simulated uniform$(-1.732, 1.732)$ random variates versus a theoretical Normal$(0, 1)$ distribution.

tiles. This time the data have the same skewness, but *lower* kurtosis or lighter tails, than the hypothetical $F(x)$. Figure 7.2.8 shows such a $Q-Q$ plot, comparing the quantiles of 100 simulated uniform$(-1.732, 1.732)$ random variables to the quantiles of a theoretical Normal$(0, 1)$ distribution. Again, the first and second moments have been matched, so that a superimposed 45° reference line through the origin is appropriate.

• In another type of picture the scatter forms a convex curve indicating that the data values are less than expected by the theoretical distribution in the lower and middle quantiles but more than expected for extreme upper quantiles. This translates into data having a long right tail, or equivalently, a positive right skew,

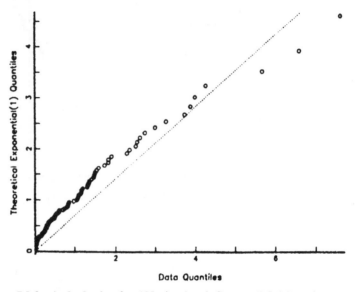

Figure 7.2.9   A *Q–Q* plot for 100 simulated Gamma(0.5, 1.0) points versus a theoretical exponential(1) distribution.

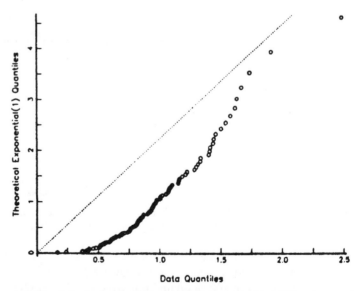

Figure 7.2.10   A *Q–Q* plot for 100 simulated Gamma(5.0, 1.0) random variables versus a theoretical exponential(1) distribution.

when compared to $F(x)$. Figure 7.2.9 illustrates this situation by comparing 100 simulated Gamma(0.5, 1) random variables to an exponential(1) distribution. Since the ratio of the exponential(1) standard deviation to the Gamma (0.5, 1) standard deviation is $1/\sqrt{2}$, the line $y = (1/\sqrt{2})x$ has been superimposed through the scatter.

- Again, the scatter may form a concave curve with data values higher than predicted by $F(x)$ in the middle and upper ranges. Hence the data have a long left tail or, equivalently, are negatively skewed in comparison to $F(x)$. Figure 7.2.10 compares 100 simulated Gamma(5, 10) random variables to an exponential(1) distribution. This time the ratio of the exponential(1) standard deviation to the Gamma(5, 1) standard deviation is $1/\sqrt{1/5} = \sqrt{5}$, so that the line $y = \sqrt{5}x$ is the appropriate reference.

- Other departures from linearity include outliers, gaps, and plateaus (Chambers et al., 1983, Chap. VI). Structures of these types are tied to "problems" in the data that occur in simulation only when modeling physical input or output. Study of them is left as outside reading.

EXAMPLE   **7.2.4 *Q–Q* Plot Analysis of Simulated Estimates of the Polarization Parameter in a Muon Decay Distribution**

Through previous examples in Chapters 6 and 7, we have seen that using the moment estimator (Equation 6.0.3) of the polarization parameter δ based on samples of size 20 from the muon decay model with $\delta = 0.8$ results in estimates that according to moment summaries (Table 6.1.6), the Kolmogorov–Smirnov test, and an ecdf plot (Figure 6.2.8b) have approximately a Normal(0.8, 0.344) distribution. Recall however, that the Shapiro–Wilk test (Example 6.2.6) rejected Normality.

A *Q–Q* plot for the same 1000 moment estimates of δ versus a Normal(0, 1) distribution is presented in Figure 7.2.11. Notice that the distribution is very Normal

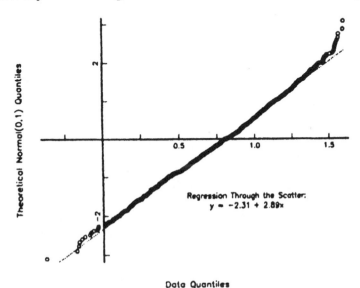

Figure 7.2.11 A *Q–Q* plot for 1000 moment estimates of δ versus a theoretical Normal(0, 1) distribution. The true parameter value δ is 0.8 and $m = 20$.

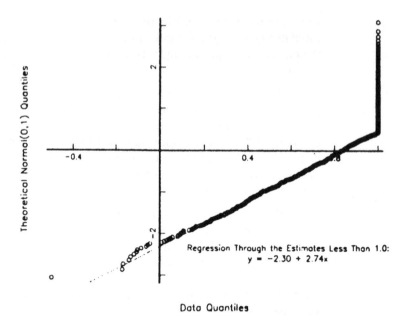

Figure 7.2.12 A Q–Q plot for 1000 mle estimates of δ versus a theoretical Normal(0, 1) distribution. The true parameter value δ is 0.8 and $m = 20$.

in the body but has extreme tails, particularly the upper one, which are shorter than predicted for truly Normally distributed data.

A least-squares regression line has been placed through the scatter. This can be used to estimate the mean and standard deviation of the data as 0.799 and 0.346, respectively (see Exercise 7.14 and compare to the results in Example 6.1.4).

A similar Q–Q plot for 1000 mle estimates (Equation 6.0.6) of δ is given in Figure 7.2.12. Notice how strongly the truncation at $\hat{\delta} = 1$ shows up, as well as the slightly non-Normal left tail. In this case, a least-squares regression has been run through only the estimates less than 1. The regression equation suggests that if not for the truncation rule, the mle estimates would follow a Normal(0.84, 0.365) distribution (see Exercise 7.14).

7.2.2 Percentile Plots

Percentile or P–P plots are often mentioned as alternatives to Q–Q plots. In fact, because of the duality between percentiles and quantiles, the two types of plots offer nearly identical information.

For the Q–Q plot, we started with the ordered random variables, $X_{(1)}$, $X_{(2)}$, $\ldots$, $X_{(m)}$ and a prespecified distribution $F(x)$. We then plotted the $k/(m + 1)$ quantile of the ecdf, namely $X_{(k)}$, versus the $k/(m + 1)$ quantile of $F(x)$: that is,

$$X_{(k)} \quad \text{versus} \quad F^{-1}\left(\frac{k}{m + 1}\right) \quad k = 1, 2, \ldots, m$$

For a $P-P$ plot, we simply plot

$$\frac{k}{m+1} \qquad \text{versus} \qquad F(X_{(k)}) \qquad k = 1, 2, \ldots, m$$

This is nothing more than applying the monotone probability integral transformation $F(\cdot)$ to both variables in the $Q-Q$ plot. From another viewpoint, if F is the correct distribution for X, then the $F(X_{(k)})$'s are order statistics from a uniform$(0, 1)$ sample for which $E(F(X_{(k)})) = k/(m+1)$. It is for this reason that $k/(m+1)$ is used instead of k/m in $Q-Q$ and $P-P$ plots.

If we have guessed the correct distribution for our data so that

$$X_{(k)} \cong F^{-1}\left(\frac{k}{(m+1)}\right)$$

it follows, by applying the monotone transformation $F(\cdot)$ to each side, that

$$\frac{k}{m+1} \cong F(X_{(k)})$$

It does not follow, however, that if we specify the correct distributional form but with the wrong mean and standard deviation, that the $P-P$ plot will still fall on a straight line (as it did for $Q-Q$ plots). Figure 7.2.13 demonstrates this point with a $P-P$ plot of the 1000 moment estimates of the polarization parameter (true value

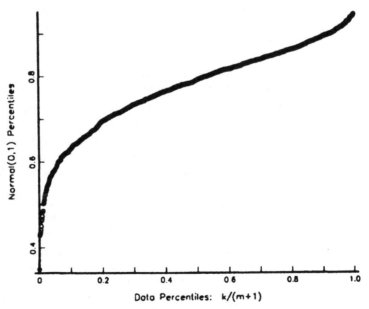

Figure 7.2.13 A $P-P$ plot for 1000 moment estimates of δ versus a theoretical Normal$(0, 1)$.

0.8), versus a Normal(0, 1) distribution. Although the true data distribution is very close to Normal(0.797, 0.343), the curve is far from linear (compare this with the corresponding Q–Q plot in Figure 7.2.11).

Thus Q–Q plots may be preferable to P–P plots because of invariance in the linearity of the plot to shifting and scaling. The Q–Q and P–P plots also differ in their emphasis. Because the more extreme order statistics, those close to $X_{(1)}$ and $X_{(m)}$, tend to be separate from the rest of the data, as well as being highly variable, they receive much visual emphasis in Q–Q plots. On the other hand, P–P plots treat each data point essentially equally, since the function $F(X_{(k)})$ is bounded between 0 and 1 and is plotted at equal spacings of width $1/(m + 1)$ on the ordinate. The value of emphasizing the tails of the empirical distribution where unusual features of the data often appear must be balanced against the high variability of those points in the extreme tails. Finally, there is one last consideration in choosing between P–P and Q–Q plots: computational feasibility. Computing $F(X_{(k)})$ for P–P plots may be easier than computing $F^{-1}[k/(m + 1)]$ for Q–Q plots. For example, $F(\cdot)$ is much easier to compute than $F^{-1}(\cdot)$ for Gamma(β, k) variables (see Table 6.1.2). Approximations to $F^{-1}(\cdot)$ for Gamma(β, k) variables are given in Chambers et al. (1983, p. 227).

Exercises   **7.1.** Compute the hazard rate for exponential(λ) and Weibull(k, β) distributions. Give conditions for the Weibull parameters k and β under which the hazard is increasing or decreasing.

7.2. Verify Equations 7.1.7, 7.1.8, and 7.1.12, giving the properties of the nonparametric probability function estimate for discrete data.

^T**7.3.** If we know that the data $X_i, X_2, \ldots, X_n$ are iid Poisson(λ), then the maximum likelihood estimate of the probability function $p_k = \lambda^k \exp(-\lambda)/(k!)$ is given by $\tilde{p}_k = \bar{X}^k \exp(-\bar{X})/k!$.

Use the delta method as described, for example, in Bishop, Fienberg, and Holland (1975), to calculate an approximate variance for $\tilde{p}_k$. Compare this result to the variance of the nonparametric estimate of p_k as given in Equation 7.1.8.

7.4. Verify the relationship $f(x) = dF(x)/dx$ for the logistic, Laplace, Cauchy, and Weibull distributions as given in Table 6.1.2.

^T**7.5.** We have the following result from Feller (1966, Volume II, p. 19): If $X_1, X_2, \ldots, X_m$ are iid exponential(λ), and $X_{(1)}, X_{(2)}, \ldots, X_{(m)}$ are the order statistics, then $X_{(1)}$, $X_{(2)} - X_{(1)}, \ldots, X_{(m)} - X_{(m-1)}$ are independent with the density of $X_{(k+1)} - X_{(k)}$ given by $(m - k)\lambda \exp[-(m - k)\lambda x]$.

Use this fact and the delta method (Bishop, Fienberg, and Holland, 1975) to compute and compare the bias and variance of the nearest-neighbor density estimate (Equation 7.1.14) to that of the estimate given by Equation 7.1.15 with $k = 1$, for exponential data.

^C**7.6.** A random variable X has a triangular distribution if it has the same distribution as the sum of two uniform$(0, 1)$ random variables. Generate a sample of 100 triangularly distributed random variables and compare graphically the density estimates resulting from the boxcar (Equation 7.1.21) and Epanechnikov (Equation 7.1.25) kernels, using $b = 0.32$ in both cases. Repeat the comparison using 500 triangular random variables.

^T**7.7.** Asymptotically with sample size m, what would happen to the mean and variance of the boxcar kernel density estimate in the following cases?

    a. b stays constant while $m \to \infty$

    b. $b \to 0$, $m \to \infty$, but $bm \to 0$

    c. $b \to 0$, $m \to \infty$, but $bm \to c$ with $0 < c < \infty$

    Illustrate your calculations using increasingly large simulated samples of triangular data (see Exercise 7.6).

^{C}7.8. Generate 50 variables from a Laplace$(1, 1)$ distribution and 50 variables from a Laplace$(-1, 1)$ distribution. For the combined sample of 100 points, use the Epanechnikov kernel and various values of b to assess the effect of bandwidth on estimating the density of this mixture of densities. Repeat the experiment using 500 Laplace$(1, 1)$ and 500 Laplace$(-1, 1)$ variables.

^{C}7.9. Repeat the simulation of the mixture of Laplace variables in Exercise 7.8, but present the data with a histogram. Vary the bin widths to investigate the various amounts of smoothing. Compare these results to the results for the Epanechnikov kernel in Exercise 7.8.

^{C}7.10. Investigate in Hoaglin, Mosteller, and Tukey (1983) the techniques for producing a stem-leaf plot. Generate 25 Laplace$(0, 1)$ variables and create a stem-leaf plot. What are the comparative advantages and disadvantages of stem-leaf plots versus histograms?

7.11. Verify the statement in subsection iii of Section 7.1.4 that we would expect about 5 outliers plus far outliers in a boxplot of 100 Normal data points. How many far outliers would we expect?

^{T}7.12. Use the results on the asymptotic distribution of data quantiles in Section 6.3 to suggest a method for producing asymptotic, pointwise confidence bounds for a Q–Q plot.

^{T}7.13. Assume that a Q–Q plot of an ordered data sample of size m versus theoretical Normal$(0, 1)$ quantiles shows a relationship of approximately the form

$$\Phi^{-1}\left[\frac{k}{(m + 1)}\right] = a + bX_{(k)}$$

What would be a reasonable guess as to the distribution of the data?

^{T}7.14. We often draw a regression line through a Q–Q plot to help us see departures from linearity. The estimated coefficients in the regression equation can also be shown to give information about the mean and variance of the data.

    a. Assume we are comparing our data to the quantiles of a theoretical distribution $F(x)$ with mean θ and variance σ^2 and observe the relationship

$$F^{-1}\left[\frac{k}{(m + 1)}\right] = a + X_{(k)}$$

    If we use linear regression to find an estimate $\hat{a}$ for a, show that $\theta - \hat{a}$ estimates the mean of the data. If $F(x)$ is symmetric, show that $\theta - \hat{a} = \bar{X}$.

    b. Assume that we observe the relationship

$$F^{-1}\left[\frac{k}{(m + 1)}\right] = bX_{(k)}$$

    and calculate the regression estimate $\hat{b}$ of b. Suggest and justify an estimate of the standard deviation of the data based on $\hat{b}$.

c. Assume that we observe the relationship

$$F^{-1}\left[\frac{k}{(m+1)}\right] = a + bX_{(k)}$$

and calculate the regression estimates $\hat{a}$ and $\hat{b}$ of a and b. Suggest and justify estimates of the mean and standard deviation of the data based on $\hat{a}$ and $\hat{b}$. Do your estimates necessarily equal $\bar{X}$ and S?

T**7.15.** If you produced a Q–Q plot using theoretical uniform$(0, 1)$ quantiles and recognized in the scatter the relationship

$$F^{-1}\left[\frac{k}{(m+1)}\right] \cong \exp(-X_{(k)})$$

what distribution would you think your data came from?

C**7.16.** Based on the general ideas presented in subsection iii of Section 7.2.1, what would you expect the Q–Q plot for bimodal data to look like? Check your intuition by producing a Q–Q plot for a mixture of two Laplace densities: 500 points from a Laplace$(-2, 1)$ distribution and 500 points from a Laplace$(2, 1)$ distribution.

7.17. Discuss the comparative sensitivity of P–P and Q–Q plots to departures from the theoretical distribution in the tails of the distribution.

T**7.18.** In a P–P plot, how would the following differences from the theoretical distribution manifest themselves in the graphics: (a) different mean, (b) different variance, (c) skewness, (d) shorter tails, and (e) longer tails?

7.19. Give examples of distributions in which it would be simpler to produce P–P plots, as opposed to Q–Q plots.

References

Barlow, R. E., and Proschan, F. (1975). *Statistical Theory of Reliability and Life Testing.* Holt, Rinehart & Winston: New York.

Bishop, Y. M. M., Fienberg, S. E., and Holland, P. W. (1975). *Discrete Multivariate Data Analysis.* MIT Press: Cambridge, MA.

Chambers, J. M., Cleveland, W. S., Kleiner, B., and Tukey, P. A. (1983). *Graphical Methods for Data Analysis.* Wadsworth & Brooks/Cole: Pacific Grove, CA.

Conover, W. J. (1980). *Practical Nonparametric Statistics,* 2nd ed. Wiley: New York.

Cox, D. R., and Oakes, D. (1984). *Analysis of Survival Data.* Chapman & Hall: London.

Epanechnikov, V. A. (1969). Nonparametric Estimates of a Multivariate Probability Density. *Theory of Probability and Its Applications, 14,* 153–153.

Feller, W. (1966). *An Introduction to Probability Theory and Its Applications,* Vol. II, 3rd ed. Wiley: New York.

Freedman, D., and Diaconis, P. (1981a). On the Histogram as a Density Estimator: L_2 Theory. *Zeitschrift für Wahrscheinlichkeitstheorie, 57,* 453–476.

Freedman, D., and Diaconis, P. (1981b). On the Maximum Deviation Between the Histogram and the Underlying Density. *Zeitschift für Wahrscheinlichkeitstheorie, 58,* 139–168.

Heidelberger, P., and Lewis, P. A. W. (1984). Quantile Estimation in Dependent Sequences. *Operations Research, 32*(1), 185–209.

Hoaglin, D. C., Mosteller, F., and Tukey, J. W. (1983). *Understanding Robust and Exploratory Data Analysis.* Wiley: New York.

Lawless, J. F. (1982). *Statistical Models and Methods for Lifetime Data.* Wiley: New York.

Lewis, P. A. W. (1971). Quantile Estimation. In *Proceedings of the Fifth Symposium on Computer Science and Statistics,* pp. 75–77. Oklahoma State University: Oklahoma City.

Parzen, M. (1962). On Estimation of a Probability Density Function and Its Mode. *Annals of Mathematical Statistics, 33,* 1065–1076.

Robertson, T. (1967). On Estimating a Density Which Is Measurable with Respect to a Sigma Lattice. *Annals of Mathematical Statistics, 27,* 482–493.

Robinson, D. W. (1975). Nonparametric Quantile Estimation Through Stochastic Approximation. Ph.D. thesis, Naval Postgraduate School, Monterey, CA.

Rosenblatt, M. (1956). Remarks on Some Nonparametric Estimates of a Density Function. *Annals of Mathematical Statistics, 27,* 832–837.

Rosenblatt, M. (1971). Curve Estimates. *Annals of Mathematical Statistics, 42,* 1815–1842.

Scott, D. W. (1979). On Optimal and Data-based Histograms. *Biometrika, 66,* 605–610.

Scott, D. W., and Factor, L. E. (1981). Nonparametric Probability Density Estimators. *Journal of the American Statistical Association, 76,* 9–15.

Tapia, R. A., and Thompson, J. R. (1978). *Nonparametric Probability Density Estimation.* Johns Hopkins University Press: Baltimore.

Tukey, J. W. (1977). *Exploratory Data Analysis.* Addison-Wesley: Reading, MA.

Van Ryzin, J. (1973). A Histogram Method of Density Estimation. *Communications in Statistics, 2,* 493–506.

Wahba, G. (1975). Optimal Convergence Properties of Variable Knot, Kernel and Orthogonal Series Methods for Density Estimation. *Annals of Statistics, 3,* 15–29.

Watson, G. S., and Leadbetter, M. R. (1963). On the Estimation of the Probability Density, I. *Annals of Mathematical Statistics, 34,* 480–491.

Wegman, E. J. (1970). Maximum Likelihood Estimation of a Unimodal Density Function. *Annals of Mathematical Statistics, 41,* 2169–2174.

Comparisons in Multifactor Simulations: Graphical and Formal Methods

8.0   Introduction

From the survey in Chapters 6 and 7, it should be clear that there are many statistical tools available to summarize the behavior of a single sample of real or simulated univariate data. Each technique is best suited to highlight structures of certain types and to answer questions of particular types about the distribution of the data. Only by trying a combination of graphical and numerical summaries can you hope to have a reasonable understanding of the data distribution.

When only one sample is to be analyzed—for instance in modeling physical input (perhaps the distribution of service to a queue)—then interactive exploration is a good approach: start with standard summaries (moments, quantiles, and a histogram or boxplot) and then use more specialized techniques (density estimates, Q–Q plots, statistical tests) to pursue interesting details and to check for a hypothesized distribution. Many of the programs in the software supplement, *Enhanced Simulation and Statistics Package*, (HISTPC, LISTPC, SWPC, KSPC, and PLOTPC) were designed to help in such an analysis.

More often though, there will be *many* samples to analyze, each the output of a similar simulation in which only certain input parameters have been changed. The goal then, as outlined in Chapter 1, is to compare the multiple outputs and to decide (a) which set of parameter inputs mimics reality or (b) how the system responds over this range of parameter values and, in particular, if it changes and how it changes. As one example of a multifactor *systems simulation*, we can think about the following situation.

EXAMPLE   **8.0.1 Using Simulation to Determine the Best Queueing Discipline in a Multiserver Queue**
The examples in Sections 3.2 and 3.3 introduced the idea of a queue, and in Section 3.3 we were presented with two ways in which customers could be served in a multiple-server queue. Customers could either enter a single line and then be served as any teller became free, or they could join one of multiple lines and wait in front of their chosen server. The distribution of *stationary* waiting times in the multiple-

server queue, which is approximated by the waiting time of say, the 5000th customer, will vary with the queueing discipline, and also with factors such as the number of servers, the interarrival time distribution, the service time distribution, and the traffic intensity. This example will be pursued in Example 8.1.1; the main question to be answered is, Does one of the two queueing disciplines, single-line or multiple-line, provide "better" service? The answer will depend on levels of the other factors and the definition of "better." This type of queue occurs everywhere, in particular in banks, and both queueing disciplines are used.

Of course some queueing situations are nonhomogeneous, so that the transient behavior of the system and the effect of the time parameter are also important. This is true, for instance, in studying the two foregoing queueing disciplines in a bank, where there is strictly no stationary waiting time. Analysis of this nonhomogeneous situation, which is more difficult because of the addition of the time factor, is not considered in this chapter or this volume.

Another broad family of simulations that produce multiple outputs are *statistical simulations*, where we wish to explore the behavior of statistical estimates or procedures as a function of sample size, population distribution, and other factors. As an example, consider the following situation:

EXAMPLE **8.0.2 Using Simulation to Determine the Best Estimator of the Center of a Distribution**
Given a sample of data from a symmetric distribution, the sample mean (Equation 6.1.1), median (Equation 6.3.4), and trimmed mean (Section 3.4) are all possible estimators for the center of the distribution. Statistical theory gives us guidance in choosing among these estimators in some cases: for example, Pitman's theorem (Mood, Graybill, and Boes, 1974, Chap. 7.6) implies that the sample mean has smaller variance than the sample median when the data are from a Normal distribution. Simulation, however, allows us to start with data from any distribution, *symmetric or nonsymmetric*, and compare the distributions of the three estimators on any criteria we choose. This example is carried through in subsection ii of Section 8.2.2.

Setting up either systems simulations or statistical simulations becomes a problem in experimental design. We must identify the critical assumptions in our simulation model and produce all the output necessary to isolate the effects of each factor and of combinations of factors. These topics are discussed for general simulations in Section 8.1 and for the special class of statistical simulations in Section 8.2.

In addition, organizing and analyzing such large quantities of output is a problem in itself. Standard analysis of variance techniques and a more flexible graphical approach using arrays of multifactor quantile plots are presented in Section 8.1.1. These techniques are illustrated through application to queueing data. Another systematic graphical method ideally suited to statistical simulations is presented in Section 8.2.1 and implemented by SMTBPC, the simulation test bed program from the software supplement. Such automation of the output analysis of the simulation encourages exploration and the wider use of statistical simulation. The use of SMTBPC is illustrated by three examples in Section 8.2.2.

8.1   Graphical and Numerical Representation of Multifactor Simulation Experiments

A simulation model can be divided into two major components. First, there are the parameters and structural assumptions that are built into the model to describe or model the situation we wish to simulate. Such specifications are illustrated by examples in Chapter 3, with details laid out in Section 3.7.3. Often a particular specification is simply a trial value, which can represent a guess as to the true value in the physical system, or alternatively, a variable parameter whose influence in the system we wish to evaluate. In the latter case, that specification is commonly called a *factor* in the simulation, and each value that we wish to evaluate in a separate run of the simulation is a *level* of that factor. (Identify for yourself the factors and possible levels involved in the examples of Chapter 3.)

The second component of the simulation model is the *response* or output from the simulation, as discussed in item (2) of Chapter 2. Note that a single simulation will give us many (related) response variables, which we may wish to evaluate jointly or singly with various quantifications. For example, the mean, variance, and extreme upper quantiles of the waiting time distribution may all be relevant in evaluating the output from a queue.

Even if we focus on a single response from a simulation, even models that have been simplified as much as possible without losing realism may have so many factors at so many levels that we must pay attention to four questions.

1. How do we choose which combinations of levels of factors to evaluate in simulation runs? Note that if we have N factors with K levels each, then the simulation must be run for K^N combinations of factors and levels. If K or N is large, it may not be possible to do all runs and still have enough replications within each run to achieve sufficient precision in each response estimate. This is called a supersaturated situation; some techniques that may be suitable for efficient identification of important factors in this situation are given in Mauro (1986).

2. How do we organize and display the output from a large number of related simulation experiments?

3. How do we model and assess the influence of the different factors on the response?

4. How do we use the control we have in simulations over the input random variables to make our assessments in Question 3 more precise?

Questions 1, 2, and 3 are really more general questions in experimental design and data analysis, not specifically tied to simulation; they are taken up in Section 8.1.1. Question 4 is directly related and unique to simulation; it is considered separately in subsection i of Section 8.1.2 in connection with common random numbers. This topic is taken up again in Chapter 11 when we discuss the general topic of variance reduction. Finally, when one of the factors in the simulation is sample size, as is often the case in what we define in Section 8.2 as a statistical simulation, more specialized techniques can be employed, and these are described in Section 8.2.1, where we discuss the simulation test bed program SMTBPC.

8.1.1 Experimental Design and Output Display and Analysis

When the effects of combinations of many different input specifications to a simulation are to be evaluated, we must first decide on which combinations of factors we wish to spend our (limited) simulation resources. If the factors being varied all have a finite number of discrete levels, then choosing the combinations to evaluate (i.e., simulate) is a standard question of *experimental design*, which is explored in general statistical terms in Cochran and Cox (1957) or Box, Hunter, and Hunter (1978). A detailed discussion with reference to simulation is given by Kleijnen (1975), and a brief overview of the more important points can be found in Law and Kelton (1982), or many of the other simulation textbooks cited in previous chapters. More recent references are Smith and Mauro (1982), Mauro and Smith (1984), and Mauro (1986).

Analyzing the output from such multiple, related experiments might also require the standard statistical tool of *analysis of variance*, as described in the references above or, for instance, in Scheffe (1964).

When one or more of the factors assume continuous values (e.g., the traffic intensity in a queue), the simulations are still performed at discretely chosen levels, but *response surface methodology* is often employed to calculate and display interpolated values for the response at intermediate values of the factors. Law and Kelton (1982) review briefly the idea of a response surface, Kleijnen (1975) provides a comprehensive bibliography for the subject, Gruber and Freimann (1986) extend the idea to the comparison of estimates, and Heidelberger, Nelson, and Welch (1985) implement the ideas graphically. The question of choosing "optimal" levels of the factors falls into the discipline of optimal experimental design, and can be explored by reading Chapter 38 of Kendall, Stuart, and Ord (1983).

We have chosen not to present details of any of these topics for two reasons:

• Experimental design, optimal design, analysis of variance, and response surface methodology are wide topics, independent of simulation, which are already well covered by other textbooks and articles.

• Many types of simulation produce results that are not well described by a single quantization of the output, such as the mean value, and in addition, produce results that are not homoscedastic (constant variance), Normally distributed, or independent. Hence, standard methods in design and analysis of variance become difficult to apply and justify. For example, waiting times in queues are positive random variables with a nonzero probability of being equal to zero.

Thus instead of presenting details of often inapplicable methodology, in this section we offer a more general graphical approach to the organization and presentation of multifactor simulation results. Conceptually, the entire distribution (from multiple replications) of the output of a simulation is presented through a quantile plot. This is an alternative to a boxplot (Section 7.1); either plot may be suitable, depending on taste and circumstance. The quantile plot is essentially a scaled vertical line with symbols marked at prespecified quantiles estimated from the data sample (notice the similarity to a boxplot). In this way, we get a picture of the center, spread, symmetry, and tails of the distribution of the output, not just one quantization, such as the average, which would normally act as the dependent variable in an analysis of variance. Each combination of levels of factors will result in one quantile plot, and comparisons between logically coherent combinations of factors are facilitated by line type, spacing, and, if available, color or motion. These ideas are explained most easily by expanding Example 8.0.1.

EXAMPLE

8.1.1 Output Display and Analysis for Multifactor Queueing Data

Our primary goal is to examine the stationary "waiting time" W in a multiserver queue to determine whether it is best to use a single-line or multiple-line queueing discipline. As an approximation to the stationary waiting time W, we use the waiting time of the 5000th customer in the queue. Notice that the waiting time W is a random output from the simulation and that many quantizations of this random variable may be relevant. For instance, we may consider the mean or median waiting time, the variance of the waiting time, or perhaps the upper extreme quantiles (0.95 or 0.99) of the waiting time distribution. Another relevant output quantization is the probability of not having to wait: that is, $P(W = 0)$. Clearly, the waiting time distribution, and therefore the comparison between the two disciplines, will be affected by a number of factors. In particular, the queueing discipline, the number of tellers, the service and interarrival time distributions, and the traffic intensity are all possible factors. Hence, for this example, we have a total of four factors, which we divide, for the purpose of controlling the size of the simulation, into the following levels:

1. Queueing discipline (main factor)
   a. Multiple line
   b. Single line

2. Number of tellers
   a. Two
   b. Five
   c. Ten

3. Service and interarrival time distributions, where "skewed" means Gamma(0.5, 1) and "regular" means Gamma(5.0, 1)
   a. Skewed service and skewed interarrival times
   b. Regular service and skewed interarrival times
   c. Skewed service and regular interarrival times
   d. Regular service and regular interarrival times

4. Traffic intensity
   a. $t = 0.3$
   b. $t = 0.6$
   c. $t = 0.75$

This $2 \times 3 \times 4 \times 3$ design gives us 72 different combinations of levels of the four factors and, for each combination, we will perform 440 independent replications of the simulation, on each replication retaining the waiting time of the 5000th customer. For each experiment, the 440 waiting times are used to form a quantile plot with symbols designating the estimated 0.25, 0.50, 0.75, 0.95, and 0.99 quantiles. Since 72 quantile plots are too many to be effectively displayed in one figure, they have been stratified by Factor 3, the service and interarrival time distributions, into four figures with 18 quantile plots in each.

The main question to be answered is, Which of the two queueing disciplines, multiple-line or single-line, is "better" when the answer may depend on the levels of the three other factors. Notice that some analytical results are available in the single-line case if the service and interarrival times have Gamma(1, 1)—that is, exponential—distributions. These results could be used to augment and validate the simulation (see, e.g., Gaver and Thompson, 1973, p. 486).

Our first figure, representing the results for all combinations with skewed service and skewed interarrival times, is presented on the front cover of this book, to make use of the added dimension of color. (Figure 8.1.1a is a copy of this plot without the benefit of color.) All the quantile plots representing the data from the multiple-line discipline are pictured to the left in blue, while those from the single-line discipline are pictured to the right in red. The traffic intensity, a continuous factor, is represented in green on the x-axis with all six experiments at a given traffic intensity graphed together. Finally, the number of servers, labeled in purple, is represented by plotting successive quantile plots for 2, 5, and 10 servers close together, for each traffic intensity and service discipline, using different line types. Hence, it is relatively easy to inspect the figures visually to determine the effect of the number of servers (by looking at each subgroup of three), the effect of service discipline (by comparing pairs of triplets), and the effect of traffic intensity (by comparing the three large clusters).

However, to observe the effect of the fourth factor, namely service and interarrival time distribution, we need to see the four figures for this factor (number 3) side by side. This is made possible in Figures 8.1.1a through d, where Figure 8.1.1a is, again, a copy of the front cover figure without the benefit of color.

It is clear immediately that traffic intensity t has a major effect on the waiting time distribution and, in fact, it is known from theory that the mean waiting time in many situations is proportional to the reciprocal of $(1 - t)$.

Note too that the vertical scales in Figures 8.1.1 are different. It would be preferable to use the same scale to give a graphical display of the relative difference of the location and spread of waiting times as the distribution–shape factor changes. However, this effect is too large to plot the data this way. The range of data in Figure 8.1.1a is 10 times the range in Figure 8.1.1d.

Beyond this, it appears to be generally true that the multiple-line discipline has larger extreme upper quantiles than the corresponding single-line discipline quantiles. The effect of the number of servers on extreme quantiles depends on both the traffic intensity and the service and interarrival time distributions, and is not consistent between the two disciplines. Note that these quantile plots must be used with care, since they do not contain any direct information about the statistical significance of, say, differences of sample means at different factor levels. For instance, in Figure 8.1.1a the mean waiting times at $t = 0.60$ appear to decrease with the number of servers for both queuing disciplines, but it is not clear from the figure whether the effect is statistically significant. It is in fact true that in all these plots (Figures 8.1.1), increasing the number of servers while keeping the traffic

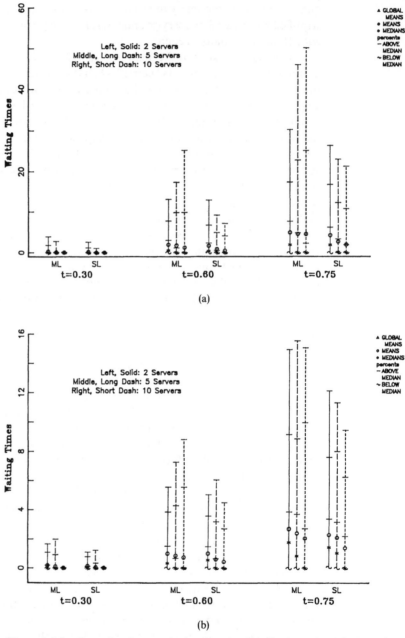

Figure 8.1.1 Quantile plots, each showing the distribution of 440 independent waiting times for the 5000th customer in a queue. The symbols ML and SL represent multiple-line and single-line, respectively. (a) Service and interarrival times skewed; other factors varied as shown. (b) Service times regular and interarrival times skewed; other factors varied as shown. (c) Service times skewed and interarrival times regular; other factors varied as shown. (d) Service and interarrival times regular; other factors varied as shown.

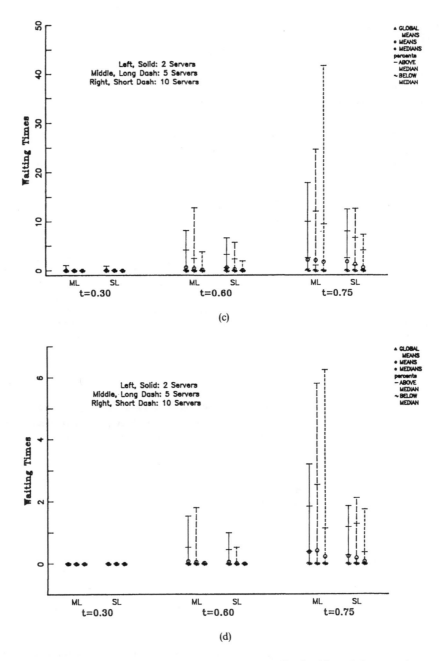

(c)

(d)

intensity constant produces a (possibly not statistically significant) decrease in mean waiting times.

The means and standard deviations calculated from the data in the quantile plots are given in Table 8.1.1, so that direct comparisons of average waiting times for the multiple- and single-line discipline are possible on a pairwise basis. If a more sophisticated analysis of variance is desired—for instance, to see if there still remains

Table 8.1.1　Estimated means (standard deviations) for the distributions of waiting times (each estimate is based on 440 independent replications)

Service, interarrival distributions t = traffic intensity	Multiple-line discipline			Single-line discipline		
	2 server	5 server	10 server	2 server	5 server	10 server
(skew, skew)						
$t = 0.30$	0.28(0.82)	0.08(0.50)	0.0008(0.02)	0.20(0.52)	0.03(0.22)	0.005(0.08)
$t = 0.60$	1.9(2.9)	1.7(3.8)	1.2(4.6)	1.6(2.6)	0.25(1.9)	0.46(1.4)
$t = 0.75$	4.9(6.9)	4.5(9.3)	4.5(11.3)	4.2(5.8)	2.6(4.6)	1.9(4.2)
(regular, skew)						
$t = 0.30$	0.18(0.37)	0.10(0.42)	0.0(0)	0.13(0.27)	0.05(0.22)	0.002(0.03)
$t = 0.60$	0.97(1.4)	0.82(1.7)	0.18(1.9)	0.97(1.2)	0.58(1.2)	0.42(0.98)
$t = 0.75$	2.7(3.3)	2.4(3.5)	1.7(3.8)	2.3(2.7)	2.1(2.8)	1.4(2.2)
(skew, regular)						
$t = 0.30$	0.03(0.21)	0.01(0.17)	0.0(0)	0.03(0.18)	0.0(0)	0.0(0)
$t = 0.60$	0.65(1.6)	0.45(2.3)	0.18(1.8)	0.62(1.4)	0.30(1.2)	0.07(0.45)
$t = 0.75$	2.2(3.9)	2.1(5.4)	1.7(7.8)	1.8(3.2)	1.2(2.8)	0.59(1.7)
(regular, regular)						
$t = 0.30$	0.001(0.1)	0.0(0)	0.0(0)	0.0002(0.004)	0.0(0)	0.0(0)
$t = 0.60$	0.08(0.30)	0.06(0.34)	0.02(0.29)	0.06(0.19)	0.01(0.08)	0.003(0.03)
$t = 0.75$	0.37(0.70)	0.41(1.2)	0.22(1.1)	0.23(0.46)	0.18(0.47)	0.06(0.27)

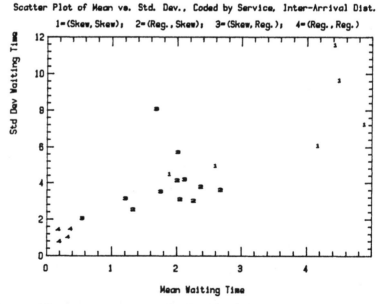

Figure 8.1.2　Coded scatter plot of the mean versus standard deviation for the case of traffic intensity $t = 0.75$. The means and standard deviations are approximately linearly related, with the main factor affecting the mean waiting time being the service and interarrival time distribution (1, 2, 3, and 4).

a difference in average waiting times due to queueing discipline after adjusting for the other three factors simultaneously—we must look at the ANOVA assumptions carefully. Although the means in Table 8.1.1 can be taken to be reasonably Normal, having been calculated from 440 values, they do not have equal variances, so that straightforward ANOVA is not possible. This fact of unequal variance is emphasized in the coded scatter plot of mean versus standard deviation in Figure 8.1.2, for the case of traffic intensity equal to 0.75. The standard deviations are increasing approximately linearly with the means.

Note that using all 440 values from each simulation in an ANOVA with 440 replications in each cell is inappropriate, since the distributions shown by the quantile plots are very skewed and highly non-Normal. Indeed, as we saw in Examples 6.2.2, 6.3.4, and 8.1.1, a significant proportion (depending on t) of the waiting times are exactly zero, so that even a log transform to induce Normality would be impossible. Instead, to do ANOVA we have sectioned each set of 440 waiting times into 8 subgroups of 55 waiting times each and calculated the 8 averages. The logs of these averages were then averaged (to induce Normality), and these averages served as the dependent variables in a three-way ANOVA with 1 replicate per cell. The three effects included in the ANOVA model were number of servers, queueing discipline, and service and interarrival time distributions.

Because the traffic intensity had such an obvious marked effect, it was used as a stratification variable, leading to the two ANOVA tables shown as Tables 8.1.2a and b. The third ANOVA corresponding to traffic intensity equal to 0.3 could not be performed because many cells had observed average values of 0, which were not amenable to log transform. The results for the two traffic intensities, 0.60 and 0.75, are very similar in showing a highly significant influence from all three of the factors

Table 8.1.2 ANOVA results using the average of the 8 log averages of 55 waiting times as the response.

(a) Traffic intensity $t = 0.60$

Source of variation	Degrees of freedom	Sum of squares	F-ratio	Significance level
Service/arrival distribution	3	53.6	80.0	0.0000
Number of servers	2	7.2	16.1	0.0001
Multiple/single discipline	1	2.5	11.2	0.0038
Error	17	3.8		
Total (mean corrected)	23	67.0		

(b) Traffic intensity $t = 0.75$

Source of variation	Degrees of freedom	Sum of squares	F-ratio	Significance level
Service/arrival distribution	3	27.5	147.3	0.0000
Number of servers	2	1.6	12.6	0.0005
Multiple/single discipline	1	1.8	28.6	0.0001
Error	17	1.1		
Total (mean corrected)	23	31.9		

considered: the service and interarrival time distributions, the number of servers, and the type of queueing discipline.

The three factors can also be viewed individually graphically, as a supplement to the ANOVA. For the case of traffic intensity equal to 0.75, the array of graphs in Figure 8.1.3 show the average sectioned log mean waiting times (i.e., the data that went into the ANOVA) for each of the factors: service and interarrival time distributions, number of servers, and type of queueing discipline. The standard error bars in those figures allow us to do pairwise comparisons between levels of each factor, but they do not allow us to make the omnibus statements about a factor, or about one factor after adjustment for the other two factors, which can be made from the analysis of variance table.

Figure 8.1.3a shows clearly the strong effect of interarrival and service distribution noted in connection with the different vertical scales in Figures 8.1.1. Based on these analyses, then, we could conclude that queueing discipline does make a difference to the mean waiting time with the single-line discipline producing a shorter expected wait. This last conclusion was drawn after calculating the ANOVA effect estimates for multiple- and single-line disciplines to be 0.32 and -0.32, respectively, when traffic intensity is $t = 0.60$, and 0.27 and -0.27, respectively, when traffic intensity is $t = 0.75$. These analyses, of course, presume that mean waiting time is the desired response variable and that a linear model with no interactions is sufficient to describe the data. A more detailed analysis would take into account possible interactions in the factors.

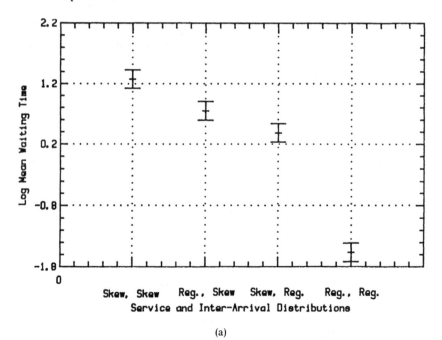

(a)

Figure 8.1.3   The 95% confidence intervals for the sectioned log mean data stratified by (a) service and interval time distributions, (b) number of servers, and (c) service discipline. Traffic intensity is fixed at $t = 0.75$ in each case.

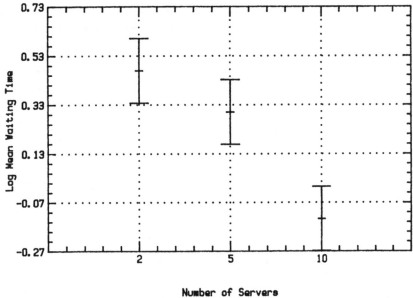

(b)

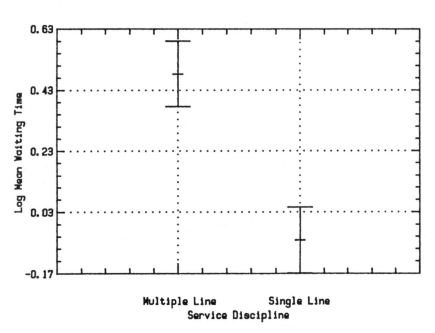

(c)

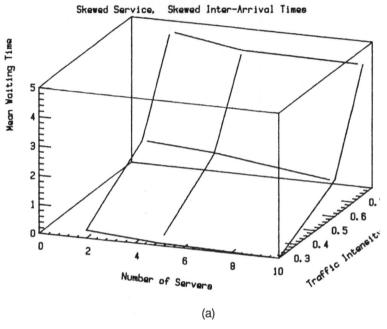

(a)

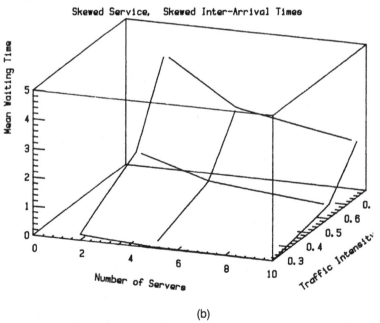

(b)

Figure 8.1.4 Mean waiting time as a function of number of servers and intensity for skewed service and interarrival times: (a) multiple-line que discipline and (b) single-line queueing discipline.

Another approach to the output analysis problem we are discussing in this section is response surface methodology. This allows us to look at a particular response and see how it varies as two factors change. As an example, we will again focus on average waiting time and use three-dimensional graphics to plot the change in average waiting time as traffic intensity and the number of servers change. For illustration, we focus on the skewed service, skewed interarrival time case, and present the response surfaces for the multiple-line and single-line disciplines in Figures 8.1.4a and b, respectively. Since the average waiting time is calculated for only 9 points on each response surface, all the other points must be interpolated—in this case, by a quadratic function built into the graphics routine we used. Choosing the "right" function and "optimal" points from which to interpolate brings us back to the areas of response surface methodology and optimal design mentioned at the beginning of this section.

8.1.2 Experimental Design and Variance Reduction

Experimental design in simulation is distinguished from experimental design in real data work by two things. First, in simulation the statistician is involved from the beginning and can specify the input factors and levels required to achieve an optimal design, as mentioned in Section 8.1.1. (Note that in well-planned data studies this can also be the case.) Second, the input random variables for each simulation experiment are in the control of the statistician, and this circumstance can sometimes be exploited to make comparisons between experiments more precise, in particular by using *common random numbers* instead of independent random numbers for the different runs of the experiment. Related ideas for increasing the precision of the estimate calculated from one simulation experiment (i.e., for obtaining variance smaller than could be achieved through crude simulation) are described in Chapter 11, "Variance Reduction."

Of course another distinction between the simulation and the real-world case is that in simulation one needs to model the real-world situation, as discussed in Chapter 3, and this can produce additional problems of validation and model detail. We ignore these questions here.

i. Pairwise Comparisons and Common Random Numbers

One basic idea for variance reduction in experimental design rests on the following idea: assume that the iid output random variables from m replications of a simulation under one set of levels of factors are $X_1, X_2, \ldots, X_m$ and that the output under a different set of levels of factors are $Y_1, Y_2, \ldots, Y_m$. Also assume that the mean response is the quantity of interest, so that we wish to test $H_0: E(X) = E(Y)$, and, to do so, we perform a paired sample test on $\bar{X} - \bar{Y}$. Then, if we can design the simulation so that $\text{cov}(X_i, Y_i) > 0$ for each i, $\text{var}(\bar{X} - \bar{Y}) = \text{var}(\bar{X}) + \text{var}(\bar{Y}) - 2\,\text{cov}(\bar{X}, \bar{Y})$ will be smaller than if each pair (X_i, Y_i) were independent (see Chapter 10 for details). It follows that $\bar{X} - \bar{Y}$ will be a better (smaller variance) estimate of $E(X) - E(Y)$, and that tests based on $\bar{X} - \bar{Y}$ will be more sensitive to departures from H_0.

To try to create the condition $\text{cov}(X_i, Y_i) > 0$, we use the method of *common random numbers*, where the input stream of random variables that produces the output X_i is identical to the input stream of random variables that produces Y_i, for each i. For example, if the X_i's are exponential(1) and the Y_i's are exponential(2), then the Y_i's are obtained as scaled values of the X_i's.

It is a general belief and empirical observation, though not a proven fact (Schruben and Margolin, 1978), that using identical input streams will create positive correlation in the outputs. (Conditions for this are given in Bratley, Fox, and Schrage, 1983: Chap. 2, Sect. 2.1.) Likewise, it is believed that if many *streams* of inputs are needed in an experiment to produce X_i, then the identical inputs, up to some parametric or structural change, should be used to produce Y_i. For example, in the multiple-server queue, we may want to use exactly the same interarrival times in the multi-line discipline as in the single-line discipline, and the same service times for both as well. The only change is the structural one involved in the queueing discipline. Notice that if X_i and Y_i are correlated through the use of common random variates, then so are $\bar{X}$ and $\bar{Y}$, so that var($\bar{X} - \bar{Y}$) is no longer equal to var($\bar{X}$) + var($\bar{Y}$), nor can it be estimated by $S_{\bar{X}}^2 + S_{\bar{Y}}^2$. Instead, we must use the paired sample variance

$$\sum_{i=1}^{m} \frac{[(X_i - Y_i) - (\bar{X} - \bar{Y})]^2}{m(m-1)}$$

to estimate var($\bar{X} - \bar{Y}$) and perform hypothesis tests (see Chapter 10 for details). In this equation $\bar{X} - \bar{Y}$ is the mean of the differences of the pairs. Individual tests of $\bar{X}$ and $\bar{Y}$ are not affected by the common random numbers since the X_i are independent over i, as are the Y_i.

One drawback to this scheme occurs if we decide, for instance, that $E(X) = E(Y)$ and try to combine all of the output, using $(\bar{X} + \bar{Y})/2$, to estimate the common mean. In this case, the variance of our estimate

$$\text{var}\left(\frac{\bar{X} + \bar{Y}}{2}\right) = \tfrac{1}{4}[\text{var}(\bar{X}) + \text{var}(\bar{Y}) + 2\,\text{cov}(\bar{X}, \bar{Y})]$$

is higher (since cov($\bar{X}, \bar{Y}$) > 0) than if independent input random number streams were used to generate the (X_i, Y_i) pairs.

ii. Multiple Comparisons

The variance reduction technique above refers to a simulation with one factor having only two levels. In this case, using common random variates usually will reduce the variance of the contrast $\bar{X} - \bar{Y}$, without affecting the variance of the individual effects, $\bar{X}$ and $\bar{Y}$, but increasing the variance of the overall mean $\bar{X} + \bar{Y}$. In more complicated experimental designs with more factors and more levels, we may wish to choose input streams that are either:

- Common to all experiments
- Independent among all experiments
- Antithetic (i.e., creating negative dependence between outputs; see examples in Section 4.2.5 or read ahead in Chapter 11) between some experiments
- A mixture of the three types just mentioned

Which scheme we choose will depend on what effects or contrasts we wish to estimate, what model for the output we wish to analyze, and what criteria (i.e., what tradeoffs) in variance reduction we desire. No design will reduce the variance of all possible effects.

Significant work in designing input streams for simulation experiments was done by Schruben and Margolin (1978). In using a definition of optimality that minimizes the determinant of the dispersion matrix for the estimators in a linear model, Schruben and Margolin find optimal designs that use common variates and antithetic variates in orthogonal blocks of the design matrix. For simple designs, it is simple and important to implement their results. For example, in a 2^n factorial design (an n-factor experiment, with two levels for each factor and observations made for all 2^n combinations of levels of factors) with design matrix (X_{ij}), common inputs are used for any experiment i such that $\Pi_j X_{ij} = -1$. More complicated designs, as well as the assumptions used to derive their results, are given by Schruben and Margolin, and we urge readers to use their paper if a large, multifactor simulation is to be run.

The amount of savings in computing time that is achieved by careful design work will depend on the amount of correlation that can be created, as well as the effects being examined. Unfortunately, even if common variates are used as input, very large, complicated simulations may show little correlation in their outputs because of the common inputs. One must always consider the amount of time spent in designing and programming a simulation, and balance this against the amount of variance reduction one can hope to achieve.

8.2 Specific Considerations for Statistical Simulation

All statistical estimators and statistical procedures depend to some degree on the sample size and the distribution of the data that are being analyzed. Even for nonparametric and distribution-free statistics, the power to reject an incorrect null hypothesis depends on sample size and on the particular alternative distribution. Hence, *it is important to determine what statistic or estimator should be used in what sort of situation, where sample size is an ever-present factor in the determination.* In particular, a statistical procedure that is "best" asymptotically (i.e., for large samples) may not be best for small samples.

EXAMPLE   ### 8.2.1 Robust and Resistant Statistics
Recently, a great deal of attention has been focused on the topics of statistical estimation and testing through research in *robust* statistics. A statistic (i.e., estimator or test statistic from some procedure) is said to be robust if it performs well—for instance, has low bias or variance, or high power—across a broad range of distributional assumptions for the data. A related category, *resistant* statistics, is designed to be effective if the data contain a certain, usually small, percentage of values not from the distribution we are trying to examine (i.e., contamination, or outliers in the sense of false values). Since simulation output will never contain outliers, we need not use resistant methods to analyze simulation results; instead the focus here

is on the use of simulation to evaluate resistant methods for use in the analysis of real data.

We cannot use real data to identify robust and/or resistant statistics because we presumably do not know the correct answer, hence cannot decide which statistic produces results closest to the truth. A second approach, theoretical analysis (see, e.g., Hoaglin, Mosteller, and Tukey, 1983), allows us to compare statistics, but usually for their asymptotic properties only. For small sample sizes, we face the problem that robust and resistant statistics are often very nonlinear: they cannot be expressed as linear combinations of the data and sometimes they are the result of iterative calculations (an example is the biweight mean: Hoaglin, Mosteller, and Tukey, 1983, Chap. 11). In many cases the statistics are linear combinations of order statistics, which are themselves nonlinear functions of the data (see Section 3.4). This nonlinearity makes finite sample calculations virtually impossible except in the simplest cases and leaves simulation as the most viable tool. A book by Andrews et al. (1972) documents a large-scale simulation study of robust, resistant estimators of location.

Of course, robustness is not the only statistical property that is of interest. In many problems we wish to compare statistical procedures or estimators for data whose distribution, up to certain parameters, is known.

EXAMPLE   **8.2.2 Estimating the Shape Parameter of a Gamma Distribution**
Given a sample of data whose distribution appears very much like a Gamma distribution, we may be interested in estimating the shape parameter, k, of that hypothesized Gamma(k, β) distribution whose distributional form is given in Table 6.1.2. This problem occurs often in reliability, where the value of the shape parameter determines the behavior of the failure rate (Barlow and Proschan, 1975, Chap. 5). Many estimates of k have been suggested. These estimators, when the rate parameter β or, alternatively, the mean $\mu = k/\beta$ is unknown and is also being estimated, include:

- The moment estimator: $\tilde{k} = \bar{X}^2/S^2$
- The maximum likelihood estimator (mle): $\hat{k}$ is the solution of

$$n[\ln(\hat{k}) - \psi(\hat{k})] = n \ln \bar{X} - \sum_i \ln X_i$$

     where $\psi(\cdot)$ is the digamma function (see, e.g., Cox and Lewis, 1966)
- Jackknifed versions of both the moment estimator and mle (see Chapter 9 for a discussion of jackknifing). Jackknifing is applied to these statistics to eliminate the term of order $1/n$ in the bias (see Equation 6.1.38) in these statistics.

It is difficult to calculate any theoretical properties that allow comparisons between the competing estimators except for asymptotic comparisons of the moment and mle estimates (Cox and Lewis, 1966, Chap. III; Johnson and Kotz, 1970, Chap. 17), but simulation results (Lewis et al., 1985) are very enlightening. They show that for $k = 5.0$ both the moment and maximum likelihood estimators have positive skewness and kurtosis for small n (10–100) and that, although both estimators are quite biased, the maximum likelihood estimator has slightly smaller standard deviation than the moment estimator. At $n = 33$, the ratio is 1.425 to 1.481, slightly higher than the asymptotic ratio of 0.78. Jackknifing appears to help by removing much of the bias in the two estimators. However, the inflation in the standard deviation of the estimator due to jackknifing almost offsets the bias deflation; mean square error is about the same for the two estimators with or without jackknifing.

Other examples of situations in which we want to compare the performance of different statistics include the following.

- The t test versus the trimmed t test (see Section 3.4)
- Estimating the center of a symmetric distribution by either the mean, the median, or the trimmed mean (subsection ii of Section 8.2.2)
- Estimating serial correlation by either the biased, unbiased, or two-sample estimator (subsection iii of Section 8.2.2)
- Robust regression versus least-squares regression
- Nonparametric versus parametric estimation

Performance of the competing statistics can be measured by looking at questions such as the following.

- What bias does each statistic have as a function of sample size?
- How do the variances (standard deviations) of the statistics compare, again as a function of sample size?
- How do the asymptotic variances of the competing statistics compare?
- What are the small-sample distributions of the statistics?
- Are the statistics asymptotically Normal, and with what means and variances?
- How fast do the distributions approach asymptotic Normality?

From the list of questions, it should be clear that sample size is a critical factor, affecting bias, variance, and distribution. The sample size in statistical simulation is identical to the time parameter n in the general framework for simulation in Section 3.7. Recognizing this, the procedure for running a *single* statistical simulation can be outlined as follows (it is basically the prescription for crude simulation given in Chapter 4).

1. Generate the simulated process until time n; in the case of a statistical simulation, generate a statistic based on a sample of n random variables from a given distribution.
2. Compute and record the result; in this case, calculate the desired statistic based on the n data points.
3. Replicate the process to get m observations of the statistic.
4. Analyze the empirical distribution of the statistic.

To perform an entire statistical simulation experiment, we must go through the procedure above for all the following factors:

1. All competing statistics that are to be compared
2. All sample sizes that can give information about small-sample and asymptotic behavior

3. Different distributions for the underlying data sample, including:
   a. The various families of distributions (i.e., Normal, exponential)
   b. The different parameters within each family
   c. Possible dependence among the observations

It is not possible to look at all combinations of all these factors (i.e., you cannot try the t test on Gamma(k, β) data for *all* values of k and β), but you should try out a "grid" of parameter values. For instance, it would be good to know how the t test performs on one-sided data that are very skewed [Gamma$(0.5, 1)$]; moderately skewed [Gamma$(1, 1)$, which is equivalent to exponential(1)]; and slightly skewed [Gamma$(5, 1)$]. This still requires generating many simulation output sets, whose organization and analysis are considered in the next section, where we present a graphical method ideally suited to displaying the changes that occur when sample size is allowed to vary, as is often the case in statistical simulations. The more general techniques already presented in Section 8.1 can, of course, also be applied.

8.2.1   Organizing and Analyzing the Results of Statistical Simulations: The SMTBPC Program

In this section, we show how to use simulation to investigate a statistical procedure or to compare competing estimators. We also introduce the basic ideas of the SMTBPC program, which is a simulation testbed for organizing the outputs of statistical simulations. The program is contained in the software supplement *Enhanced Simulation and Statistics Package*, along with random-number generators, which are needed for the input to the program.

As an example of a statistical simulation that can be analyzed with SMTBPC, we can ask: How robust is the t statistic? That is, will a t test for a hypothesized mean value still be valid if performed on non-Normal data? For the t test to be valid, the t statistic T_n (Equation 3.4.1), computed from n data points, must have a t distribution with $n - 1$ degrees of freedom or, if n is large (say, greater than 40), the t statistic should have approximately a Normal$(0, 1)$ distribution. Since sample size and the underlying distribution of the data affect the distribution of the t statistic, these two factors must be built into our simulation design. Competing procedures, such as a test based on the trimmed t statistic (Section 3.4) are not considered here, nor do we address questions of the power of the t test. However, investigation of the power of the t test when testing a location parameter in non-Normal situations is a typical case in which simulation is needed to augment theoretical results. For analytical results see Yoav (1983).

According to Section 8.2, we should start the investigation of the robustness of the t statistic by picking a particular sample size n, and a fixed distribution for the underlying data. We then generate n points from that distribution, calculate the statistic T_n, and repeat this until we have, say, m observations of T_n. (The total number of random variables generated, nm, will be denoted N.) To look at the empirical distribution of the statistic, we need to be thorough, yet our presentation of the simulation data must be compact enough to allow comparisons with other factors in the design. For this purpose the boxplot (subsection iii of Section 7.1.4) is nearly one-dimensional, but still gives a good graphical view of the distribution of the statistic, provided the data are continuous. In addition, basic numerical summaries [Section 6.1: mean, standard deviation, standard deviation of the mean (to show the precision of the estimated mean), coefficient of skewness, and coefficient of kurtosis] quantify some of the qualitative features of the boxplot and make comparisons to standard distributions easier.

To make this example of the t statistic concrete, we generate 500 samples of 10 simulated Normal(0, 1) random variables each [i.e., $n = 10$, $m = 500$, and $N = 5000$ total Normal(0, 1) random variables] and compute 500 values of the t statistic T_{10}. The boxplot and summary statistics in Figure 8.2.1 show an empirical distribution for the t statistic that is close to the theoretical t distribution with 9 degrees of freedom that 10 Normal(0, 1) points should produce (see Table 8.2.1 below for the theoretical values of the summary statistics).

Since the effects of sample size are of particular importance in evaluating the bias, variability, and asymptotic distribution of the t statistic, we should repeat the simulation with independent data using a different sample size, n_2. The number of replications m_2 is adjusted so that the product $n_2 m_2 = nm = N$ remains constant. This keeps the variances of the estimated means of the two empirical distributions approximately equal (see Exercise 8.1 and comments below on regressions for the mean), although the boxplots represent samples of different sizes (m and m_2). For comparisons, the boxplot and summary statistics for the empirical distribution of the m_2 simulated values of the statistic based on n_2 simulated data points should be placed next to the previous one. The x-axis in the graph can then be interpreted as data sample size.

Figure 8.2.2 illustrates this comparison with the first boxplot representing 500 estimates of the t statistic based on samples of 10 points (i.e., T_{10}) and the second representing 250 estimates of the t statistic based on samples of 20 points (i.e., T_{20}). *Note that both boxplots are plotted on the same vertical scale.*

It is clear from Figure 8.2.2 that the t statistic has longer tails at $n = 10$ than at $n = 20$. Also, the percentage of outliers in the boxplot at $n = 10$ indicates definite non-Normality.

We can keep changing the sample size until we run out of room on the paper for more boxplots. The SMTBPC program from the software supplement, which produced Figures 8.2.1 and 8.2.2, automates the division of the data and can accommodate up to eight different sample sizes, $n_1, n_2, \ldots, n_8$.

Once we have run the simulation eight times with different input data [here Normal(0, 1) random variables] and sample sizes, we will have eight estimated means and eight estimated variances of the empirical distributions, one for each of those sample sizes. Working with these estimates, we can quantify and extrapolate the changes in mean and variance with sample size by assuming that the asymptotic expansions in Equations 6.1.38 and 6.1.39 hold. The calculated means will be used as the dependent variables, and the inverse powers of the respective sample sizes will be used as the independent variables to estimate the coefficients $\theta, a_1, a_2, \ldots$ in Equation 6.1.38. Since the estimated means are equivariant, as discussed in Exercise 8.1, simple least-squares regression can be used. Similarly, we can use the estimated variances of our statistic as the dependent variables and solve for estimates of

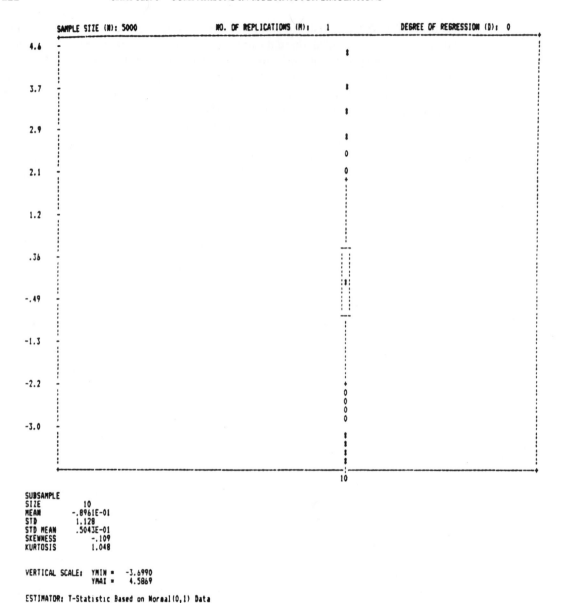

Figure 8.2.1   A boxplot and summary statistics for 500 values of the *t* statistic, each computed from 10 simulated independent Normal(0, 1) random variables.

b_1, b_2, b_3, ... in Equation 6.1.39. This time, the variances are *not* equivariant (see Exercise 8.2) and we must use weighted regression, the weight being equal to $n_i^{1/2}$.

Since we have only a small number of estimates of the mean or variance, we can reasonably solve for only the first few coefficients in either asymptotic expansion. (SMTBPC limits you to 6, although 3 or 4 is usually sufficient.) Hence, our regression model may be truncated and false, creating misleading coefficients. In addition, *it is not necessarily true that either Equation 6.1.38 or 6.1.39 holds*, so that

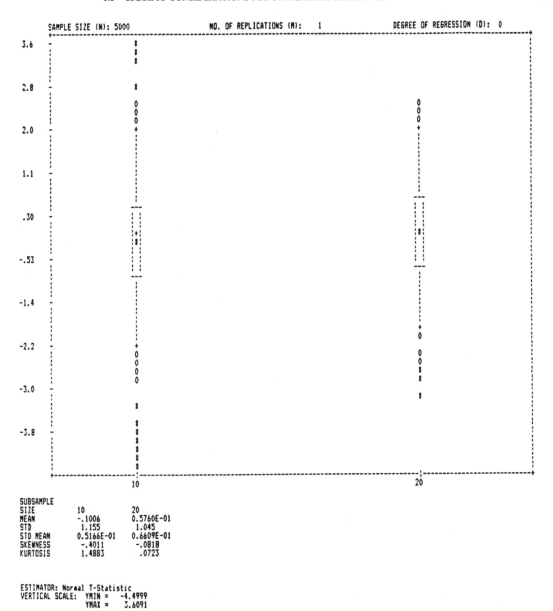

Figure 8.2.2 The simulation experiment in Figure 8.2.1 repeated with two sample sizes, $n_1 = 10$ and $n_2 = 20$.

none of the coefficients may make sense. As an example, the variance of the t statistic approaches 1 for most distributions of data, thereby invalidating Equation 6.1.39 (which is true for many statistical estimators, but not Normalized test statistics). It is up to you to decide whether such regression models are appropriate for the simulation you are working on. Alternatively, one could look at $T_n/(n^{1/2})$, whose moments will satisfy the asymptotic expansions of Equations 6.1.38 and 6.1.39.

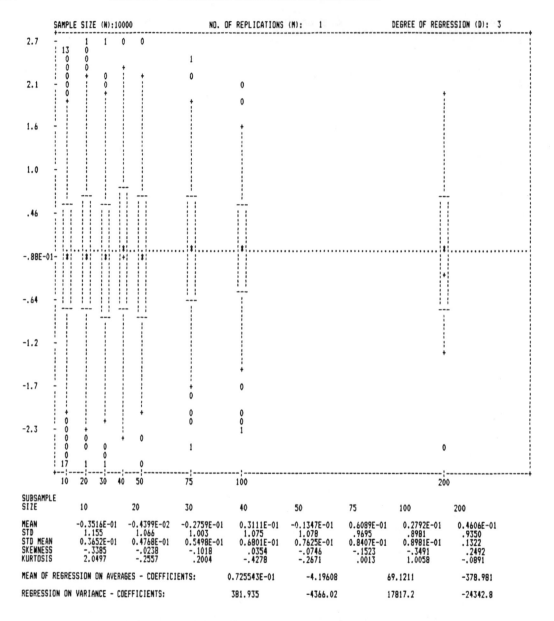

Figure 8.2.3 A one-factor simulation experiment exploring the distribution of the t statistic T_n with Normal(0, 1) data for eight sample sizes n_i. In addition, a regression analysis of the mean and variance of T_n has been performed by the SMTBPC program by setting the "degree of regression" parameter to 3.

If the asymptotic expansions are appropriate, the regression estimate of θ estimates the asymptotic mean of our statistic and the estimate of a_1 gives a value for the bias term of order $(1/n)$. From the variance equation, the term b_1 shows up in calculating asymptotic relative efficiency (Cox and Hinkley, 1974, Chap. 9), so that our estimate of b_1 can be used in estimating asymptotic relative efficiency. Higher order terms in either the mean or variance expansion are usually estimated with high variability and are difficult to interpret.

The SMTBPC output given in Figure 8.2.3 shows the empirical distribution of the t statistic T_n when it is calculated from independent Normal samples of size 10, 20, 30, 40, 50, 75, 100, and 200. In each case, $N = 10,000$ independent simulated Normal$(0, 1)$ random variables were divided into appropriate sized samples for calculation of the t statistic. The regression estimates of θ and a_1, at the bottom of the figure, show that the estimate of θ (0.072) is close to the known true mean (0.0). Also, since the t statistic has mean 0 when computed from data from a symmetric distribution, the estimate of a_1 (-4.20), the coefficient of the $1/n$ term in the asymptotic expansion, seems to be far from its true value of 0. However the standard deviation of the estimated coefficient is unknown, so precise conclusions cannot be drawn. Similar observations hold for the estimates of a_2 and a_3, whose true values are also zero.

To find out how precise the estimate of θ (or any other parameter in the regression on the means) is, all the work represented in Figure 8.2.3 can be repeated M independent times. From those M replications, we will get M estimates of θ,

$$\bar{\theta}(1), \bar{\theta}(2), \ldots, \bar{\theta}(M)$$

which can be averaged together to obtain a lower variance (more precise) estimate of θ,

$$\bar{\bar{\theta}} = \sum_{i=1}^{M} \frac{\bar{\theta}(i)}{M}$$

as well as an estimated variance of $\bar{\bar{\theta}}$, (see Equation 4.1.2),

$$\text{vâr}(\bar{\bar{\theta}}) = \frac{\sum (\bar{\theta}(i) - \bar{\bar{\theta}})^2}{M(M - 1)}$$

Notice that M times as much data must be generated and M times as many estimates of the statistic will be calculated at each sample size. The estimates from all M replications are combined and used as one large sample to produce the boxplots and summaries.

Figure 8.2.4 repeats the experiment in Figure 8.2.3 $M = 15$ times, so that $NM = 150,000$ independent simulated Normal$(0, 1)$ random variables are used at each sample size. Estimates of the parameters in the regression on the means, and their estimated variances, can be found below the moment summaries. In addition, a dashed line through the boxplots shows the estimated asymptotic mean while a dotted line shows the estimated regression equation for the mean. In the present

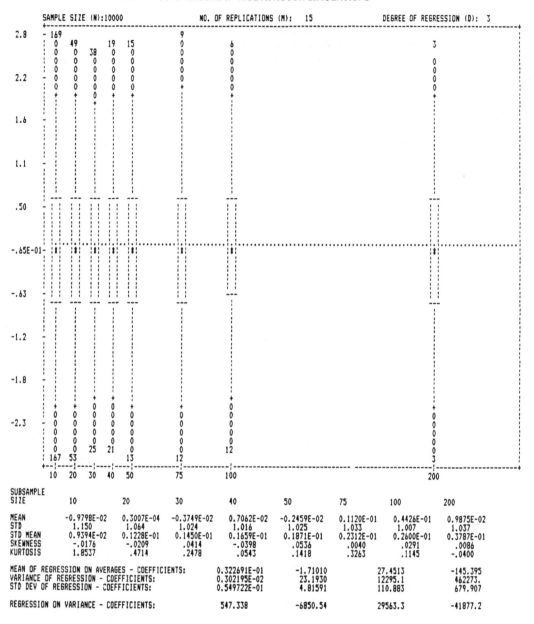

Figure 8.2.4 This figure is similar to Figure 8.2.3 except that the number of replications (*M*) has been set to 15. This gives standard deviation estimates for the parameters in the regression on the mean. Note the use of the reduced graphics option, which counts the number of extreme outliers and is a useful feature when it is necessary to emphasize the body of the boxplot.

case these are identical and are shown as a dotted line. These lines, when distinct, give a visual presentation of how the bias in the statistic is changing with sample size (see Figure 8.2.7).

Note that θ, with true value 0, is now estimated as 0.032 with estimated standard deviation 0.055. Thus this estimate is within one sample standard deviation of its true value. Similarly a_1 is now estimated as -1.71 where its true value is again zero. This estimate of a_1 has an estimated standard deviation of 4.81. A figure such as Figure 8.2.4 does an excellent job of displaying and quantifying the distribution of a statistic and the changes in that distribution as sample size increases. To study the *added factor* of underlying data distribution, similar experiments must be run and figures produced when we change the assumption of Normal$(0, 1)$ data. Such a study is pursued in Section 8.2.2i. If competing procedures or estimators exist, such as the trimmed t statistic, analogous figures must also be produced for each of these.

The process of organizing multiple simulation outputs is made easy by the SMTBPC program which, given subroutines that calculate the desired statistics and sources of data, will output up to three figures using optionally the *same scaling*. Having the same scaling is imperative for proper comparison. The flexibility and simplicity of use of the SMTBPC program are documented in the software supplement, while further comments about the design of the output are given in Lewis et al. (1985).

Finally, note that an automated output display procedure is useful not only for statistical simulation output analysis, but also for any simulation output where sample size or time is an important factor in the design. An example of this is the sequential design of regenerative simulations (see Heidelberger and Lewis, 1981; this will be expanded on in Volume II).

8.2.2 Examples of Statistical Simulation

The procedure from Section 8.2.1 for organizing and presenting the output from multiple statistical simulations is now implemented using the SMTBPC program to look at:

- The robustness of the t statistic T_n defined at Equation 3.4.1
- Three estimates of the center of two symmetric distributions
- Three estimates of serial correlation in small samples

Further examples of statistical simulations using SMTBPC—for instance, to demonstrate the effectiveness of jackknifing and bootstrapping—appear later (Chapter 9).

i. Robustness and the *t* Statistic

The t statistic (Section 3.4) is widely used—often too widely—to test hypotheses and construct confidence intervals for the mean of a distribution. In fact, under certain provisos, the t statistic has been suggested for use in analyzing simulation output (see Section 6.1.4). Those provisos reduce to two rules:

- If sample size is small, say $n \leqslant 40$, then we want the t statistic to have a distribution that is close to a t distribution with $n - 1$ degrees of freedom.

- If sample size is large, say $n > 40$, then the t distribution with $n - 1$ degrees of freedom is very hard to distinguish from a Normal$(0, 1)$ distribution, so we demand that the t statistic have a distribution close to a Normal$(0, 1)$ distribution.

These rules are satisfied fully if the data that go into calculating the t statistic come from a Normal(μ, σ) distribution, a fact that is illustrated by the simulation experiment in Figure 8.2.4. It is suggested that the reader use Table 8.2.1 to calculate and compare the relevant moments of the t distribution with the simulated values in Figure 8.2.4.

Notice in particular in Figure 8.2.4 that the heavy tails of the t distribution melt into Normal$(0, 1)$ tails as sample size increases to 40. *One way* of seeing this is from the kurtosis at $n_1, n_2, \ldots, n_8$. Note, however, that the estimates of kurtosis (or any moment) will not keep getting closer to 0 indefinitely as n increases; in fact, the

Table 8.2.1 Theoretical moments of the t distribution

Moments	Degrees of freedom	
	$n - 1$	∞
$E(T_n)$	0.0	0.0
Standard deviation, T_n	$\left[\dfrac{(n - 1)}{(n - 3)}\right]^{1/2}$	1.0
Coefficient of skewness, T_n	0.0	0.0
Coefficient of kurtosis, T_n	$\left[\dfrac{3(n - 3)}{(n - 5)}\right] - 3$	0.0

Table 8.2.2 Estimated kurtosis $\hat{\gamma}_2$ for the t statistic with several types of distribution for the data

The variability in the estimates can be approximated by the Normal theory approximation to the standard deviation of the kurtosis estimate $\hat{\gamma}_2$ and is given in the first row. The theoretical kurtosis for T_n computed with Normal data is given in the last row.

n_i	10	20	30	40	50	75	100	200
$m_i = NM/n_i$	15,000	7,500	5,000	3,750	3,000	2,000	1,500	750
s.d. of $\hat{\gamma}_2$ [a] $(24/m)^{1/2}$	0.040	0.057	0.069	0.080	0.089	0.110	0.126	0.179
$\hat{\gamma}_2$: Normal data Figure 8.2.4	1.85	0.47	0.25	0.05	0.14	0.33	0.11	0.00
$\hat{\gamma}_2$: Laplace data Figure 8.2.5	0.42	0.08	0.03	0.10	-0.05	0.29	-0.04	0.08
$\hat{\gamma}_2$: Uniform data Figure 8.2.6	2.20	0.62	0.33	0.03	0.32	0.28	-0.07	0.27
$\hat{\gamma}_2$: Gamma(0.5, 1) data Figure 8.2.7	62.99	13.39	4.13	2.69	2.44	0.60	0.57	0.22
Theoretical kurtosis for (Normal data) T_n	1.200	0.400	0.240	0.171	0.133	0.086	0.063	0.031

[a] Presumes normality of the simulated T_n.

estimates of kurtosis are becoming more variable as the sample size n grows larger (see Exercise 8.4). Hence, you should look for trends in the estimated parameters, not complete convergence to fixed values. Table 8.2.2 illustrates this, giving in the third row the estimated value of the kurtosis of the t statistic for the sample sizes n used in Figure 8.2.4. These should be compared to the theoretical values in the last line of Table 8.2.2.

To determine whether the t statistic still follows the bulleted rules above if the data are not Normal, the experiment was repeated using Laplace$(0, 1)$ data. Since the Laplace distribution is symmetric, the distribution of the t statistic must still be symmetric, but the high kurtosis of the Laplace distribution may affect the distribution of the t statistic. The SMTBPC output in Figure 8.2.5a bears out this intuition. When compared to either Figure 8.2.4 or to theoretical t distribution moments, we see that the t statistic based on Laplace data has the correct mean, standard deviation, and skewness, but the tails are too short. This fact is evident in both the boxplot (consider the number of far outliers) and the estimated kurtosis, listed on the third row of results in Table 8.2.2.

It appears in fact, that from sample size 20 on, the Normal$(0, 1)$ distribution may be a better approximation to the distribution of the Laplace data t statistic than the t distribution. This observation is supported by Figure 8.2.5b (output optionally from the SMTBPC program), which shows numerically, estimates of selected quantiles of the estimated distribution of the t statistic in Figure 8.2.5a, and graphically, the quantiles adjusted for mean and variance (Section 6.3.3). The graphics make it easy to compare the empirical quantiles to the standard Normal$(0, 1)$ quantiles, which are represented by the tick marks on the right-hand side of the plot.

Alternatively, if the experiment is run on uniform$(0, 1)$ data, which has a very small kurtosis, the distribution of the t statistic, shown in Figure 8.2.6, has too high a kurtosis (row 4 of results in Table 8.2.2) at small sample sizes. Hence, if we performed t tests on small samples of uniform data, we would reject a correct null hypothesis too often.

Finally, if the underlying data is from a highly non-Normal, very skewed Gamma$(0.5, 1)$ distribution, then Figure 8.2.7 is produced. The high kurtosis and negative skewness are clearly evident in the boxplots out to sample size 200. The conclusion is that we should not feel comfortable using a t test on highly skewed data even if we have a relatively large sample.

ii. Using the Mean, Median, or Trimmed Mean to Estimate Location

A great deal of theoretical work has gone into comparing the sample mean, median, and trimmed mean as estimators of the center of a symmetric distribution. Since all these estimators have expectation equal to the mean in symmetric populations (i.e., are unbiased), the quantity of interest is the variance of each estimator.

For this example, we will be using a version of the trimmed mean (Section 3.4), namely the 5% trimmed mean so that $((0.05)n)$ of the ordered observations are removed from each end of the sample. Asymptotic calculations, summarized in Table 8.2.3 (Gastwirth and Cohen, 1970; Lehmann, 1983, Chap. 5), show that the sample mean has minimum variance when the data come from a Normal$(0, 1)$

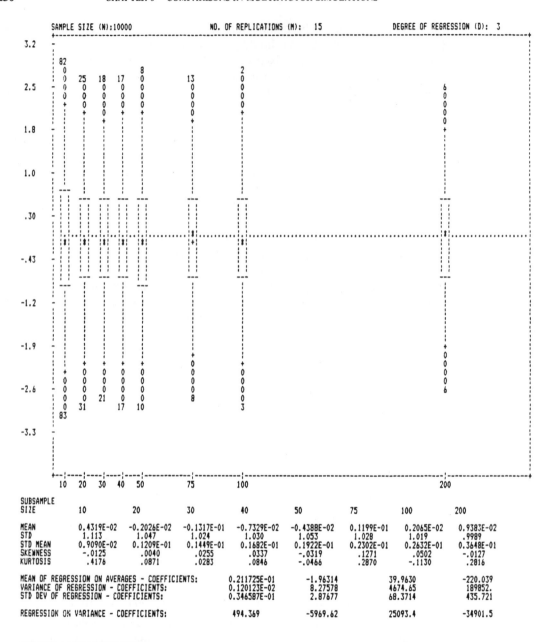

SAMPLE SIZE (N):10000   NO. OF REPLICATIONS (M):   15   DEGREE OF REGRESSION (D): 3

SUBSAMPLE SIZE	10	20	30	40	50	75	100	200
MEAN	0.4319E-02	-0.2026E-02	-0.1317E-01	-0.7329E-02	-0.4388E-02	0.1199E-01	0.2065E-02	0.9383E-02
STD	1.113	1.047	1.024	1.030	1.053	1.028	1.019	.9989
STD MEAN	0.9090E-02	0.1209E-01	0.1449E-01	0.1682E-01	0.1922E-01	0.2302E-01	0.2632E-01	0.3648E-01
SKEWNESS	-.0125	.0040	.0255	.0337	-.0319	.1271	.0502	-.0127
KURTOSIS	.4176	.0871	.0283	.0846	-.0466	.2870	-.1130	.2816

MEAN OF REGRESSION ON AVERAGES - COEFFICIENTS:	0.211725E-01	-1.96314	39.9630	-220.039
VARIANCE OF REGRESSION - COEFFICIENTS:	0.120123E-02	8.27578	4674.65	189852.
STD DEV OF REGRESSION - COEFFICIENTS:	0.346587E-01	2.87677	68.3714	435.721
REGRESSION ON VARIANCE - COEFFICIENTS:	494.369	-5969.62	25093.4	-34901.5

ESTIMATOR: LAPLACE T-STATISTIC
VERTICAL SCALE: YMIN =   -3.9251
                  YMAX =   3.2116

Figure 8.2.5a   An SMTBPC run to investigate the distribution of the *t* statistic computed from iid Laplace-distributed data, showing a boxplot and summary statistics for each of eight sample sizes, $n_i = 10, 20, 30, 40, 50, 75, 100,$ and 200. Reduced graphics have been used and the vertical scaling is the same as that in Figures 8.2.6 and 8.2.7. Note that there are fewer outliers at $n = 10$ than there are in Figure 8.2.4 at $n = 10$ for Normal data.

SAMPLE SIZE (N):10000   NO. OF REPLICATIONS (M): 15   DEGREE OF REGRESSION (D): 3

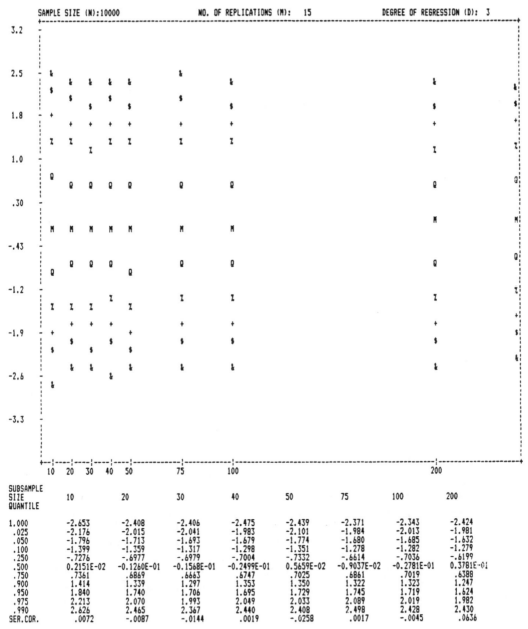

SUBSAMPLE SIZE	10	20	30	40	50	75	100	200
QUANTILE								
1.000	-2.653	-2.408	-2.406	-2.475	-2.439	-2.371	-2.343	-2.424
.025	-2.176	-2.015	-2.041	-1.983	-2.101	-1.984	-2.013	-1.981
.050	-1.796	-1.713	-1.693	-1.679	-1.774	-1.680	-1.685	-1.632
.100	-1.399	-1.359	-1.317	-1.298	-1.351	-1.278	-1.282	-1.279
.250	-.7276	-.6977	-.6979	-.7004	-.7332	-.6614	-.7036	-.6199
.500	0.2151E-02	-0.1260E-01	-0.1568E-01	-0.2499E-01	0.5659E-02	-0.9037E-02	-0.2781E-01	0.3781E-01
.750	.7361	.6869	.6663	.6747	.7025	.6861	.7019	.6388
.900	1.414	1.339	1.297	1.353	1.350	1.322	1.323	1.247
.950	1.840	1.740	1.706	1.695	1.729	1.745	1.719	1.624
.975	2.213	2.070	1.993	2.049	2.033	2.089	2.019	1.982
.990	2.626	2.465	2.367	2.440	2.408	2.498	2.428	2.430
SER.COR.	.0072	-.0087	-.0144	.0019	-.0258	.0017	-.0045	.0636

ESTIMATOR: LAPLACE T-STATISTIC

Figure 8.2.5b Plot and tabulation of standardized quantiles $((\hat{x}_\alpha - \bar{x})/s)$ for each sample size for 11 values of α. These can be compared to the quantiles of the Normal(0, 1) distribution on the right-hand side of the plot.

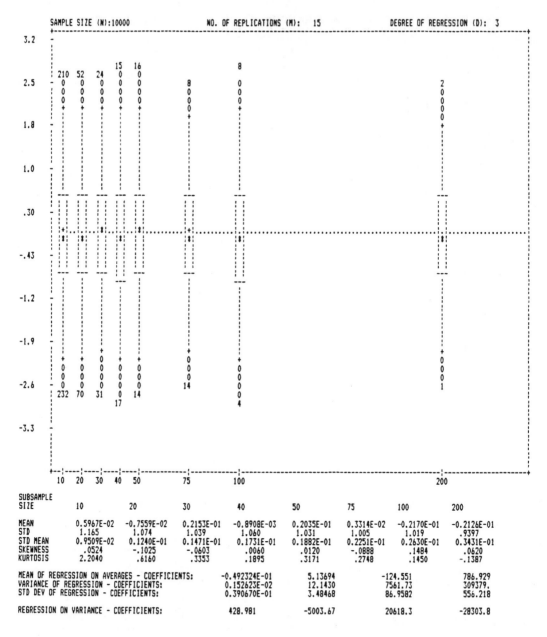

Figure 8.2.6 Uniformly distributed data used to produce a figure analogous to Figure 8.2.5a, to which this figure should be compared.

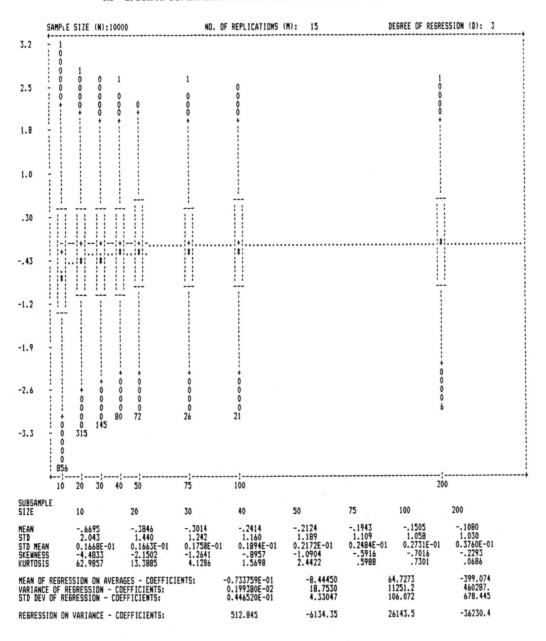

Figure 8.2.7 Gamma(0.5, 1)-distributed data, used to produce a figure analogous to Figures 8.2.5a and 8.2.6. Note that the *t* statistic in this case is negatively skewed and extremely non-Normal up to sample size 100.

Table 8.2.3 Asymptotic variances of the mean, median, and 5% trimmed mean

Data distribution	Estimator		
	Mean	Median	5% trimmed mean
Normal(0, 1)	1.00/n	1.57/n	1.01/n
Laplace(0, 1)	2.00/n	1.00/n	1.65/n

distribution and that the sample median has minimum variance for Laplace(0, 1) data. The 5% trimmed mean represents a general-purpose (i.e., robust) compromise between the two.

Small-sample calculations have also been done, with much effort (Gastwirth and Cohen, 1970), which show that the asymptotic results hold fairly well in small Normal samples ($n = 20$), and, for the mean and trimmed mean, in small Laplace samples. In Laplace samples of size 20, the median is 33% more variable than predicted by the asymptotics.

Hence, Figures 8.2.8 through 8.2.13, which show the simulated distributions of the mean, median, and 5% trimmed mean based on Normal(0, 1) and Laplace(0, 1) data, contain no surprises. They do, however, provide three things that are quite valuable:

• *A very simple example of the use of statistical simulation that could serve to teach, very vividly and graphically, an important lesson in point estimation.* This lesson is that the mean, although a very popular estimator, is a good estimator (i.e., has low variance) only when the data have tails about as long as those of a Normal distribution. If the tails are heavy like a Laplace, then the median is a better estimator (i.e., is a lower variance estimator than the mean). Finally, if the tails are somewhere between those of a Normal and a Laplace distribution, or if we really don't know what the tails are, this limited simulation experiment indicates that we would be better off using the 5% trimmed mean. This is because it has only a slightly larger variance than the mean in the Normal(0, 1) case and does better than the mean in the Laplace case.

• *A look at the small-sample distributions of the mean, median, and 5% trimmed mean.* All are known to be asymptotically Normal (Lehmann, 1983, Chap. 5) but, in general, their small-sample distributions (hence, testing theory) are not theoretically tractable.

• *A framework that can be extended easily to include any desired sample size, any underlying distribution for the data, and any other estimator of location for comparison.*

Since we have known theoretical results for purposes of comparisons, we can also evaluate the precision of our simulation estimates. Comparison to the small-sample ($n = 20$) results in Gastwirth and Cohen (1970) shows that the standard deviations of the estimators are good to the second decimal place—more than sufficient for practical comparison and choice. Be careful though, about the esti-

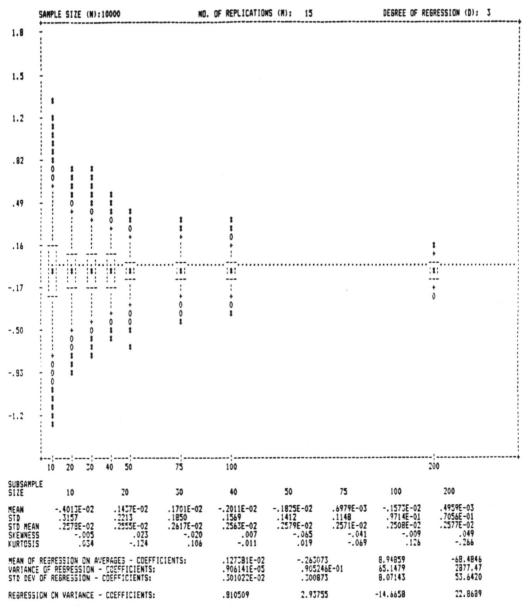

Figure 8.2.8 Mean–Normal(0, 1) data. The mean from a Normal(0, 1) sample of size n is Normal(0, $n^{-1/2}$) and this is reflected in the symmetry and number of outliers in the boxplots at the different sample sizes. Note, too, that in the moment summaries the true values of the coefficients of skewness and kurtosis are zero.

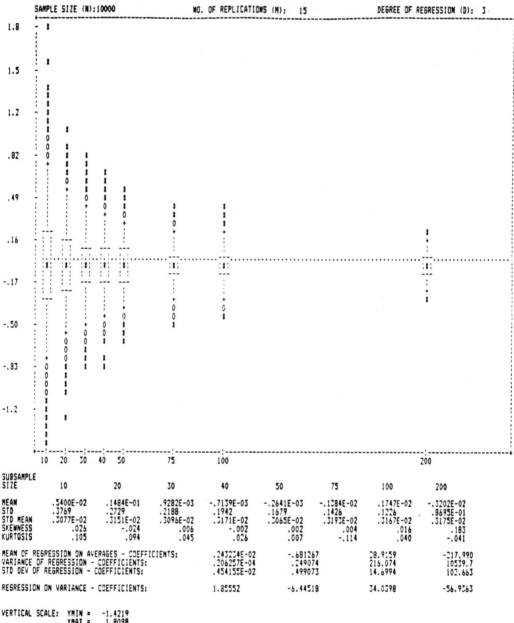

Figure 8.2.9 Median–Normal(0, 1) data. The median is a highly nonlinear function of the sample data and therefore, even for small sample sizes from Normal data, is not necessarily Normally distributed. There tend to be more outliers in the median than in the mean, and the estimated standard deviation of the median is 0.3769, which is to be compared to the estimated standard deviation of the mean, 0.3157, at sample size 10 (in Figure 8.2.8). (True value is $10^{-1/2}$.) This reflects the greater efficiency of the mean as a location estimator in Normal samples.

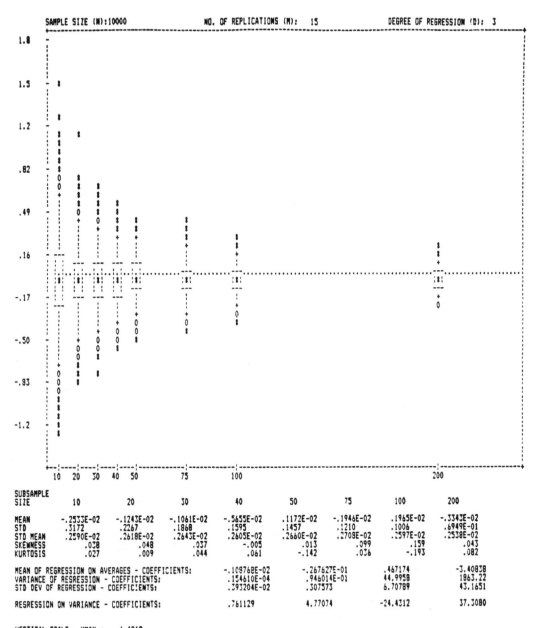

SUBSAMPLE SIZE	10	20	30	40	50	75	100	200
MEAN	-.2533E-02	-.1243E-02	-.1061E-02	-.5655E-02	.1172E-02	-.1946E-02	.1965E-02	-.3343E-02
STD	.3172	.2267	.1868	.1595	.1457	.1210	.1006	.6949E-01
STD MEAN	.2590E-02	.2618E-02	.2643E-02	.2605E-02	.2660E-02	.2709E-02	.2597E-02	.2538E-02
SKEWNESS	.038	.048	.037	-.005	.013	.099	.159	.043
KURTOSIS	.027	.009	.044	.061	-.142	.036	-.193	.082

MEAN OF REGRESSION ON AVERAGES - COEFFICIENTS:	-.109768E-02	-.267627E-01	.467174	-3.40838
VARIANCE OF REGRESSION - COEFFICIENTS:	.154610E-04	.946014E-01	44.9958	1863.22
STD DEV OF REGRESSION - COEFFICIENTS:	.393204E-02	.307573	6.70789	43.1651
REGRESSION ON VARIANCE - COEFFICIENTS:	.761129	4.77074	-24.4312	37.3080

VERTICAL SCALE: YMIN = -1.4219
                 YMAX = 1.8098

ESTIMATOR: Distribution of the Trimmed Mean Based on Normal(0,1) Data

Figure 8.2.10 Trimmed mean–Normal(0, 1) data. Here the mean has been trimmed; note that the estimated standard deviation is only slightly greater than it is for the mean: 0.3172 versus 0.3157 (in Figure 8.2.8).

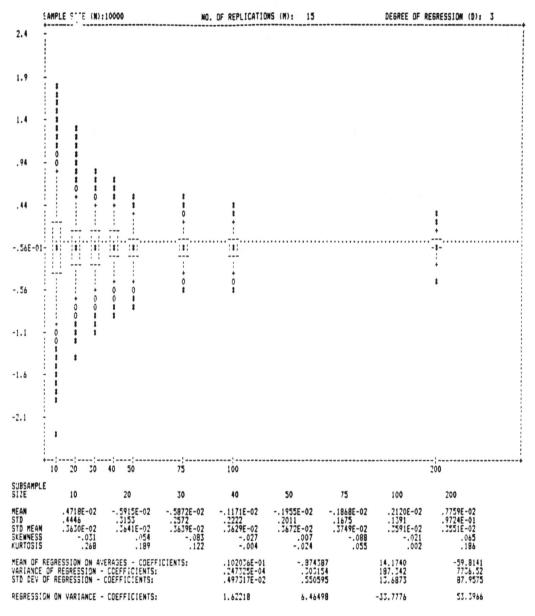

SAMPLE SIZE (N):10000   NO. OF REPLICATIONS (M): 15   DEGREE OF REGRESSION (D): 3

SUBSAMPLE SIZE	10	20	30	40	50	75	100	200
MEAN	.4718E-02	-.5915E-02	-.5872E-02	-.1171E-02	-.1955E-02	-.1868E-02	.2120E-02	.7759E-02
STD	.4446	.3153	.2572	.2222	.2011	.1675	.1391	.9724E-01
STD MEAN	.3630E-02	.3641E-02	.3639E-02	.3629E-02	.3672E-02	.3749E-02	.3591E-02	.3551E-02
SKEWNESS	-.031	.054	-.083	-.027	.007	-.088	-.021	.065
KURTOSIS	.268	.189	.122	-.004	-.024	.055	.002	.186

MEAN OF REGRESSION ON AVERAGES - COEFFICIENTS:	.102026E-01	-.974387		14.1740	-59.8141
VARIANCE OF REGRESSION - COEFFICIENTS:	.247325E-04	.303154		197.342	7736.52
STD DEV OF REGRESSION - COEFFICIENTS:	.497317E-02	.550595		13.6873	87.9575

REGRESSION ON VARIANCE - COEFFICIENTS:	1.62218	6.46498	-33.7776	53.3966

VERTICAL SCALE: YMIN =   -2.4555
                  YMAX =   2.4433

ESTIMATOR: Distribution of the Mean Based on Laplace(1) Data

Figure 8.2.11   Mean–Laplace(0, 1) data. The simulation experiment of Figures 8.2.8 to 8.2.10 is repeated for data having a Laplace(0, 1) distribution. Convergence to Normality is quite rapid, as can be seen from the shape of the boxplots or the small values of the (sample) skewness and kurtosis.

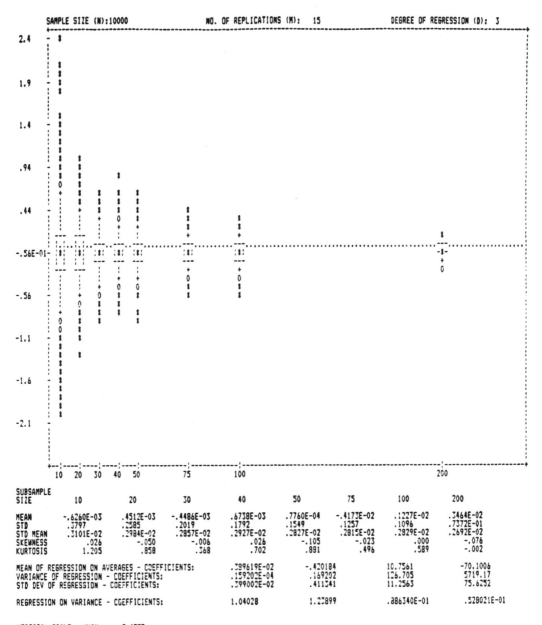

Figure 8.2.12 Median–Laplace(0, 1) data. The median in a Laplace(0, 1) sample is the mle and the minimum variance estimate of the mean. Note that the estimated standard deviation for the median at $n = 10$ is 0.3797, compared to the value for the mean (from Figure 8.2.11) of 0.4446. The corresponding numbers at $n = 200$ are 0.0737 versus 0.0972. From Table 8.2.1 the ratio of these quantities should be 0.5. The deviation is because of the small number of replications (750) obtained in the simulation at $n = 200$.

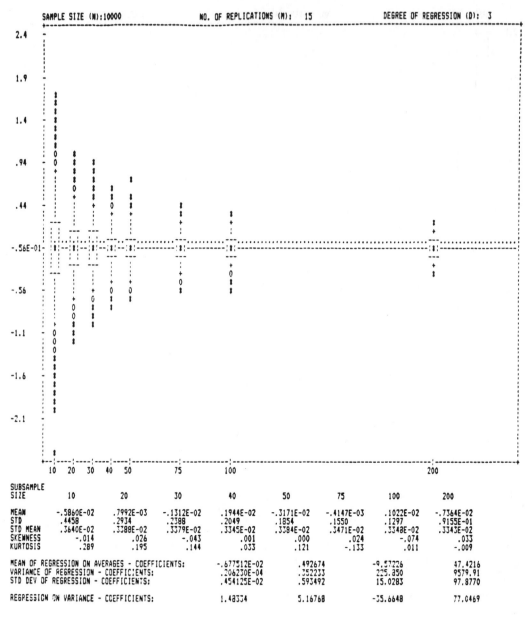

Figure 8.2.13   Trimmed mean–Laplace(0, 1) data. The trimmed mean is, again, a less efficient estimator of the center of the distribution than the median. The estimated standard deviations at $n = 10$ are 0.4458 (trimmed mean) and 0.3797 (median; from Figure 8.2.12) and at $n = 200$ are 0.0916 and 0.0737 (from Figure 8.2.12).

mated coefficients for the regressions on the means and variances. For instance, the estimate of b_1, the coefficient of the leading term in the asymptotic expansion of the variance of the median of Normal data, from Figure 8.2.9, is 1.856, quite far from the true value of 1.57, in Table 8.2.3. To make the asymptotics more precise, we would need to run the simulation at larger sample sizes (up to now we have been interested only in small-sample performance of estimators), and larger values of N and M (this will mean going to larger, faster computers than the IBM/PC).

iii. Estimating Serial Correlation

One topic in simulation that we have skirted up to now is serial covariance and serial correlation (sometimes called autocorrelation), which measure dependence in a *sequence* of random variables. Only in discussing uniform random variable generation in Chapter 5 have we worried about dependence from one data point to the next. Otherwise, in generating or analyzing simulation output, we have designed the simulation to use and produce *independent* observations. We introduce an alternative type of simulation in Volume II when we discuss systems simulation. In systems simulation, dependence is measured in part by the lag one serial correlation, defined for a stationary sequence $X_1, X_2, \ldots, X_n$ as

$$
\begin{aligned}
\rho(1) &= \frac{\text{cov}(X_i, X_{i+1})}{\{\text{var}(X_i)\text{var}(X_{i+1})\}^{1/2}} \\
&= \frac{E[\{X_i - E(X_i)\}\{X_{i+1} - E(X_{i+1})\}]}{\{\text{var}(X_i)\text{var}(X_{i+1})\}^{1/2}}
\end{aligned}
$$

[For stationary sequences, $\rho(1)$ does not change with i.]

There are three well-known estimators for $\rho(1)$:

1. The so-called "unbiased" estimator:

$$
\rho^*(1) = \frac{\dfrac{1}{n-1}\sum_{i=1}^{n-1}(X_i - \bar{X})(X_{i+1} - \bar{X})}{\dfrac{1}{n}\sum_{i=1}^{n}(X_i - \bar{X})^2}
$$

This estimator is not actually unbiased, but its numerator is a less biased estimator of $\text{cov}(X_i, X_{i+1})$ than is the numerator in estimator 2.

2. The "biased" estimator (also called the *Yule–Walker estimator*):

$$
\hat{\rho}(1) = \frac{\dfrac{1}{n}\sum_{i=1}^{n-1}(X_i - \bar{X})(X_{i+1} - \bar{X})}{\dfrac{1}{n}\sum_{i=1}^{n}(X_i - \bar{X})^2} = \frac{n-1}{n} \cdot \rho^*(1)
$$

The numerator, because of the factor $1/n$, has a larger bias, as an estimator of

$cov(X_i, X_{i+1})$, than the numerator in estimator 1. However, in some cases (Parzen, 1962), $\hat{\rho}(1)$ has a smaller mean squared error than the estimator $\rho^*(1)$.

3. The "two-sample" estimator:

$$\tilde{\rho}(1) = \frac{\dfrac{1}{n-1} \sum\limits_{i=1}^{n-1} (X_i - \bar{X}_1)(X_{i+1} - \bar{X}_2)}{\dfrac{1}{n-1} \left\{ \sum\limits_{i=1}^{n-1} (X_i - \bar{X}_1)^2 \sum\limits_{i=1}^{n-1} (X_{i+1} - \bar{X}_2)^2 \right\}^{1/2}}$$

where

$$\bar{X}_1 = \sum_{i=1}^{n-1} \frac{X_i}{n-1} \qquad \text{and} \qquad \bar{X}_2 = \sum_{i=2}^{n} \frac{X_i}{n-1}$$

Because we are assuming that the sample is stationary, and hence has only one mean and variance to estimate, having two estimators of the mean and variance seems to waste information and $\tilde{\rho}(1)$ is not often used.

If we assume that the X_i are iid so that $\rho(1) = 0$ then, with a few, quite general assumptions, theory tells us (Priestley, 1981, Chap. 5), that

$$E\{\rho^*(1)\} = -1/(n-1)$$
$$var\{\rho^*(1)\} \sim 1/n \text{ as } n \to \infty$$
$$\rho^*(1) \sim \text{Normal}(0, n^{-1/2}) \text{ as } n \to \infty$$

and

$$E\{\hat{\rho}(1)\} = -1/n;$$
$$var\{\hat{\rho}(1)\} \sim 1/n \text{ as } n \to \infty$$
$$\hat{\rho}(1) \sim \text{Normal}(0, n^{-1/2}) \text{ as } n \to \infty$$

Since the form of $\tilde{\rho}(1)$ is so close to the other two, we would expect it to have similar properties to those of $\hat{\rho}(1)$ and $\rho^*(1)$.

Other than the equations above for the means, there are no general small-sample results we can turn to for finding out which estimator to use.

Instead, we turn to simulation and Figures 8.2.14, 8.2.15, and 8.2.16, which show the empirical distributions of $\rho^*(1)$, $\hat{\rho}(1)$, and $\tilde{\rho}(1)$, respectively, for small samples of iid simulated Normal(0, 1) data. The most striking feature in all the plots is the bias of the estimators. Since the bias has known form, at least for $\rho^*(1)$ and $\hat{\rho}(1)$, we need not worry about it; we simply adjust our estimate or confidence interval to take account of it. This will not be the case if the true correlation $\rho(1)$ is nonzero. Then we have available only approximate asymptotic expressions for the bias term (Lomnicki and Zaremba, 1957; Priestley, 1981, Chap. 5) or special results such as those for first-order autoregressive processes [AR(1)].

Moving on to the standard deviations of the serial correlation estimates, we

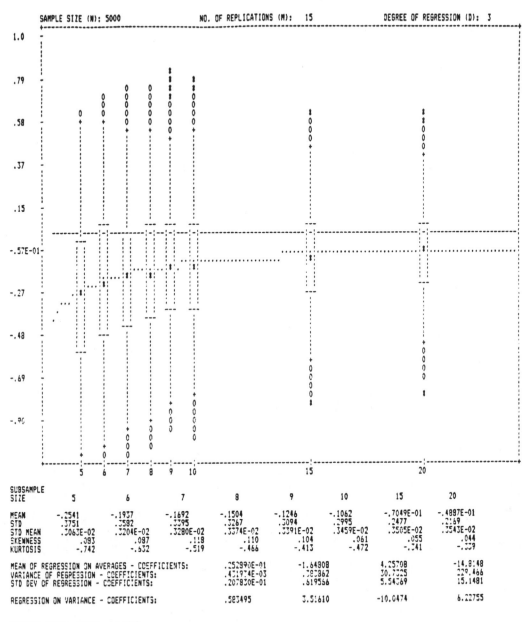

Figure 8.2.14   Simulation results from SMTBPC for the "unbiased" estimator $\rho^*(1)$ of the first serial correlation coefficient for n iid Normal$(0, 1)$ data points. The unbiasedness refers only to the fact that the estimator of the first serial covariance in the numerator of the estimator is unbiased. We see that this ratio estimator is severely biased at small samples; its bias is known to be exactly $-1/(n-1)$. Note that the distribution of $\rho^*(1)$ is very non-Normal at small sample sizes. In fact the kurtosis is negative, and there is only one observation outside of the whiskers of the boxplot at $n = 5$.

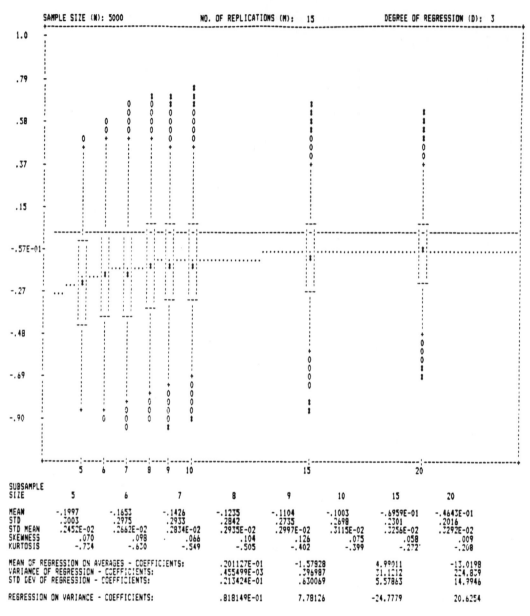

Figure 8.2.15 The estimator $\hat{\rho}(1)$ of the first serial correlation coefficient, which is actually equal to $((n-1)/n)\rho^*(1)$. Thus the variance of $\hat{\rho}(1)$ is less than that of $\rho^*(1)$, as shown by the simulation.

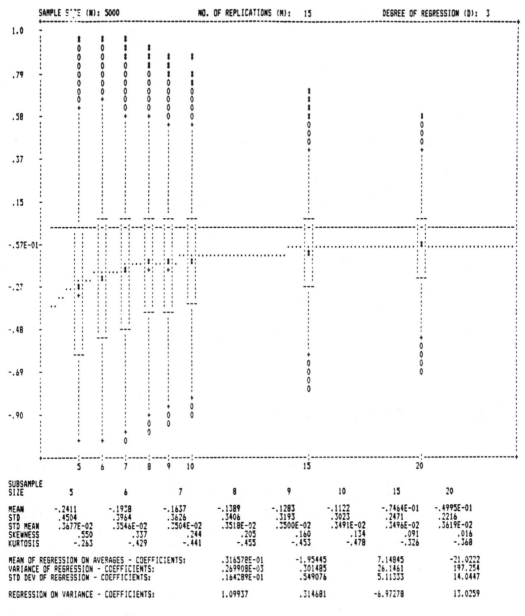

SUBSAMPLE SIZE	5	6	7	8	9	10	15	20
MEAN	-.2411	-.1938	-.1637	-.1389	-.1283	-.1122	-.7464E-01	-.4995E-01
STD	.4504	.3964	.3626	.3406	.3193	.3023	.2471	.2216
STD MEAN	.3677E-02	.3546E-02	.3504E-02	.3518E-02	.3500E-02	.3491E-02	.3496E-02	.3619E-02
SKEWNESS	.550	.337	.244	.205	.160	.134	.091	.016
KURTOSIS	-.263	-.429	-.441	-.455	-.453	-.478	-.326	-.368

				8	9		15	20
MEAN OF REGRESSION ON AVERAGES - COEFFICIENTS:				.316578E-01	-1.95445		7.14845	-21.0222
VARIANCE OF REGRESSION - COEFFICIENTS:				.269908E-03	.301485		26.1461	197.254
STD DEV OF REGRESSION - COEFFICIENTS:				.164289E-01	.549076		5.11333	14.0447
REGRESSION ON VARIANCE - COEFFICIENTS:				1.09937	.314681		-6.97278	13.0259

VERTICAL SCALE: YMIN = -1.0712
                 YMAX = .9992

ESTIMATOR: Lag One Serial Correlation: Two Sample Estimate
*** WIDEST Y VALUES FOUND: YMIN=-1.073   YMAX= .9992

Figure 8.2.16 The two-sample estimator $\tilde{\rho}(1)$ has a higher (estimated) sampling standard deviation than the estimator $\rho^*(1)$ of Figure 8.2.14. Thus at $n = 5$ the numbers are 0.4504 versus 0.375 and, at $n = 20$, they are 0.2216 versus 0.2169.

see immediately that the "biased" estimate $\hat{\rho}(1)$ has a smaller small-sample standard deviation than either competitor. For Normal data, this is the estimator (of the ones considered) we would want to use. Notice also that the boxplots show very symmetric but short-tailed distributions for all the estimators. This is supported by the estimated coefficients of skewness and kurtosis ($\hat{\gamma}_1 = 0.083$ and $\hat{\gamma}_2 = -0.742$ at $n = 5$ in Figure 8.2.14) and indicates that tests of the correlation coefficient based on asymptotic Normality of the serial correlation coefficient estimates would be too conservative. Not only would we accept the correct null hypothesis too often but, we suspect, we would also not accept an incorrect null hypothesis. This aspect, the *power* of the test, would have to be examined through another simulation using dependent data.

Finally, note that the regression estimates are fairly accurate in many cases: that is, in Figure 8.2.16 the leading term in the expansion for the mean has value 0.0317 with estimated standard deviation 0.0164. The true value is zero. At other times the estimates may not be sufficiently precise and may require larger samples. As an example, we have mentioned that the two estimates of serial correlation, $\rho^*(1)$ and $\hat{\rho}(1)$, differ only by the factor $n/(n - 1)$, so that asymptotically their variances are equal. However, the leading coefficients for the regression on the variances in Figures 8.2.14 and 8.2.15 differ by nearly an order of magnitude, their values being 0.583 and 0.082, respectively. This would lead to false conclusions about the asymptotic relative efficiency of $\rho^*(1)$ and $\hat{\rho}(1)$ in tests of $H_0: \rho(1) = 0$, unless larger samples were used to obtain more accurate estimates of the coefficients in the regression on the variance.

8.3   Summary and Computing Resources

We would emphasize two points made throughout this chapter. The first is that classical experimental design and analysis of variance techniques are not always suitable for comparisons of multifactor simulations. These techniques assume that the data are independent and Normally distributed, that the mean of the data is the quantity of interest, and that the mean value of the response is independent of its variance. The effects of the factors are also assumed to be additive. Especially in systems simulations, these assumptions may not be true, although we saw in Example 8.1.1 that batching and averaging of data, and data transformations, can sometimes alleviate the situation. Alternatively, more sophisticated approaches specific to simulation are being developed; see, for instance, the work of Schruben (1986).

The second point is that graphics—looking at outputs—are essential. Of course, this depends on the sophistication of the user's computing facility. The FORTRAN subroutines in the software supplement to the book can be used, for example, for a simplified version of Figure 8.1.2. For more sophisticated response surface plots in this chapter (e.g., Figure 8.1.3a), the microcomputer package STATGRAPHICS from STSC, Inc. was used. The very detailed multifactor quantile plots as in, for example, Figure 8.1.1a, were produced on an IBM mainframe using the experimental APL-based GRAFSTAT program from IBM Research. The plots were produced in color on a Hewlett-Packard 7275 plotter driven from the mainframe via an IBM 3179G2 terminal.

Exercises T8.1. Show that if the variance of an estimator $\tilde{\theta}_n$ has an asymptotic expansion given by Equation 6.1.39,

$$\text{var}(\tilde{\theta}_n) = \frac{b_1}{n} + \frac{b_2}{n^{3/2}} + \frac{b_3}{n^2} + \cdots$$

then the variance of the average estimator

$$\bar{\theta}_n = \sum_{i=1}^{m} \frac{\tilde{\theta}_n(i)}{m}$$

is approximately the same, as n gets large, for all pairs m and n such that $mn = N$ is a constant. In other words, show that

$$\text{var}(\bar{\theta}_{n_1}) \simeq \text{var}(\bar{\theta}_{n_2})$$

as long as $\bar{\theta}_{n_1}$ is an average over m_1 estimators, $\bar{\theta}_{n_2}$ is an average over m_2 estimators, and $n_1 m_1 = n_2 m_2$. Relate this result to the comments in the text concerning equivariant regression in SMTBPC.

T8.2. In computing the regression on the variance of an estimator $\tilde{\theta}_n$, we use as the dependent variable

$$S^2(n, m) = \sum_{i=1}^{m} \frac{(\tilde{\theta}_n(i) - \bar{\theta}_n)^2}{m - 1}$$

where $\bar{\theta}_n$ is defined in Exercise 8.1. In what sorts of situations can we assume that $\tilde{\theta}_n$ is approximately Normally distributed? Under this Normality assumption and under the expansion for the variance of $\tilde{\theta}_n$ given in Exercise 8.1, compute an approximate variance for $S^2(n, m)$. Use your result to justify the comment in the text that the regression on the variances in SMTBPC must be weighted by factors $n_i^{1/2}$, which are proportional to the inverse of the standard deviations of the variance estimators.

8.3. Contrast and compare the theoretical moments of the t distribution given in Table 8.2.1 with the simulated values obtained in the simulation reported in Figure 8.2.4.

8.4. The estimates of kurtosis in the SMTBPC program, as in Figure 8.2.4, come from smaller and smaller samples, of size m_i, as n_i increases, since $n_i m_i = N$ is fixed. If the data were Normally distributed, use the results of Table 6.1.5 (last row) to find the variances of the kurtosis estimates. Comment on the estimates of kurtosis in Table 8.2.2.

C8.5. Repeat the simulations of subsection iii of Section 8.2.2 with $\rho \neq 0$, say $\rho = 0.9$ and $\rho = -0.9$. Thus, we have discretized the continuous valued factor ρ, with values in $(-1, 1)$, into three representative levels.

^T**8.6.** Paired comparisons and common random numbers would be useful in comparing the means of waiting times in the GI|GI|k queue of Examples 8.0.1 and 8.1.1 with single-line or multiple-line queueing disciplines. In those examples independent replications were used. Describe how you would use paired comparisons and common random numbers, noting that there are the other auxiliary factors such as the number of servers k and the traffic intensity t.

^T**8.7.** (Continuation) A further extension in the use of common random variates can be implemented through the SMTBPC program. Assume that we wish to compare two estimators but both are unbiased and the true mean is known. Such a comparison should be made on the basis of the variances of the two estimators. Hence, rather than simulate each estimator separately and compare their independently estimated variances, we would be better off attacking the problem directly. To do this we need to simulate and analyze the difference in the variance of the two estimators.

To be concrete, we consider comparing the variances of the mean $\bar{X}_n$ and the trimmed mean $\bar{X}_{n,1}$ as competing estimators of location in symmetric contaminated populations centered at 0. The difference in the variances of these estimators can be examined through the statistic

$$D2(n, i) = [\bar{X}_n(i)]^2 - [\bar{X}_{n,1}(i)]^2$$

where $\bar{X}_n(i)$ and $\bar{X}_{n,1}(i)$ are the mean and trimmed mean for the ith sample of size n. We assume that the samples are independent over i, but that we use common random samples—that is, the same data—to compute both $\bar{X}_n(i)$ and $\bar{X}_{n,1}(i)$ for each i.

a. Argue why the use of common random samples should be more efficient than using independent samples.

b. Carry out the proposed study using SMTBPC, generating data samples from a contaminated Normal distribution

$$f_X(x) = p\phi(x; 0; 1) + (1 - p)\phi(x; 0; h) \qquad -\infty < x < \infty; 0 < p < 1; h > 1$$

where $\phi(x; \mu; \sigma)$ is the Normal density function given in Table 6.1.2, for several values of p and h. Comment on the results. Note that SMTBPC gives not only an estimate of the expected value of the difference in the variances, but also an estimate of the precision of this estimate.

^T**8.8.** Consider the problem posed in Exercise 8.7, but modify the assumption that the means of the two estimators are known: assume now that the means are equal (which is already implied by both being unbiased) but unknown. We are still interested in comparing the variances of the two estimators. Show that the statistic $D2(n, i)$ is still an unbiased estimate for the difference of the variances. Should the statistic $D2(n, i)$ be modified for the current situation in which the means are not assumed known and equal to zero? If so, how?

References

Andrews, D. F., Bickel, P. J., Hample, F. R., Huber, P. J., Rogers, W. H., and Tukey, J. W. (1972). *Robust Estimates of Location: Survey and Advances.* Princeton University Press: Princeton, NJ.

Barlow, R., and Proschan, F. (1975). *Statistical Theory of Reliability and Life Testing: Probability Models.* Holt, Rinehart & Winston: New York.

Box, G. E. P., Hunter, W. G., and Hunter, J. S. (1978). *Statistics for Experimenters: An Introduction to Design, Data Analysis and Model Building*. Wiley: New York.

Bratley, P., Fox, B. L., and Schrage, L. E. (1983). *A Guide to Simulation*. Springer: Berlin.

Cochran, W. G., and Cox, G. M. (1957). *Experimental Designs*, 2nd ed. Wiley: New York.

Cox, D. R., and Hinkley, D. B. (1974). *Theoretical Statistics*. Chapman and Hall: London.

Cox, D. R., and Lewis, P. A. W. (1966). *The Statistical Analysis of Series of Events*. Methuen: London; Wiley: New York.

Gastwirth, J., and Cohen, M. (1970). Small Sample Behavior of Some Robust Linear Estimates of Location. *Journal of the American Statistical Association, 65*, 946–973.

Gaver, D. P., and Thompson, G. L. (1973). *Programming and Probability Models in Operations Research*. Wadsworth & Brooks/Cole: Pacific Grove, CA.

Gruber, J., and Freimann, K. (1986). Combined Response Surface Regressions in Monte Carlo Studies of Small Sample Properties of Estimators: Theory and an Application. Discussion paper no. 108, Department of Economics, University of Hogen, West Germany, presented at the 1986 ASA Meetings.

Heidelberger, P., and Lewis, P. A. W. (1981). Regression-adjusted Estimates for Regenerative Simulations, with Graphics. *Communications of the Association for Computing Machinery, 24*, 260–273.

Heidelberger, P., Nelson, R. D., and Welch, P. D. (1985). A Future Direction in Stochastic Modeling: The Application of Interactive Analysis and Computer Graphics. Research report AC 11088 (no. 49780), Computer Science, IBM Thomas J. Watson Research Center, Yorktown Heights, New York.

Hoaglin, D. C., Mosteller, F., and Tukey, J. W. (1983). *Understanding Robust and Exploratory Data Analysis*. Wiley: New York.

Johnson, N. L., and Kotz, S. (1970). *Distributions in Statistics: Continuous Univariate Distributions*, Vol. 1. Wiley: New York.

Kendall, M., Stuart, A., and Ord, K. (1983). *The Advanced Theory of Statistics*, Vol. 3, 4th ed. Macmillan: New York.

Kleijnen, J. P. C. (1975). *Statistical Techniques in Simulation*, Part II. Dekker: New York.

Law, A. L., and Kelton, W. D. (1982). *Simulation Modeling and Analysis*. McGraw-Hill: New York.

Lehmann, E. L. (1983). *Theory of Point Estimation*. Wiley: New York.

Lewis, P. A. W., Orav, E. J., Drueg, H. W., Linnebur, D. G., and Uribe, L. (1985). An Implementation of Graphical Analysis in Statistical Simulations. *International Statistical Review, 53*, 69–90.

Lomnicki, Z. A., and Zaremba, S. K. (1957). On the Estimation of Autocorrelation in Time Series. *Annals of Mathematical Statistics, 28*, 140–158.

Mauro, C. A. (1986). Efficient Identification of Important Factors in Large Scale Simulations. In *Proceedings of the 1986 Winter Simulation Conference*, J. Wilson, J. Hendriksen, and S. Roberts, Eds. Institute of Electrical and Electronics Engineers: Piscataway, NJ, pp. 296–305.

Mauro, C. A., and Smith, D. E. (1984). Factor Screening in Simulation: Evaluation of Two Strategies Based on Random Balance Sampling. *Management Science, 30*, 209–221.

Mood, A. M., Graybill, F. A., and Boes, D. C. (1974). *Introduction to the Theory of Statistics*. McGraw-Hill: New York.

Parzen, E. (1962). *Stochastic Processes*. Holden-Day: San Francisco.

Priestley, M. B. (1981). *Spectral Analysis and Time Series*, Vol. 1, *Univariate Series*. Academic Press: New York.

Scheffe, H. (1964). *The Analysis of Variance*. Wiley: New York.

Schruben, L. W. (1986). Simulation Optimization Using Frequency Domain Methods. In *Proceedings of the 1986 Winter Simulation Conference*, J. Wilson, J. Hendriksen, and

S. Roberts, Eds. Institute of Electrical and Electronics Engineers: Piscataway, NJ, 336–369.

Schruben, L. W., and Margolin, B. H. (1978). Pseudorandom Number Assignment in Statistically Designed Simulation and Distribution Sampling Experiments. *Journal of the American Statistical Association, 73*, 504–524.

Smith, D. E., and Mauro, C. A. (1982). Factor Screening in Computer Simulation. *Simulation,* 49–54.

Yoav, B. (1983). Is the t-Test Really Conservative When the Parent Distribution Is Long-Tailed? *Journal of the American Statistical Association, 78*, 645–654.

9

Assessing Variability in Univariate Samples: Sectioning, Jackknifing, and Bootstrapping

9.0 Introduction

In Chapter 4, we stressed that it is essential to assess the sampling variability of a point estimate, $\tilde{\theta}$, of a characterization θ of a simulated random variable. To assess variability, our approach has been to first calculate the theoretical variance of our estimator, $\mathrm{var}(\tilde{\theta})$, and then to use our simulated data to estimate this quantity. For a sample average, $\tilde{\theta} = \bar{X}$, we showed in Chapter 4 that $\mathrm{var}(\tilde{\theta}) = \sigma_X^2/m$, where m is the sample size, and we then estimated $\mathrm{var}(\tilde{\theta})$ by S_X^2/m, where S_X^2 is the sample variance. For more complicated estimators such as sample quantiles or ratio estimators, say of the coefficient of variation, the results in Chapter 6 and particularly Sections 6.1.2 and 6.3.2 allow calculation of $\mathrm{var}(\tilde{\theta})$. However, the estimates of variability that follow from these calculations rely on unstable estimates of the probability density function (Equation 6.3.14) or on estimates of higher distributional moments. Hence, in this chapter, we present three alternative methods for estimating $\mathrm{var}(\tilde{\theta})$ that can be used for any estimate $\tilde{\theta}$, regardless of its complexity and regardless of whether $\mathrm{var}(\tilde{\theta})$ can be evaluated exactly or approximately, in closed form. We also show how each of these three methods can be used to produce confidence intervals around the true value θ. Since confidence intervals are sometimes preferred to point estimates, we conclude this chapter with a brief, general discussion of the way simulations can be used to evaluate and compare competing methods for computing confidence intervals.

When working with large simulation samples, the most basic and straightforward variance assessment technique is *sectioning* or *batching*. In small and moderate size samples, the use of sectioning is limited for many estimators by problems of bias, and, when considering confidence intervals, by non-Normality.

The use of *pilot samples*, and of a technique called *jackknifing* (Miller, 1974), is introduced as a supplement to sectioning to assess and relieve the problem of bias. Unfortunately, in using the jackknife methodology, we are forced to use a different estimator of θ called the jackknifed estimator, which may have a larger variance than the original unjackknifed estimator. Thus, the jackknifed estimator may be less desirable since it may have larger mean square error (mse)—variance plus bias squared—than the unjackknifed estimator. The existence and magnitude of this problem is not analytically predictable. Jackknifing can also be used for assessing variability and calculating confidence intervals. As with sectioning, confidence intervals using jackknifing are based on the Normality of large sample averages.

A third technique for assessing variability, which is useful for working with small samples that result from complicated simulation studies, is *bootstrapping* (Efron, 1981). This is a resampling scheme, somewhat related to the jackknife, which allows estimation of variance and calculation of confidence intervals while making no distributional assumptions. Both jackknifing and bootstrapping are computer-intensive methods, and bootstrapping is in fact a simulation procedure that "generates" data from the empirical distribution of the data sample. As such, bootstrapping is useful only in very complicated simulation studies where it takes much longer to generate simulated output from replications than to resample the existing output data sample. (Note that in working with "real" data where we may not be able to collect more observations, bootstrapping is a very powerful tool for assessing variability and constructing confidence intervals.)

Finally, note that we have chosen to present this material after Chapter 8, although we have returned to the analysis of univariate samples from simple experiments in which one factor is considered at a single level. This was done to make use of the SMTBPC program. The all-important small-sample properties of jackknifing and bootstrapping are best obtained by simulation studies and are most easily displayed by a program such as SMTBPC. In this chapter, SMTBPC is used to study variance estimation with several bootstrapping and jackknifing schemes when estimating the correlation coefficient in bivariate Normal samples.

9.1   Preliminaries

As in Chapters 6 and 7, we focus on the analysis of random samples of independent, identically distributed observations of univariate random variables. For example, we have the waiting time W_n of the nth customer in a queue, for a fixed value of n. (Analysis of bivariate random variables such as W_n and W_{n+1} is more complicated and is deferred until Chapter 10, although estimation of the correlation coefficient for bivariate samples is used for illustration in the last example in this chapter.) For simplicity, we drop the subscript n and let X denote a generic univariate random variable. A simulated iid sample from m independent replications of the simulation will be denoted by $X_1, \ldots, X_j, \ldots, X_m$.

The object in this chapter is to go beyond obtaining a point estimate $\tilde{\theta}$ for some characterization θ of X, to presenting information about the precision of $\tilde{\theta}$ from the data. This could be accomplished by finding an estimate of the variance of the estimator (since the estimator is a random variable) or a $100(1 - \alpha)\%$ confidence interval estimate for the unknown value of θ. The latter takes the form of a random interval that covers the true (but unknown) value θ with probability $(1 - \alpha)$. From the frequentist point of view, this means that, in the long run, an average of $100(1 - \alpha)\%$ of such observed random intervals would contain the true value θ.

It is important to note that the width of a $100(1 - \alpha)\%$ confidence interval measures the precision with which we know θ. At the same time, a narrow confidence interval is not useful if it does not cover the true value, θ, the prescribed $100(1 - \alpha)\%$ of the time. Hence, we not only need to discuss methods for computing confidence intervals, but we will actually look at the properties of the confidence intervals themselves. As an example, we can think of comparing a confidence interval for the mean of a Normal(μ, σ) population based on $\bar{X}$ and $S_{\bar{X}}^2$, as in Equation 6.1.40, versus an interval based on order statistics, as in Equation 6.3.8. The last section of this chapter considers a comparison of confidence intervals for the correlation coefficient based on Normal theory results, jackknifing, and bootstrapping.

9.2 Assessing Variability of Sample Means and Percentiles

9.2.1 Variance Estimates for Means and Percentiles

Assume that we have a random sample $X_1, \ldots, X_j, \ldots, X_m$ generated by simulation from an unknown distribution $F_X(x)$ and that we wish to estimate $E(X)$. A standard approach is to work with the sample average

$$\bar{X} = \frac{1}{m} \sum_{j=1}^{m} X_j \tag{9.2.1}$$

which is an unbiased estimate of $E(X)$ and has variance

$$\text{var}(\bar{X}) = \frac{\sigma_X^2}{m}$$

where σ_X^2 is unknown. We then estimate the quantity σ_X^2 from the data and, upon dividing by m, get an *estimate of the variance of the estimate of the mean*. This procedure was discussed briefly in Chapter 4, "Crude Simulation," but we reiterate it in more detail here because of its importance.

Thus, we estimate $\text{var}(X)$ in the usual way by the sample variance

$$S_X^2 = \frac{1}{m - 1} \sum_{j=1}^{m} (X_j - \bar{X})^2 \tag{9.2.2}$$

which is an unbiased estimate of $\text{var}(X)$, then estimate $\text{var}(\bar{X})$ by

$$S_{\bar{X}}^2 = \text{vâr}(\bar{X}) = \frac{S_X^2}{m} \tag{9.2.3}$$

By analogy, the estimated standard deviation of the mean, often called the standard error, is

$$S_{\bar{X}} = \frac{S_X}{\sqrt{m}} = \sqrt{(S_{\bar{X}}^2)} \tag{9.2.4}$$

Note that $S_{\bar{X}}^2$ is an unbiased estimate of var($\bar{X}$), since S_X^2 is an unbiased estimate of var(X). However, neither S_X nor $S_{\bar{X}}$ is unbiased (see Chapter 6), since the expectation of the square root of a random variable is not equal to the square root of the expectation.

The same steps, Equations 9.2.2 and 9.2.3, hold for estimating the variance of a percentile estimate $\tilde{p}$, the usual estimate of $P(X \leqslant x) = p$ (Conover, 1980, p. 95; Chap. 6.2). In this case, however, we work with binary indicator functions (Equation 6.2.5), which behave as Bernoulli(p) random variables. The average of these Bernoulli variables equals $\tilde{p}$:

$$\tilde{p} = \frac{1}{m} \sum_{j=1}^{m} I(X_j; x)$$

so that

$$S_{\tilde{p}}^2 = \frac{1}{m(m-1)} \sum_{j=1}^{m} \{I(X_j; x) - \tilde{p}\}^2 = \frac{\tilde{p}(1 - \tilde{p})}{m - 1} \tag{9.2.5}$$

Alternatively, the mle estimate of the variance of $\tilde{p}$ can be obtained as follows:

$$\text{vâr}(\tilde{p}) = \frac{\tilde{p}(1 - \tilde{p})}{m} \tag{9.2.6}$$

The rationale for this second variance estimate is that, since there is only one parameter p in the distribution of the population, we know that the variance of $\tilde{p}$ must be a function of that parameter. Therefore we substitute the mle estimate $\tilde{p}$ of the parameter into the expression for the variance (Equation 6.2.7) to obtain the mle estimate of the variance of the estimator $\tilde{p}$. The estimated standard deviation of $\tilde{p}$ can be the square root of either Equation 9.2.5 or 9.2.6.

9.2.2 Confidence Intervals for the Means of Normal Data

From the variance estimate in Section 9.2.1, we could use the guidelines from subsection ii of Section 6.1.1 to interpret the precision with which we know the original quantity of interest, $E(X)$. However we often also want a *confidence interval*: that is, a statement that a random set (L, U) covers the true mean $100(1 - \alpha)\%$ of the time, where $1 - \alpha$ is typically 0.95, and L and U are functions of the m observations, with $L < U$. To obtain a confidence interval for $E(X)$, we must know something about the distribution, $F_X(x)$. At a minimum, we require that $F_X(x)$ have a given parametric form (i.e., Normal, exponential). Unfortunately, in most situations this is precisely what we do not know. For the case in which the unknown distribution is symmetric, so that the mean and the median are the same, nonparametric methods based on order statistics are available (Conover, 1980, p. 122). In particular, for *symmetric* populations, confidence intervals can be based on the sign test for location or on the Wilcoxon signed rank test (Conover, 1980, p. 280). The need for distributional assumptions for confidence interval estimates contrasts with the completely nonparametric procedure of Equation 9.2.4 for estimating the standard deviation of $\bar{X}$.

We review here briefly the confidence intervals derived from the t statistic in the case of iid Normal random variables. If the X_j are a sample of m Normal random variables with mean $E(X)$, then the t statistic

$$T_m = \frac{\bar{X} - E(X)}{S_{\bar{X}}}$$

has a t distribution, with $m - 1$ degrees of freedom: that is,

$$P\left[t_{(m-1),\alpha/2} \leqslant \frac{\{\bar{X} - E(X)\}}{S_{\bar{X}}} \leqslant t_{(m-1),(1-\alpha/2)} \right] = 1 - \alpha \tag{9.2.7}$$

Note that $t_{(m-1),(1-\alpha/2)} = -t_{(m-1),\alpha/2}$ is defined as the $(1 - \alpha/2)$ quantile of the t distribution with $m - 1$ degrees of freedom.

Inverting Equation 9.2.7 to obtain a confidence interval estimate for the unknown mean $E(X)$, we get, using the symmetry of the t distribution,

$$\begin{aligned} P\{-t_{(m-1),(1-\alpha/2)}S_{\bar{X}} &\leqslant E(X) - \bar{X} \leqslant t_{(m-1),(1-\alpha/2)}S_{\bar{X}}\} \\ &= P\{\bar{X} - t_{(m-1),(1-\alpha/2)}S_{\bar{X}} \leqslant E(X) \leqslant \bar{X} + t_{(m-1),(1-\alpha/2)}S_{\bar{X}}\} \\ &= P\{L \leqslant E(X) \leqslant U\} = 1 - \alpha \end{aligned} \tag{9.2.8}$$

Thus, we not only have a method for estimating the sampling standard deviation $\sigma_{\bar{X}}$ of a sample mean, but also, when the random variable X is Normally distributed, a method for generating a confidence interval for the true mean.

9.3 Sectioning to Assess Variability: Arbitrary Estimates from Non-Normal Samples

9.3.1 Variance Estimates from Sectioning

Clearly, in many simulations the X_j's are not going to be Normally distributed. For example, in the MI|MI|1 queue, the distribution of the stationary waiting time has a jump at zero and, conditional on the variable being greater than zero, has an exponential distribution. Thus, the method for obtaining confidence intervals given in the preceding section is not valid.

A more general technique for obtaining variance estimates and confidence intervals, applicable *even if the estimate of θ is not a sample average and even if the data are not Normally distributed*, is to section the data into, say, v nonoverlapping subsections, each of size l. We assume for simplicity that v divides the sample size m evenly. We then compute v point estimates of the quantity of interest from the subsamples and average them to form a new estimate. The variability of the new estimate can be estimated from the sample variance of the v point estimates, divided by v. (If the v point estimates are approximately Normal, we can also use the Normal theory t statistic, as described in the next section, to obtain a confidence interval for θ.)

Illustrating with $\theta = E(X)$, we assume the X_j's are independent, but not necessarily Normally distributed. Let

$$\bar{X}_1 = \frac{1}{l} \sum_{j=1}^{l} X_j \quad \text{and} \quad \bar{X}_2 = \frac{1}{l} \sum_{j=l+1}^{2l} X_j$$

and so forth up to $\bar{X}_v$.

Let

$$X^* = \frac{1}{v} \sum_{i=1}^{v} \bar{X}_i$$

be the new sectioned estimate of $E(X)$, and let

$$S_{X^*}^2 = \frac{1}{v} \left\{ \frac{1}{v-1} \sum_{i=1}^{v} (\bar{X}_i - X^*)^2 \right\} \tag{9.3.1}$$

be the new estimate of the variance of X^*. Since the estimate under consideration is a sample average, it is easily shown that the new sectioned estimate of $E(X)$, namely X^*, is exactly equal to the sample average $\bar{X}$. The new estimate of variance, $S_{X^*}^2$, however, is *not* equal to $S_{\bar{X}}^2$, although both are unbiased estimates of $\mathrm{var}(X^*) = \mathrm{var}(\bar{X})$. The difference in the variability of $S_{X^*}^2$ and $S_{\bar{X}}^2$ is considered in Section 9.3.3.

This computationally simple sectioning technique is most useful for estimates that are more complex than sample averages. Through sectioning we can obtain estimates of variance for quantile estimates based on order statistics or for estimates based on sample moment ratios such as the sample coefficients of variation, skewness, and kurtosis, as defined at Equations 6.1.19, 6.1.13, and 6.1.15, respectively.

The more general formulation of sectioning is as follows. It is given in terms of a univariate sample, but the technique applies equally well to the bivariate samples considered in Chapter 10.

Again suppose we have a sample of size m from a population with unknown distribution function $F_X(x)$, that is, $X_1, \ldots, X_j, \ldots, X_m$, which are assumed to be independent and identically distributed.

Let θ be a parameter or quantification of $F_X(x)$—for example, the coefficient of variation, $\theta = C(X) = \sigma(X)/E(X)$, or any other quantification discussed in Chapter 6.

Let $\tilde{\theta}(m) = \tilde{\theta}(X_1, \ldots, X_m)$ be a statistic (i.e., a function of $X_1, \ldots, X_m$), which is used to estimate θ. For example, for $\tilde{C}(X)$, from Equation 6.1.19:

$$\tilde{\theta}(m) = \tilde{C}(X) = \frac{S_X}{\bar{X}} = \frac{\left\{ \dfrac{1}{m-1} \sum_{j=1}^{m} (X_j - \bar{X})^2 \right\}^{1/2}}{\dfrac{1}{m} \sum_{j=1}^{m} X_j}$$

This point estimate is known to be asymptotically unbiased, consistent, and Normally distributed by the results of Section 6.1.2.

Although we cannot directly assess the sampling variability of $\tilde{\theta}(m)$, we will construct a new sectioned estimate based on $\tilde{\theta}(m)$, whose variability can be estimated as follows.

1. Divide the sample into v sections each of size l, assuming for simplicity that v divides m exactly: that is,

$$\frac{m}{v} = l \quad \text{or} \quad m = lv$$

(If this were not true, some of the data would have to be discarded. The appropriate division to minimize this data loss is an added factor in the choice of v, a topic that will be discussed later.)

Note that the v sections are disjoint and the sectioning is done in such a way as to produce "random" subsamples from $F_X(x)$ in each section. In particular one would *not* order the data first and then pick the largest l X_j's for the first section, and so on!!

2. Form the *same* estimate $\tilde{\theta}$ from each section, calling the estimates $\tilde{\theta}_i(l)$, $i = 1, \ldots, v$.

Thus if θ were the population coefficient of variation and we were going to estimate it with $\tilde{C}(X)$ as above, then

$$\tilde{\theta}_1(l) = \frac{\left\{\frac{1}{l-1} \sum_{j=1}^{l} (X_j - \bar{X}_1)^2\right\}^{1/2}}{\bar{X}_1} \tag{9.3.2}$$

$$\tilde{\theta}_2(l) = \frac{\left\{\frac{1}{l-1} \sum_{j=l+1}^{2l} (X_j - \bar{X}_2)^2\right\}^{1/2}}{\bar{X}_2} \tag{9.3.3}$$

$$\vdots \qquad\qquad \vdots$$

$$\tilde{\theta}_v(l) = \frac{\left\{\frac{1}{l-1} \sum_{j=vl-l+1}^{vl} (X_j - \bar{X}_v)^2\right\}^{1/2}}{\bar{X}_v} \tag{9.3.4}$$

where

$$\bar{X}_i = \frac{1}{l} \sum_{j=(i-1)l+1}^{il} X_j \quad i = 1, \ldots, v$$

3. Form a new sectioned estimator

$$\bar{\theta}(m) = \frac{1}{v} \sum_{i=1}^{v} \tilde{\theta}_i(l) \tag{9.3.5}$$

the average of the v estimates. Note that $\bar{\theta}(m)$ is also a function of $X_1, \ldots, X_j, \ldots,$ X_m but is in general a function of the X_j's that is *different* from $\tilde{\theta}(m)$. (In the case of the sample mean we have argued that $X^* = \bar{\theta}(m)$ and $\bar{X} = \tilde{\theta}(m)$ are the same; what about the case of the sample variance?) In general, the means, variances, and distributions of $\bar{\theta}(m)$ and $\tilde{\theta}(m)$ will be different.

4. If we use $\bar{\theta}(m)$ to estimate θ, then, since $\bar{\theta}(m)$ is simply the average of the iid random variables $\tilde{\theta}_1(l), \tilde{\theta}_2(l), \ldots, \tilde{\theta}_v(l)$, we have that

$$E\{\bar{\theta}(m)\} = E\{\tilde{\theta}(l)\} \tag{9.3.6}$$

and

$$\operatorname{var}\{\bar{\theta}(m)\} = \frac{\operatorname{var}\{\tilde{\theta}(l)\}}{v} \tag{9.3.7}$$

Now, the advantage and purpose of sectioning is that $\operatorname{var}\{\tilde{\theta}(l)\}$ *can be estimated (unbiasedly) from the sample variance of the* $\tilde{\theta}_i(l)$ *as follows:*

$$S^2_{\tilde{\theta}(l)} = \hat{\operatorname{var}}\{\tilde{\theta}(l)\} = \frac{1}{v-1} \sum_{i=1}^{v} \{\tilde{\theta}_i(l) - \bar{\theta}(m)\}^2 \tag{9.3.8}$$

and

$$S^2_{\bar{\theta}(m)} = \hat{\operatorname{var}}\{\bar{\theta}(m)\} = \frac{S^2_{\tilde{\theta}(l)}}{v} \tag{9.3.9}$$

Thus, we have an estimate, $\hat{\operatorname{var}}\{\bar{\theta}(m)\}$ of the variance of $\bar{\theta}(m)$, which is something we usually do not have for $\tilde{\theta}(m)$ directly. Note that sectioning can be done on most pocket calculators that incorporate routines for averaging a set of v numbers, computing S^2, S, and the sample standard deviation of the average of the set of numbers, usually called the standard error and denoted $S_{\bar{X}}$.

9.3.2 Confidence Intervals from Sectioning

From the prescription for sectioning given in Section 9.3.1, we have defined the v estimates from the sections,

$$\tilde{\theta}_1(l), \ldots, \tilde{\theta}_i(l), \ldots, \tilde{\theta}_v(l)$$

the new sectioned estimate,

$$\bar{\theta}(m) = \frac{1}{v} \sum_{i=1}^{v} \tilde{\theta}_i(l)$$

and the new sectioned estimate of the variance of $\bar{\theta}(m)$,

$$S^2_{\bar{\theta}(m)} = \frac{1}{v(v-1)} \sum_{i=1}^{v} \{\tilde{\theta}_i(l) - \bar{\theta}(m)\}^2$$

We can now proceed to construct a confidence interval for θ by following the development in Section 9.2.2. Specifically, if the $\tilde{\theta}_i(l)$ can be assumed to be Normally distributed, then direct use of the t statistic leads to the $100(1 - \alpha)\%$ confidence interval:

$$(\bar{\theta}(m) - t_{(v-1),(1-\alpha/2)}S_{\bar{\theta}(m)}, \quad \bar{\theta}(m) + t_{(v-1),(1-\alpha/2)}S_{\bar{\theta}(m)}) \qquad (9.3.10)$$

Again, as in Section 9.2.2, this result is valid only if the $\tilde{\theta}_i(l)$ are Normally distributed. However, as we have seen in Chapter 6, most of the estimators we work with are asymptotically Normally distributed when the sample size, in this case l, is large enough. Hence, we need not require Normality of the data $X_1, \ldots, X_j, \ldots, X_m$. Instead we may be able to choose l large enough (see Section 9.3.3) so that the $\tilde{\theta}_i(l)$ are approximately Normally distributed and the use of the sectioned confidence interval is approximately valid.

Moreover, the Normality of the $\tilde{\theta}_i(l)$ can be checked by probability plotting methods (Section 7.2) or by more formal methods such as the Shapiro–Wilk test for Normality (subsection ii of Section 6.2.4).

The advantage of confidence intervals based on sectioning is that we no longer require Normality of the data but can rely instead on data analysis and large-sample theory to check and ensure the Normality of the $\tilde{\theta}_i(l)$. The disadvantage to confidence intervals from sectioning is expressed in the fact that the intervals are based on the quantiles of a t distribution with $v - 1$ degrees of freedom. This contrasts with the $m - 1$ degrees of freedom that would be appropriate if we could assume the X_i to be Normal and were using $\bar{X}$ to estimate the true mean (recall the example in Section 9.3.1). Since smaller degrees of freedom correspond to wider confidence intervals through the coefficient $t_{(v-1),(1-\alpha/2)}$, our goal is to keep v as large as possible. The issue of choosing v is discussed in detail in the next section. For now, we can look at Table 9.3.1, where we see that the 0.975 quantile of T_{v-1}, for large v, rapidly approaches the value of this quantile for Normal variables ($t_{\infty,0.975} = 1.960$). In fact, at $v - 1 = 20$, there is seen to be very little difference between the

Table 9.3.1 Table of 0.975 and 0.995 quantiles of the t distribution with $v - 1$ degrees of freedom

$v - 1$	$t_{v-1,0.975}$	$t_{v-1,0.995}$
2	4.303	9.925
5	2.571	4.032
10	2.228	3.169
15	2.131	2.947
20	2.086	2.845
30	2.042	2.750
40	2.021	2.704
∞	1.960	2.576

two quantiles. Thus the confidence intervals will be approximately the same width when $v = 21$ as when $v > 21$. There remain, however, questions about how to obtain confidence interval estimates from sectioning when the $\tilde{\theta}_i(l)$ are not Normally distributed, and about the limitations of sectioning in general. Both these considerations are intimately tied up with the choice of sample size l, or equivalently the number of sections v. As we will see in the next section, although choosing v too large *may* lead to a poor estimator $\bar{\theta}(m)$ for θ (possibly biased, etc.), the procedure in Section 9.3.1 for estimating the variance of $\bar{\theta}(m)$ is valid, although perhaps not very useful. This is true no matter how bad an estimator $\bar{\theta}(m)$ is, or how non-Normal the underlying random variables $X_1, \ldots, X_j, \ldots, X_m$ may be. Also, the variances of $\tilde{\theta}(m)$ and $\bar{\theta}(m)$ may be different because of small-sample effects, as illustrated in the next section.

9.3.3 Limitations of Sectioning: Normality and Bias Considerations

When sectioning the sample mean, the original variance and confidence interval can be recovered by taking the number of sections v equal to m. This follows from the substitution of m for v in the formulas in Sections 9.3.1 and 9.3.2. This gives a (slightly) narrower confidence interval than that obtained by sectioning, where by sectioning we mean that $v < m$. So what is to stop us from taking $v = m$ in general?

1. First, taking $v = m$ is impractical for most estimators $\tilde{\theta}(m)$ of interest. For example, if the estimate, $\tilde{\theta}(m)$ is the sample variance, we must have $l \geqslant 2$ because of computational considerations: we cannot compute the sample variance of a sample of size $l = 1$. Going the other way, even taking $l = m/2$ may not be sufficient when we take into account the more important limitations on sectioning due to bias and Normality considerations.

2. Sectioning gives a *variance estimate* that is independent of Normality assumptions. But, if we want to use that variance estimate to obtain *confidence intervals* based on the t statistic, then each of the estimates $\tilde{\theta}_i(l)$ that is averaged to give the overall estimate $\bar{\theta}(m)$ must be approximately Normally distributed. This dictates that we make l large.

3. Perhaps the greatest limitation to sectioning, especially with large v, is the consideration that estimates of higher moments and of quantiles, which are generally biased estimates, will have more bias in them if the size of each section is too small. This again dictates that we make l large.

We can quantify this third point as follows, using quantile estimation as an example.

We noted in Chapter 6 that, for example, the quantile estimator $\hat{x}_\alpha$ given at Equation 6.3.2, has bias given by

$$E(\hat{x}_\alpha) = x_\alpha + \frac{a}{m} + O(m^{-2})$$

where a is a constant depending on the distribution of X and $O(m^{-2})$ is a term that is of order m^{-2}. If we section into v subsections, each of size l, then after averaging the v estimates from the subsections we get

$$E(\bar{x}_\alpha) = x_\alpha + \frac{a}{l} + O(l^{-2}) = x_\alpha + \frac{va}{m} + O(m^{-2}) \qquad (9.3.11)$$

The bias term, to the order $1/m$, now has coefficient va and, thus, is seen to be v times as large as the bias term for the whole sample, if $a \neq 0$.

4. A fourth reason for making l large (v small) is that the variance of the new sectioned estimate $\bar{\theta}(m)$ will usually be smaller if l is large, hence $\bar{\theta}(m)$ will be a more desirable estimate.

A specific example that illustrates this last effect of section size is the case of $\theta = \sigma^2$ and $\tilde{\theta}(m) = S^2(m)$, the sample variance for a sample of size m, which estimates the variance of the population. (An equation for the variance of $S^2(m)$ is given in Table 6.1.4. This could be used directly to estimate $\text{var}\{S^2(m)\}$, but it involves estimating μ_4, the fourth central moment of X.) For Normal samples, the variance of the sectioned estimator can be calculated and compared to the variance of the usual estimator, $S^2(m)$, as follows:

$$\text{var}\{\bar{\theta}(m)\} = \text{var}\left\{\frac{1}{v}\sum_{j=1}^{v} S_j^2(l)\right\} = \frac{2\sigma^4}{\{v(l-1)\}} = \frac{2\sigma^4}{m-v}$$

$$> \frac{2\sigma^4}{m-1} = \text{var}\{S^2(m)\} \qquad \text{for } v > 1$$

Therefore, the variance of the sectioned estimator increases as v gets large.

Note also that since $S^2(l)$ is an unbiased estimate of σ^2, the choice of v is not dependent on the bias considerations in Point 3. But, in the case of a Normal sample, $S^2(l)$ is proportional to a chi-square random variable with $l - 1$ degrees of freedom. Thus $S^2(l)$ can not be considered approximately Normal until its degrees of freedom equal about 50, meaning that l, the size of the section, must be at least 51.

In summary, to minimize bias inflation, help promote Normality, and decrease the variance of the sectioned estimator, we want to make the section size l large (or, equivalently, make v small). For a fixed sample size m, though, this conflicts with two considerations:

- The width of the sectioned confidence interval depends highly on the constant $t_{(v-1),(1-\alpha/2)}$, as argued in Section 9.3.2. To achieve a narrow interval, we need to make v large, hence l small.
- Our estimate of the variance of the sectioned estimate of θ will be more precise if v is large, hence l small.

This last point can be illustrated by assuming that the variance estimate

$$\hat{\text{var}}\{\bar{\theta}(m)\} = S^2_{\bar{\theta}(m)} = \frac{1}{v(v-1)}\sum_{i=1}^{v}\{\tilde{\theta}_i(l) - \bar{\theta}(m)\}^2 \qquad (9.3.12)$$

is proportional to a chi-square random variable with $v - 1$ degrees of freedom. This will be true if the $\tilde{\theta}_i(l)$ are Normally distributed, as often happens when l is large. We also assume that $\tilde{\theta}_i(l)$ has a variance given by the expansion in Equation 6.1.39:

$$\text{var}\{\tilde{\theta}_i(l)\} = \frac{b_1}{l} + O(l^{-3/2})$$

Then

$$\text{var}[\text{vâr}\{\bar{\theta}(m)\}] \cong \frac{2b_1^2}{\{m^2(v - 1)\}} \tag{9.3.13}$$

which decreases as v increases.

Note that while it may be desirable to estimate $\text{var}\{\bar{\theta}(m)\}$ very precisely, it is not our main goal: our objective was to estimate $\bar{\theta}(m)$ precisely, and then to assess $\text{var}\{\bar{\theta}(m)\}$. Hence, this last point may not be a very compelling one in determining the tradeoff between v and l.

Putting all the foregoing considerations together, we can prescribe a rough rule of thumb: make v at least 12, no more than 20. It can be seen from Table 9.3.1 that, for example, the t statistic 0.975 quantile at $v - 1 = 20$ is only 6% larger than the corresponding Normal quantile (2.086 versus 1.960). Hence, making v larger than 20 will not result in appreciably narrower confidence intervals but may induce non-Normality and bias into $\bar{\theta}(m)$ and may make $\text{var}\{\bar{\theta}(m)\}$ much bigger than $\text{var}\{\tilde{\theta}(m)\}$. Having v less than 12 may produce confidence intervals that are too wide to give precise information about the parameter of interest and may make the variability of $\text{vâr}\{\bar{\theta}(m)\}$ much larger than the variability of $\text{vâr}\{\tilde{\theta}(m)\}$.

If m is 1000 or greater (which is usually desirable in a simulation), the sample sizes within sections generally will be big enough to make bias small and satisfy Normality assumptions for the $\tilde{\theta}_i(l)$. Also, $\text{var}\{\bar{\theta}(m)\}$ will be approximately equal to $\text{var}\{\tilde{\theta}(m)\}$.

When m is, say, less than 1000, an alternative technique called the jackknife can be used both to eliminate bias and to give an estimate of the variance of the jackknifed estimator (see Section 9.4).

9.3.4   Example: Sample Characteristics of a Gamma$(k = 0.2, \beta = 1.0)$ Population

A sample of 1000 random variables $X_1, \ldots, X_j, \ldots, X_m$ was generated from a population with a Gamma probability density function (Table 6.1.2)

$$f_X(x) = \beta^k \frac{x^{k-1}e^{-\beta x}}{\Gamma(k)} \qquad x \geq 0; k > 0; \beta > 0$$

where k is a shape parameter and $E(X) = k/\beta$. In this example β was chosen to be 1.0 and k was chosen to be 0.2, giving a very "skewed" population. This characteristic of the distribution is easily quantified; from Table 6.1.2, $C(X) = (k)^{-1/2} = 2.236$, the coefficient of skewness is $\gamma_1 = 2C(X) = 4.472$ and the coefficient of kurtosis is $\gamma_2 = 6/k = 30$.

The sample was run through a FORTRAN program called SECTN, which is part of the software supplement, *Enhanced Simulation and Statistics Package*. The inputs to this program are

X: The data array
N: Size of the data array, denoted by m in this chapter
K: Number of sections required, denoted by v this chapter

The output of SECTN is given in Figure 9.3.1 and Figure 9.3.2 for the Gamma data for number of sections $v = 10$ and $v = 20$, respectively.

ESTIMATED SAMPLE PARAMETERS - SAMPLE SIZE= 1000   NON-JACKNIFED   (JACKNIFE K=   0)

SECTION	MEAN	MEDIAN	VARIANCE	STD. DEV.	COEF VAR	SKEWNESS	KURTOSSIS	MINIMUM	MAXIMUM
1	0.2703	1.1635E-02	0.4800	0.6928	2.563	4.903	30.23	8.3970E-08	5.413
2	0.2652	2.0771E-02	0.2336	0.4833	1.822	2.340	5.459	4.7131E-10	2.329
3	0.1908	1.0658E-02	0.1184	0.3441	1.804	2.366	5.401	1.6810E-13	1.582
4	0.2669	1.7222E-02	0.5206	0.7215	2.703	4.992	28.61	1.1787E-10	5.307
5	0.2179	1.6681E-02	0.2609	0.5108	2.344	4.293	21.03	2.4097E-08	3.366
6	0.2327	2.4237E-02	0.2275	0.4770	2.050	2.852	7.693	2.6657E-10	2.333
7	0.1953	2.0957E-02	0.1184	0.3441	1.763	2.287	5.239	8.5664E-15	1.785
8	0.1916	3.1283E-02	0.2074	0.4554	2.377	4.341	19.88	1.1891E-09	2.916
9	0.2780	3.0426E-02	0.2422	0.4921	1.770	2.363	5.332	5.1982E-10	2.305
10	0.2635	4.8733E-02	0.2720	0.5216	1.980	2.968	8.672	3.2291E-09	2.741
UNSECTIONED	0.2372	2.1816E-02	0.2668	0.5166	2.178	4.266	26.67	8.5664E-15	5.413

ESTIMATED PARAMETERS OF SAMPLE PARAMETERS

PARAMETER	MEAN	MEDIAN	VARIANCE	STD. DEV.	COEF VAR	SKEWNESS	KURTOSSIS	STD. DEV./NS**.5
MEAN	0.2372	0.2481	1.2786E-03	3.5757E-02	0.1507	-0.3221	-2.000	1.1307E-02
MEDIAN	2.3260E-02	2.0864E-02	1.2786E-04	1.1307E-02	0.4861	1.278	0.9607	3.5757E-03
VARIANCE	0.2681	0.2379	1.7855E-02	0.1336	0.4984	1.080	-0.1521	4.2255E-02
STD. DEV.	0.5043	0.4877	1.5343E-02	0.1239	0.2456	0.6280	-0.4531	3.9170E-02
COEF. VAR.	2.118	2.015	0.1239	0.3520	0.1663	0.5472	-1.545	0.1113
SKEWNESS	3.371	2.910	1.274	1.129	0.3348	0.4789	-1.921	0.3569
KURTOSSIS	13.76	8.183	103.0	10.15	0.7379	0.7493	-1.516	3.210

NS=# OF SECTS

Figure 9.3.1 Output of the SECTN program for $v = 10$, where the data consist of a random sample of size $m = 1000$ generated from a Gamma distribution with shape parameter $k = 0.2$ and mean 0.2. For this population $C(X) = 2.236$, $\gamma_1 = 4.472$, and $\gamma_2 = 30$. The lower table gives a row of summary statistics for the data, consisting of sectioned estimates, in seven columns of the upper table.

ESTIMATED SAMPLE PARAMETERS - SAMPLE SIZE=   1000   NON-JACKNIFED   (JACKNIFE K=   0)

SECTION	MEAN	MEDIAN	VARIANCE	STD. DEV.	COEF VAR	SKEWNESS	KURTOSSIS	MINIMUM	MAXIMUM
1	0.2399	1.3403E-02	0.6376	0.7985	3.328	5.866	35.31	8.3970E-08	5.413
2	0.3008	1.0056E-02	0.3303	0.5747	1.911	2.138	3.279	1.3694E-07	2.221
3	0.2741	2.7226E-02	0.2337	0.4835	1.764	2.349	5.343	2.6328E-07	2.278
4	0.2563	1.6191E-02	0.2380	0.4879	1.904	2.407	5.834	4.7131E-10	2.329
5	0.1803	1.6529E-02	0.1197	0.3459	1.919	2.456	5.454	2.1416E-09	1.582
6	0.2013	9.5594E-03	0.1193	0.3455	1.716	2.356	5.633	1.6810E-13	1.531
7	0.2263	1.9740E-02	0.1866	0.4320	1.909	1.936	2.080	1.1787E-10	1.465
8	0.3075	1.4537E-02	0.8618	0.9283	3.019	4.532	19.95	1.0174E-08	5.307
9	0.1896	1.6953E-02	0.1069	0.3269	1.725	2.439	5.985	1.0726E-07	1.502
10	0.2463	1.5454E-02	0.4186	0.6470	2.627	3.940	15.15	2.4097E-08	3.366
11	0.3120	3.2817E-02	0.3749	0.6123	1.962	2.242	3.588	4.1586E-09	2.333
12	0.1534	1.9470E-02	7.2020E-02	0.2684	1.750	2.520	6.367	2.6657E-10	1.296
13	0.1708	2.6268E-02	7.6588E-02	0.2767	1.621	1.860	2.212	8.5664E-15	0.9934
14	0.2197	1.4408E-02	0.1615	0.4018	1.829	2.255	4.546	2.0890E-10	1.785
15	0.2467	8.3980E-02	0.2365	0.4863	1.971	3.290	10.67	8.6573E-08	2.416
16	0.1364	1.7963E-02	0.1764	0.4200	3.078	6.170	38.59	1.1891E-09	2.916
17	0.3036	1.5832E-02	0.2723	0.5219	1.719	2.195	4.533	5.1982E-10	2.305
18	0.2524	4.4088E-02	0.2156	0.4643	1.840	2.640	6.599	2.2215E-08	2.138
19	0.1876	4.8733E-02	0.1727	0.4155	2.215	3.781	14.42	3.2291E-09	2.255
20	0.3393	5.9791E-02	0.3652	0.6043	1.781	2.499	5.737	3.4837E-06	2.741
UNSECTIONED	0.2372	2.1816E-02	0.2668	0.5166	2.178	4.266	26.67	8.5664E-15	5.413

ESTIMATED PARAMETERS OF SAMPLE PARAMETERS

PARAMETER	MEAN	MEDIAN	VARIANCE	STD. DEV.	COEF VAR	SKEWNESS	KURTOSSIS	STD. DEV./NS**.5
MEAN	0.2372	0.2431	3.3072E-03	5.7508E-02	0.2424	2.0902E-02	-1.074	1.2859E-02
MEDIAN	2.6150E-02	1.7458E-02	3.6699E-04	1.9157E-02	0.7326	1.905	2.726	4.2836E-03
VARIANCE	0.2688	0.2247	3.8052E-02	0.1951	0.7257	1.852	3.028	4.3619E-02
STD. DEV.	0.4921	0.4739	2.8063E-02	0.1675	0.3404	1.063	0.8162	3.7459E-02
COEF. VAR.	2.079	1.906	0.2586	0.5085	0.2446	1.583	0.7944	0.1137
SKEWNESS	2.994	2.448	1.555	1.247	0.4165	1.653	1.351	0.2788
KURTOSSIS	10.06	5.786	105.8	10.29	1.022	2.040	2.762	2.300

NS=# OF SECTS

Figure 9.3.2 Output of program SECTN using the same data as in Figure 9.3.1 except that the number of sections v is 20.

The first output from SECTN, the upper table in Figure 9.3.1, gives labels for the column contents as the mean, median, variance, standard deviation, coefficients of variation, skewness, and kurtosis, and the minimum and maximum for each section. Because the number of sections was chosen to be $v = 10$, there are $l = 100$ observations in each section. On the last line of the upper table, we have the corresponding statistics for the whole (unsectioned) sample of size $m = 1000$. Thus for the unsectioned data, the estimated mean is 0.2372, the median is 0.0218, and so forth.

The second part of the output from SECTN, the lower table in Figure 9.3.1, contains summary statistics for each of the quantifications (i.e., for each column in the upper table). Thus the "Kurtosis" row gives the sample mean, median, ..., kurtosis, and standard error (standard deviation divided by the square root of the number of sections) for the 10 kurtosis estimates in the "Kurtosis" column on the upper half of the output. Consequently, the column at the bottom headed "Mean" is the sectioned estimate $\bar{\theta}(m)$ and the last column of the bottom is the standard error estimate $S_{\bar{\theta}(1000)}$, where the θ's are the mean, median, variance, standard deviation, and coefficients of variation, skewness, and kurtosis of the population.

Some of the problems associated with the sectioning technique can be seen from this output.

The kurtosis estimate from the *whole sample* is 26.67, compared to the known value of 30. However, we cannot go on to make any inference from this apparently low estimate unless we also have an estimate of its standard deviation. Note, in addition, that all but two estimated kurtoses from the 10 sections are less than 26.67; the minimum value is 5.239 and the maximum 30.23. Thus (even without knowing the true value) the sectioning procedure, based on a rough sign test, is suspect. The mean of the 10 section values is $\bar{\theta}(m) = \bar{\gamma}_2(l) = 13.76$ with estimated standard error 3.210. This is clearly many standard deviations from the true value of 30, or from the estimated kurtosis of 26.67 for the whole sample.

Figure 9.3.2 gives the results of sectioning the same data as used in Figure 9.3.1 but now with $v = 20$. Clearly the use of less data ($l = 50$) in each section aggravates the problem of bias in the kurtosis estimate; from the last row of Figure 9.3.2 $\bar{\theta}(m)$ is now 10.06 with $S_{\bar{\theta}(m)} = 2.300$.

The other estimates (e.g., for coefficient of variation) should be similarly examined carefully for evidence of bias. It would also be clearly advantageous to do Normal Q–Q plots of the columns of sectioned estimates to determine whether confidence interval estimates can be formed with the sectioned estimates.

Figure 9.3.3 gives a Normal Q–Q probability plot for the 20 sectioned kurtoses given in Figure 9.3.2. Clearly the kurtosis estimates are very non-Normal.

Figures 9.3.4 and 9.3.5 give similar results to those just presented for sectioning the extremely skewed Gamma population; this time, however, a Normally distributed population of size $m = 1000$, with mean zero and $\sigma = 1$ was used. Note that $C(X) = \sigma/\mu$ is meaningless here. All estimates appear to be well behaved.

9.3.5 Example: Confidence Intervals for the Kurtosis Estimate in a Gamma($k = 0.2; \beta = 1.0$) Population

The assumption of Normality of the estimates $\tilde{\theta}(l)$ is critical in forming the confidence intervals given in Equation 9.3.10 and should be checked by examining the $\tilde{\theta}_i(l)$ sample. Departures from Normality not only will affect the distribution of the t statistic on which the confidence interval is based, they will also give a $\bar{\theta}(m)$ and an $S_{\bar{\theta}(m)}$ that are correlated. If the correlation is positive, small $\bar{\theta}(m)$ will be associated with small $S_{\bar{\theta}(m)}$, giving considerable bias to the confidence interval estimate.

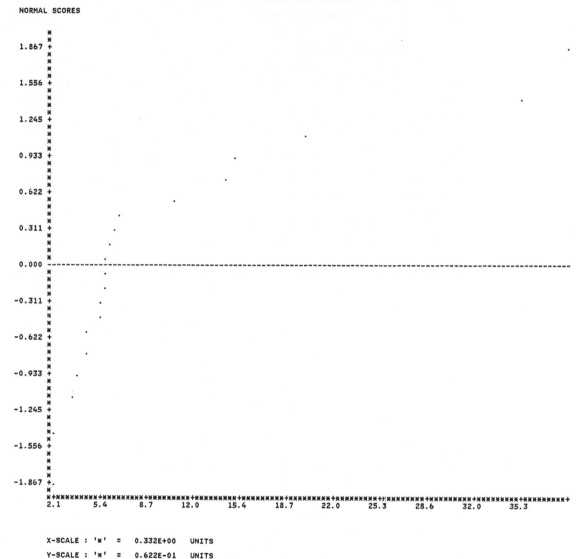

X-SCALE : 'x' =   0.332E+00   UNITS
Y-SCALE : 'x' =   0.622E-01   UNITS

Figure 9.3.3 Normal probability plot for data consisting of 20 kurtosis estimates from sections of size 50 from Gamma-distributed data with shape parameter $k = 0.2$. The data are listed in the third column from the right in the upper table in Figure 9.3.2.

Note that the estimate $S_{\bar{\theta}(m)}$ of the standard deviation of $\bar{\theta}(m)$ does not depend on Normality assumptions.

As an example of confidence intervals based on sectioning, Figure 9.3.6 gives a Normal Q–Q plot of 20 sectioned estimates of the coefficient of kurtosis from a Gamma-distributed population of size $m = 5000$. The sample size was increased from the $m = 1000$ used previously because of the obvious bias and non-Normality

ESTIMATED SAMPLE PARAMETERS - SAMPLE SIZE=   1000   NON-JACKNIFED   (JACKNIFE K=   0)

SECTION	MEAN	MEDIAN	VARIANCE	STD. DEV.	COEF VAR	SKEWNESS	KURTOSSIS	MINIMUM	MAXIMUM
1	-0.1433	-0.2500	1.019	1.009	-7.042	0.1235	4.2424E-02	-2.846	2.308
2	9.8625E-02	0.1497	0.7267	0.8524	8.643	-4.2498E-02	0.1798	-2.484	2.140
3	0.1380	0.2358	0.9540	0.9767	7.077	-0.1015	0.1273	-2.127	3.097
4	-4.0571E-02	-0.1168	0.9792	0.9895	-24.39	0.2065	0.2239	-2.741	2.787
5	2.0797E-02	1.0219E-03	0.9754	0.9876	47.49	8.4588E-02	0.5517	-2.632	2.821
6	-0.1937	-9.9085E-02	1.062	1.030	-5.318	-7.5440E-02	-0.1475	-2.914	2.136
7	2.8339E-02	-6.9447E-03	1.022	1.011	35.68	0.1609	-0.3058	-2.132	2.632
8	-8.9489E-02	-0.1216	1.141	1.068	-11.94	-6.6069E-02	0.1471	-3.328	2.618
9	-0.1004	-0.1991	0.7692	0.8771	-8.734	0.3271	-0.2936	-1.776	2.411
10	-0.1429	-0.2296	1.175	1.084	-7.585	0.4066	-0.2628	-2.538	2.656
UNSECTIONED	-4.2465E-02	-4.8700E-02	0.9844	0.9922	-23.36	8.1895E-02	4.3329E-02	-3.328	3.097

ESTIMATED PARAMETERS OF SAMPLE PARAMETERS

PARAMETER	MEAN	MEDIAN	VARIANCE	STD. DEV.	COEF VAR	SKEWNESS	KURTOSSIS	STD. DEV./NS**.5
MEAN	-4.2465E-02	-6.5030E-02	1.2245E-02	0.1107	-2.606	0.3584	-1.356	3.4993E-02
MEDIAN	-6.3673E-02	-0.1079	2.5626E-02	0.1601	-2.514	0.7821	-0.6946	5.0622E-02
VARIANCE	0.9823	0.9990	2.0324E-02	0.1426	0.1451	-0.7035	-0.4819	4.5082E-02
STD. DEV.	0.9886	0.9994	5.4639E-03	7.3918E-02	7.4771E-02	-0.8448	-0.3778	2.3375E-02
COEF. VAR.	3.389	-6.180	498.6	22.33	6.590	1.119	-0.1264	7.061
SKEWNESS	0.1024	0.1040	3.1178E-02	0.1766	1.725	0.4774	-1.242	5.5837E-02
KURTOSSIS	2.6252E-02	8.4878E-02	7.6682E-02	0.2769	10.55	0.3956	-0.7628	8.7568E-02

NS=# OF SECTS

NORMALS

Figure 9.3.4   Complete output of the SECTN program for $v = 10$, as in Figure 9.3.1, but in this case the data consist of a random sample of size $m = 1000$ generated from a Normal(0, 1) population. The lower table gives a row of summary statistics for the data in seven columns of the upper table.

that occurred when the kurtosis estimate from each section was based on only $l = 1000/20 = 50$ observations. The lack of linearity in the plot in Figure 9.3.6 is a clear indication of non-Normality in this extended sample; this was verified with the Shapiro–Wilk test statistic whose value is given at the bottom of the figure. Because of this non-Normality, confidence intervals should not be formed with the average estimate $\bar{\theta}(5000) = \bar{\gamma}_2(250) = 24.04$ and the estimated standard error of 2.645, obtained from SECTN. When we divide the 5000 sample points into 10 sections, we arrive at $\bar{\theta}(5000) = \bar{\gamma}_2(500) = 26.61$ with estimated standard error 3.95, and the estimates appear to be Normally distributed. The tables and plots are not shown. The overall indication from this example is that the bias in the estimate of kurtosis is quite severe, but it should be remembered that we are using a very skewed population and are estimating a fourth-order moment. In the Normal samples of Figures 9.3.4 and 9.3.5, extreme behavior in the kurtosis estimate is not observed.

ESTIMATED SAMPLE PARAMETERS - SAMPLE SIZE=   1000   NON-JACKNIFED   (JACKNIFE K=   0)

SECTION	MEAN	MEDIAN	VARIANCE	STD. DEV.	COEF VAR	SKEWNESS	KURTOSSIS	MINIMUM	MAXIMUM
1	-0.2303	-0.3168	1.092	1.045	-4.538	0.4743	0.1378	-2.582	2.308
2	-5.6364E-02	-3.7904E-02	0.9508	0.9751	-17.30	-0.2600	0.2024	-2.846	2.149
3	3.0363E-02	0.1910	0.6308	0.7942	26.16	-0.8156	0.5926	-2.484	1.324
4	0.1669	-3.2943E-02	0.8278	0.9099	5.452	0.4146	-0.4311	-1.690	2.140
5	0.1678	0.1893	0.9796	0.9897	5.898	-1.8170E-02	0.7633	-2.127	3.097
6	0.1082	0.2621	0.9461	0.9727	8.988	-0.1962	-0.5642	-1.956	1.867
7	1.1296E-02	-5.5306E-02	1.022	1.011	89.48	0.2200	0.7464	-2.741	2.787
8	-9.2439E-02	-0.1766	0.9513	0.9753	-10.55	0.1876	-0.3784	-2.056	1.997
9	3.0649E-02	1.8919E-02	1.125	1.061	34.61	-0.2780	0.3493	-2.632	2.470
10	1.0945E-02	-7.5733E-02	0.8452	0.9193	84.00	0.6324	0.7994	-2.044	2.821
11	-0.1484	2.6977E-02	1.077	1.038	-6.995	-0.4023	0.1735	-2.914	1.927
12	-0.2391	-0.4180	1.063	1.031	-4.313	0.2534	-0.3581	-2.384	2.136
13	0.1857	4.4792E-02	1.074	1.036	5.581	0.3667	-0.2310	-1.858	2.632
14	-0.1290	-3.3291E-02	0.9410	0.9700	-7.518	-0.1431	-0.8635	-2.132	1.560
15	-7.8727E-02	-0.1224	0.9993	0.9996	-12.70	0.2819	-0.2041	-1.994	2.618
16	-0.1003	-8.7962E-02	1.305	1.143	-11.40	-0.2932	0.2557	-3.328	2.534
17	1.2718E-02	-0.1727	0.8385	0.9157	72.00	0.4271	-0.3497	-1.681	2.411
18	-0.2136	-0.2234	0.6896	0.8304	-3.888	0.1303	-0.5539	-1.776	1.879
19	-0.1659	-0.2633	1.357	1.165	-7.021	0.4521	-0.4743	-2.538	2.656
20	-0.1199	-0.2229	1.015	1.008	-8.406	0.3691	-2.2507E-02	-2.169	2.526
UNSECTIONED	-4.2465E-02	-4.8700E-02	0.9844	0.9922	-23.36	8.1395E-02	4.3329E-02	-3.328	3.097

ESTIMATED PARAMETERS OF SAMPLE PARAMETERS

PARAMETER	MEAN	MEDIAN	VARIANCE	STD. DEV.	COEF VAR	SKEWNESS	KURTOSSIS	STD. DEV./NS**.5
MEAN	-4.2465E-02	-6.7546E-02	1.7512E-02	0.1323	-3.116	0.2837	-1.055	2.9591E-02
MEDIAN	-7.5314E-02	-6.5520E-02	3.0048E-02	0.1733	-2.302	0.1141	-0.3775	3.8761E-02
VARIANCE	0.9865	0.9894	3.0591E-02	0.1749	0.1773	8.0791E-02	0.3309	3.9109E-02
STD. DEV.	0.9894	0.9947	7.9317E-03	8.9060E-02	9.0009E-02	-0.2411	0.3571	1.9914E-02
COEF. VAR.	11.88	-4.101	1077.	32.82	2.763	1.602	0.9124	7.338
SKEWNESS	9.0136E-02	0.2038	0.1397	0.3737	4.146	-0.7348	-0.2698	8.3562E-02
KURTOSSIS	-2.0528E-02	-0.1133	0.2458	0.4958	-24.15	0.2873	-1.139	0.1109

NS=# OF SECTS

NORMALS

Figure 9.3.5 Output of the SECTN program with the number of sections v for the Normal population used in Figure 9.3.4 increased to 20 to give greater degrees of freedom for the variance estimates. The lower table gives the summary statistics for the data in seven of the columns of the upper table.

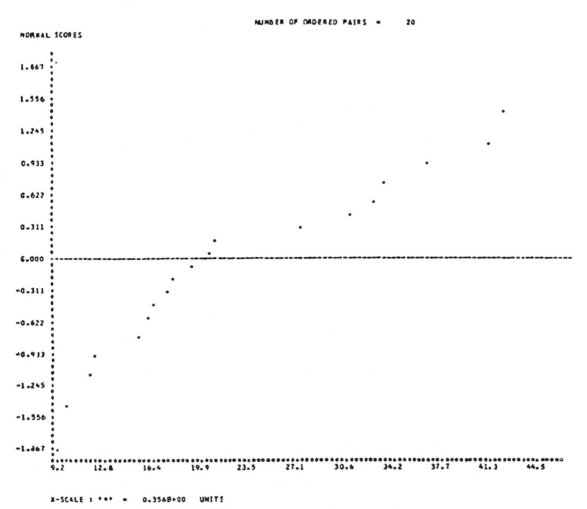

Figure 9.3.6 Normal plot of kurtosis estimates from $v = 20$ sections, each of size $l = 250$, of a sample of size $m = 5000$ from a Gamma-distributed population with shape parameter $k = 0.2$ and mean equal to 0.2. The average is 24.04 with an estimated standard deviation of 2.645. The kurtosis estimate from the whole sample is 30.83; the true value is 30.00.

9.4 Bias Elimination

9.4.1 Pilot Samples

One drawback to sectioning has been shown (Equation 9.3.11) to be the bias introduced into the section estimates $\tilde{\theta}_i(l)$ and consequently into $\bar{\theta}(m)$. There we used the fact that an estimator $\tilde{\theta}(m)$ has, in general, a known bias structure, as discussed in Chapter 6, of the form

$$E\{\tilde{\theta}(m)\} - \theta = \frac{a}{m} + O(m^{-2})$$

This analytical result suggests many ways to eliminate the a/m term in the bias, the most straightforward approach being based on regression methods. Although the impetus for these bias elimination methods is the increased bias in the section estimates $\tilde{\theta}_i(l)$, we will discuss the methods in the context of the full sample and the estimate $\tilde{\theta}(m)$.

First, divide the random sample into two disjoint subsamples of sizes m_1 and m_2, where $m_1 + m_2 = m$ and $m_1 < m_2$. We then have

$$E\{\tilde{\theta}(m_1)\} = \theta + \frac{a}{m_1} + O(m_1^{-2}) \tag{9.4.1}$$

$$E\{\tilde{\theta}(m_2)\} = \theta + \frac{a}{m_2} + O(m_2^{-2}) \tag{9.4.2}$$

Clearly, if we take a linear combination of $\tilde{\theta}(m_1)$ and $\tilde{\theta}(m_2)$ of the form $\tilde{\tilde{\theta}}(m) = \varepsilon\tilde{\theta}(m_1) + (1 - \varepsilon)\tilde{\theta}(m_2)$, we have, if we assume that the ratio of m_1 to m_2 remains constant as m increases,

$$E\{\tilde{\tilde{\theta}}(m)\} = \varepsilon\theta + \frac{\varepsilon a}{m_1} + (1 - \varepsilon)\theta + (1 - \varepsilon)\frac{a}{m_2} + O(m^{-2})$$

$$= \theta + a\left[\frac{\varepsilon}{m_1} + \frac{1 - \varepsilon}{m_2}\right] + O(m^{-2}) \tag{9.4.3}$$

To eliminate the second term in this expression, we want to set the multiplier of the constant a equal to 0. This is easily seen to be achieved by setting $\varepsilon = m_1/(m_1 - m_2)$, where, by assumption, $m_1 < m_2$. Thus

$$\tilde{\tilde{\theta}}(m) = \left[\frac{m_1}{m_1 - m_2}\right]\tilde{\theta}(m_1) + \left[\frac{-m_2}{m_1 - m_2}\right]\tilde{\theta}(m_2) \tag{9.4.4}$$

will be unbiased up to and including the term of order $1/m$.

An advantage of this method of bias elimination is that the variance of $\tilde{\tilde{\theta}}(m)$ can be obtained in terms of the variance of $\tilde{\theta}(m)$. Hence the price, in terms of variance inflation, to be paid for the bias elimination can be assessed as follows. Because the two disjoint samples of size m_1 and m_2 are independent, so are $\tilde{\theta}(m_1)$ and $\tilde{\theta}(m_2)$. In addition, we have in general the asymptotic expansion (see Equation 6.1.39),

$$\text{var}\{\tilde{\theta}(m)\} = \frac{b_1}{m} + O(m^{-3/2})$$

where b_1 is a constant that depends on the distribution of X, namely $F_X(x)$. It follows that

$$\text{var}\{\tilde{\bar{\theta}}(m)\} = \left(\frac{m_1}{m_1 - m_2}\right)^2 \frac{b_1}{m_1} + \left(\frac{m_2}{m_2 - m_1}\right)^2 \frac{b_1}{m_2} + O(m^{-3/2})$$

$$= \frac{b_1}{m} \frac{m^2}{(m_1 - m_2)^2} + O(m^{-3/2}) \tag{9.4.5}$$

Now in Equation 9.4.5 the multiplier $m^2/(m_1 - m_2)^2$ can be considered a variance inflation factor, since it multiplies the b_1/m term, which appears alone in $\text{var}\{\tilde{\theta}(m)\}$. By writing this inflation factor as $m^2/(m - 2m_1)^2$, we can see that it is slightly greater than 1 at $m_1 = 1$ and that it increases monotonically and rapidly to infinity as $m_1 \to m_2$. (This merely says that if $m_1 = m_2$, we have no basis for the comparison, and no way to eliminate the $1/m$ term in the bias.) Thus, m_1 should be small; this corresponds to the sample of size m_1 being a *pilot sample*.

Unfortunately there is one large drawback to this pilot sample method of bias elimination. If m_1 is small, then the estimate $\tilde{\theta}(m_1)$ could be *very non-Normal* and, even if $\tilde{\theta}(m_2)$ were Normal because of the large sample size m_2, the mixture $\tilde{\bar{\theta}}(m)$ could be very non-Normal. If m is fairly large this is not a problem, but then the bias is not important either. If we try to apply the pilot sample method to the estimates $\tilde{\theta}_i(l)$ in each section of a sectioned estimator $\bar{\theta}(m)$, as discussed in the preceding section, then, again, m must be fairly large.

The jackknife method of bias elimination described in the next section overcomes some of the drawbacks of the pilot sample technique, at a price of uncertainty about the variance inflation induced by the method, and also at a price of greater computation time.

9.4.2 The Jackknife for Bias Elimination

There are many ways to eliminate the $1/m$ term in the bias of $\tilde{\theta}(m)$, and the jackknife is a method that eliminates the non-Normality problems that occur with pilot samples. The price paid is uncertainty, in small to moderate samples, of the amount of possible variance inflation in the new, less biased, jackknifed estimate.

The original jackknifing proposal by Quenouille (1949), which we will call the *twofold jackknife*, combined the estimate $\tilde{\theta}(m)$ from the whole sample with estimates from two disjoint halves of the data set. Assuming that m is even, we call these estimates $\tilde{\theta}_1(m/2)$ and $\tilde{\theta}_2(m/2)$. Then the twofold jackknife estimator is based on the linear combination

$$\delta\tilde{\theta}(m) + (1 - \delta)\left\{\frac{\tilde{\theta}_1(m/2) + \tilde{\theta}_2(m/2)}{2}\right\} \tag{9.4.6}$$

Suitable choice of the parameter δ in this particular one-parameter linear combination of the three estimators makes the leading term in the expected value of Equation 9.4.6 equal to θ. Explicitly, the expectation of Equation 9.4.6 is

$$\delta\theta + \frac{\delta a}{m} + (1 - \delta)\theta + (1 - \delta)\frac{a}{m/2} + O(m^{-2}) = \theta + \frac{a}{m}\{(2 - \delta)\} + O(m^{-2})$$

so that the a/m term disappears if $\delta = 2$. Then we have the twofold jackknife estimator

$$2\tilde{\theta}(m) - \left\{ \frac{\tilde{\theta}_1(m/2) + \tilde{\theta}_2(m/2)}{2} \right\} \tag{9.4.7}$$

This twofold jackknife estimator eliminates the $1/m$ bias term in the asymptotic expansion for the expectation of $\tilde{\theta}(m)$ and avoids using a combination of a small sample and a large sample estimator, as was the case for the pilot sample estimate $\tilde{\tilde{\theta}}(m)$. Extending this idea to its limit, we combine $\tilde{\theta}(m)$ with $\tilde{\theta}_i(m - 1), i = 1, \ldots, m$, in a linear fashion, where $\tilde{\theta}_i(m - 1)$ is the estimator calculated from the complete sample with the ith data point, X_i, eliminated. Our general formula becomes

$$\theta^*(m) = \delta\tilde{\theta}(m) + (1 - \delta)\left\{ \frac{1}{m} \sum_{i=1}^{m} \tilde{\theta}_i(m - 1) \right\} \tag{9.4.8}$$

and formal manipulation shows that $\delta = m$ will eliminate the $1/m$ bias term from the expectation of $\theta^*(m)$. Thus, we have the *completely jackknifed estimate*:

$$\theta^*(m) = m\tilde{\theta}(m) - \frac{m - 1}{m} \sum_{i=1}^{m} \tilde{\theta}_i(m - 1) \tag{9.4.9}$$

This is a linear combination of $m + 1$ estimates, *all of which are from large samples*.

A number of points are of interest about this completely jackknifed estimator $\theta^*(m)$:

• It is difficult to assess the variance of $\theta^*(m)$ because it is a linear combination of *dependent* estimators. However, it can be shown that for a broad class of estimators there is asymptotically no inflation of variance in the jackknifed estimator: that is, $\text{var}\{\theta^*(m)\} \to \text{var}\{\tilde{\theta}(m)\}$ as $m \to \infty$.

• More surprisingly, $\text{var}\{\theta^*(m)\}$ can be smaller than $\text{var}\{\tilde{\theta}(m)\}$ in small samples, the result being especially true for ratio estimators involving sample moments. Examples are the estimates of the coefficient of variation $C(X)$, namely $S/\bar{X}$, and the estimates of the coefficients of skewness and kurtosis.

• The complete jackknife can be computationally prohibitive if m is large; in this case, fourfold, eightfold, ..., jackknife estimators should be used.

• If $\tilde{\theta}(m) = \bar{X}$ then it is easily seen that a twofold, fourfold, ..., or complete jackknife estimator gives back the sample average. In fact the jackknife will always work well with estimators that are linear functions, or approximately linear functions, of the observations $X_1, \ldots, X_j, \ldots, X_m$.

• Under fairly general regularity conditions—for example, that $\tilde{\theta}(m)$ be a function of the maximum likelihood estimator—it can be shown that $\theta^*(m)$ is asymptotically Normal and, when standardized by the standard deviation estimate from Section 9.5, $\theta^*(m)$ will have approximately a standardized Normal distribution. We can exploit this fact to produce confidence intervals for θ, as described in Section 9.5.

EXAMPLE **9.4.1 Jackknifing to Improve Sectioning**
An added feature of the program SECTN in the software supplement is that it will provide completely jackknifed estimates of the various quantifications *within the sections*. Thus Figure 9.4.1 gives the output of SECTN where the sample size of the Gamma population has been increased to $m = 5000$ and all estimates within sections have been subjected to complete jackknifing. The data, without jackknifing, were analyzed in Section 9.3.5, where we obtained the estimate $\bar{\theta}(5000) = \bar{\gamma}_2(250) = 24.04$ and an estimated standard error 2.645.

Now, from the last row of Figure 9.4.1, we have $\bar{\theta}(5000) = \bar{\gamma}_2(250) = 31.30$ and $S_{\bar{\theta}(5000)} = S_{\bar{\gamma}_2(250)} = 4.646$. Thus, bias appears to have been reduced by jackknifing. It is tempting to speculate that the standard deviation of the sectioned estimate is smaller than the standard deviation of the sectioned estimate with complete jackknifing in each section (2.645 versus 4.646). However, these numbers represent estimates with only 19 degrees of freedom, so the difference may be due to sampling variations alone.

9.5 Variance Assessment with the Complete Jackknife

If the complete jackknife is used *within sections* to eliminate small-sample bias, then a variance estimate for the overall average of the jackknifed estimators across sections is obtainable by the methods of Section 9.3.1.

However, an added feature of the complete jackknife that makes it valuable in small samples is that we can obtain a variance estimate directly for $\theta^*(m)$, and possibly confidence intervals (Miller, 1974; Efron, 1981). To see this, note that we can write $\theta^*(m)$ as follows:

$$\theta^*(m) = m\tilde{\theta}(m) - \frac{m-1}{m} \sum_{i=1}^{m} \tilde{\theta}_i(m-1)$$

$$= \frac{1}{m} \sum_{i=1}^{m} [m\tilde{\theta}(m) - (m-1)\tilde{\theta}_i(m-1)] = \frac{1}{m} \sum_{i=1}^{m} {}_{(i)}\tilde{\theta}(m) \tag{9.5.1}$$

where the ${}_{(i)}\tilde{\theta}(m)$ are called "pseudo-values." It was Tukey's proposal (1958) to proceed as if $\theta^*(m)$ were an average of m independent estimates ${}_{(i)}\tilde{\theta}(m)$ and then form the variance estimate

$$S^2_{\theta^*(m)} = \frac{1}{m(m-1)} \sum_{i=1}^{m} \left[{}_{(i)}\tilde{\theta}(m) - \theta^*(m) \right]^2$$

with the possibility of forming confidence intervals using the corresponding t statistic.

If this proposal seems somewhat far-fetched, note that if $\tilde{\theta}(m) = \bar{X} =$

ESTIMATED SAMPLE PARAMETERS - SAMPLE SIZE= 5000   K-FOLD JACKNIFED (JACKNIFE K= 250)

SECTION	MEAN	MEDIAN	VARIANCE	STD. DEV.	COEF VAR	SKEWNESS	KURTOSSIS	MINIMUM	MAXIMUM
1	0.2518	1.5741E-02	0.3097	0.5681	2.283	5.610	48.21	4.7131E-10	5.413
2	0.2354	1.6098E-02	0.3367	0.5920	2.547	5.584	39.42	1.6810E-13	5.307
3	0.2220	3.2381E-02	0.1861	0.4340	1.966	2.971	9.209	8.5664E-15	2.416
4	0.2452	2.9308E-02	0.2434	0.4964	2.041	3.112	10.04	5.1982E-10	2.916
5	0.1792	1.9652E-02	0.1300	0.3656	2.061	4.336	27.95	8.8001E-16	3.131
6	0.2352	3.7345E-02	0.2858	0.5523	2.397	7.533	77.34	2.6202E-19	5.726
7	0.1730	1.2142E-02	0.1799	0.4342	2.545	6.110	51.71	4.1728E-12	4.160
8	0.1966	1.6628E-02	0.1828	0.4320	2.214	3.727	16.79	2.4633E-18	3.149
9	0.1988	1.2979E-02	0.1894	0.4393	2.231	3.986	19.16	1.3122E-14	3.134
10	0.2109	1.7526E-02	0.2739	0.5379	2.596	6.865	65.50	1.2640E-15	5.398
11	0.1962	1.7225E-02	0.1666	0.4110	2.117	3.349	12.95	5.0314E-11	2.788
12	0.1755	1.7020E-02	0.1535	0.3965	2.277	3.963	18.86	1.4530E-23	2.903
13	0.2330	2.0879E-02	0.3229	0.5837	2.552	6.754	63.91	2.5714E-12	5.837
14	0.2180	2.2338E-02	0.1933	0.4445	2.054	3.772	18.40	6.9515E-18	3.372
15	0.1645	1.4973E-02	0.1367	0.3732	2.290	3.796	16.71	8.3535E-14	2.622
16	0.1396	1.5043E-02	0.1055	0.3307	2.390	4.976	29.63	1.7084E-16	2.716
17	0.2173	2.1844E-02	0.2530	0.5092	2.363	4.149	19.53	1.4091E-13	3.486
18	0.1722	1.8767E-02	0.1502	0.3953	2.332	5.744	47.76	1.7550E-11	3.749
19	0.2368	2.3199E-02	0.2442	0.4979	2.118	3.307	11.61	1.4290E-21	3.078
20	0.1689	2.0502E-02	0.1709	0.4179	2.493	4.339	21.23	7.0944E-11	2.985
UNSECTIONED	0.2022	1.8501E-02	0.2095	0.4578	2.264	4.565	30.83	1.4530E-23	5.837

ESTIMATED PARAMETERS OF SAMPLE PARAMETERS

PARAMETER	MEAN	MEDIAN	VARIANCE	STD. DEV.	COEF VAR	SKEWNESS	KURTOSSIS	STD. DEV./NS**.5
MEAN	0.2035	0.2049	1.0034E-03	3.1677E-02	0.1557	-0.2454	-1.121	7.0831E-03
MEDIAN	2.0079E-02	1.8147E-02	4.1620E-05	6.4513E-03	0.3213	1.408	1.267	1.4426E-03
VARIANCE	0.2107	0.1877	4.6577E-03	6.8247E-02	0.3239	0.4537	-1.079	1.5261E-02
STD. DEV.	0.4606	0.4368	5.9223E-03	7.6957E-02	0.1671	0.2945	-1.123	1.7208E-02
COEF. VAR.	2.293	2.286	3.7215E-02	0.1929	8.4120E-02	-1.1234E-02	-1.248	4.3137E-02
SKEWNESS	4.699	4.243	1.854	1.361	0.2897	0.6769	-0.8711	0.3044
KURTOSSIS	31.30	20.38	431.7	20.78	0.6639	0.9243	-0.5892	4.646

NS=# OF SECTS

Figure 9.4.1   Output of the SECTN program using a sample of size $m = 5000$ generated from a Gamma-distributed population for which $C(X) = 2.236$, $\gamma_1 = 4.472$, and $\gamma_2 = 30$. The number of sections is $v = 20$, and estimates are *completely jackknifed within each section*. The lower table is the summary statistics for seven of the columns in the upper table.

$(1/m)\sum X_i$, then $_{(i)}\tilde{\theta}(m) = X_i$, and the jackknife variance estimate is the same as the usual variance estimate $S_{\bar{X}}^2$. An example of the use of the complete jackknife for variance estimation is given in Section 9.6.

9.6 Variance Assessment with the Bootstrap

Another broad, small-sample method for obtaining variance estimates for $\tilde{\theta}(m)$, or confidence intervals for θ, is "the bootstrap" (Efron, 1981; Efron and Gong, 1983; Efron and Tibshirani, 1986), although there are now many variations of this procedure (see, e.g., Banks, 1988). As remarked in the introduction, bootstrapping is basically a simulation method, its application in simulation itself being limited to occasions when computing time constraints and the computing time required for a simulation may force the use of very small simulation samples.

The construction of the bootstrap estimate of the variance of $\tilde{\theta}(m)$ is as follows (Efron and Gong, 1983).

1. From the sample $X_1, \ldots, X_j, \ldots, X_m$, construct an empirical distribution function $\hat{F}_X(x)$ as at Equation 6.2.14.

2. Draw a *bootstrap sample* $X_1^*, \ldots, X_j^*, \ldots, X_m^*$ from the empirical distribution function $\hat{F}_X(x)$. That is, make m random draws from the given values $X_1, \ldots, X_j, \ldots, X_m$ with replacement. Thus X_1^* can be any one of the values $X_1, \ldots, X_j, \ldots, X_m$ with equal probability $1/m$. The rest of the bootstrap sample $X_2^*, \ldots, X_j^*, \ldots, X_m^*$ are drawn independently in the same way.

3. Compute the bootstrap estimate $\tilde{\theta}_m^*(1)$ by applying the formula for the estimator $\tilde{\theta}(m)$ to the sample $X_1^*, \ldots, X_j^*, \ldots, X_m^*$.

4. Repeat Steps 2 and 3 B times to obtain the replicated bootstrap estimates $\tilde{\theta}_m^*(1), \tilde{\theta}_m^*(2), \ldots, \tilde{\theta}_m^*(b), \ldots, \tilde{\theta}_m^*(B)$, where B is typically between 128 and 512.

5. Form the mean value of the bootstrap estimates across replications. We have

$$\tilde{\theta}_m^*(\cdot) = \frac{1}{B} \sum_{b=1}^{B} \tilde{\theta}_m^*(b) \tag{9.6.1}$$

and the sample standard deviation of the bootstrap replications:

$$S_B = \left[\sum_{b=1}^{B} \frac{\{\tilde{\theta}_m^*(b) - \tilde{\theta}_m^*(\cdot)\}^2}{B - 1} \right]^{1/2} \tag{9.6.2}$$

Then S_B is the bootstrap estimate of the standard deviation of $\tilde{\theta}(m)$.

The contention is that as $B \to \infty$, S_B approaches the standard deviation of $\tilde{\theta}(m)$ computed from a sample with distribution function $\hat{F}_X(x)$ and that this in turn will approximate the true but unknown standard deviation of $\tilde{\theta}(m)$.

Note two things here. First we are performing a simulation with B replications to estimate the standard deviation of $\tilde{\theta}(m)$ from a sample with distribution $\hat{F}_X(x)$; this estimate improves as $B \to \infty$. Second, since $\hat{F}_X(x)$ approaches $F_X(x)$ as $m \to \infty$, the procedure should get better and better as m gets larger, in the sense that the bootstrap standard deviation should approach the true standard deviation of $\tilde{\theta}(m)$.

We illustrate these ideas with an example.

9.6.1   Example: The Standard Deviation of the Correlation Coefficient in a Bivariate Normal Population

Bivariate Normal random variables are discussed in Chapter 10. Suffice it to say here that the variables X and Y in the pair $\mathbf{Z} = (X, Y)$ have marginal Normal distributions with parameters (μ_X, σ_X) and (μ_Y, σ_Y), and their dependence is specified by the correlation coefficient

$$\rho = \text{corr}(X, Y) = E\left\{\left(\frac{X - \mu_X}{\sigma_X}\right)\left(\frac{Y - \mu_Y}{\sigma_Y}\right)\right\} \tag{9.6.3}$$

This correlation coefficient is often called the Pearson product moment correlation coefficient. It can be estimated in an iid sample $\mathbf{Z}_1 = (X_1, Y_1), \ldots, \mathbf{Z}_i = (X_i, Y_i), \ldots,$ $\mathbf{Z}_m = (X_m, Y_m)$ by

$$\hat{\rho} = \frac{\sum\limits_{i=1}^{m} (X_i - \bar{X})(Y_i - \bar{Y})}{\left[\left\{\sum\limits_{i=1}^{m} (X_i - \bar{X})^2\right\}\left\{\sum\limits_{i=1}^{m} (Y_i - \bar{Y})^2\right\}\right]^{1/2}} \tag{9.6.4}$$

The distribution of $\hat{\rho}$ can be computed in bivariate Normal samples (Johnson and Kotz, 1970, Chap. 32), although not easily. A useful approximation to the standard deviation of $\hat{\rho}$ in Normal samples is (Efron and Gong, 1983)

$$\sigma_{\hat{\rho}} \approx \frac{(1 - \rho^2)}{\sqrt{m - 3}} \tag{9.6.5}$$

This is, asymptotically, the square root of the leading term in the expansion for $\sigma_{\hat{\rho}}^2$; the divisor $\sqrt{m - 3}$ makes Equation 9.6.5 a better approximation than $(1 - \rho^2)/\sqrt{m}$ in small samples without the necessity of going to terms of order $1/m^2$.

The Normal theory estimate of the standard deviation of ρ in a bivariate Normal sample of size m is thus

$$\hat{\sigma}_{\hat{\rho}} = \frac{(1 - \hat{\rho}^2)}{\sqrt{m - 3}} \tag{9.6.6}$$

The distribution of this standard error estimate is, technically, known from the distribution of $\hat{\rho}$ in Normal samples (Johnson and Kotz, 1970, Chap. 32); the formula gives little intuitive insight. Instead we will use simulation to look empirically at the distribution of $\hat{\sigma}_{\hat{\rho}}$. Figure 9.6.1 shows SMTBPC runs of the standard deviation estimator $\hat{\sigma}_{\hat{\rho}}$ for bivariate Normal ($\rho = 0.5$) samples of sizes 8, 14, 20, 26, 32, 38, 44, and 50. The distribution of $\hat{\sigma}_{\hat{\rho}}$ is negatively skewed for small m. Concentrating on the case $m = 14$, see that the mean of the sample of $14000/14 = 1000$ $\hat{\sigma}_{\hat{\rho}}$'s at $m = 14$ is 0.2156 compared to the true value (Equation 9.6.5) of $(1 - 0.5^2)/\sqrt{11} = 0.226$. Since the estimated standard deviation of the estimated mean is 0.0018, the value 0.2156 is more than 2 standard deviations from 0.226. Thus the Normal theory estimate, $\hat{\sigma}_{\hat{\rho}}$, is evidently biased at $m = 14$.

Figure 9.6.2 shows the same experiment, but using the complete jackknife estimate of $\sigma_{\hat{\rho}}$. (The extension of jackknifing to estimates computed from bivariate data requires the analogous removal of pairs of data points before computation of pseudo-values.) Figures 9.61 and 9.62 are on the same scale, and it is immediately clear that the jackknife estimate of $\sigma_{\hat{\rho}}$ is more variable than the Normal theory estimate, $\hat{\sigma}_{\hat{\rho}}$, and is positively skewed, with some rather large (positive) outliers, which are highly undesirable. Nevertheless, the jackknifing idea does provide a usable method for assessing variability in this case

Figures 9.6.3 and 9.6.4 show simulations of the properties of bootstrap estimates of $\sigma_{\hat{\rho}}$ for $B = 128$ and $B = 512$. These two figures are again on the same scale as Figures 9.6.1 and 9.6.2. A first glance shows that little is gained statistically in going from $B = 128$ to $B = 512$. The bootstrap estimates of $\sigma_{\hat{\rho}}$ are positively skewed, but they are certainly better, in terms of variability and positive outliers, than the jackknife estimates. Again, the Normal theory estimator $\hat{\sigma}_{\hat{\rho}}$ is clearly better than the bootstrap estimator in the sense that at, say, $m = 14$, its distribution is tighter than the distribution of the bootstrap estimator. The standard deviations of the estimators, at $m = 14$, are 0.0646 for the bootstrap estimator and 0.0580 for $\hat{\sigma}_{\hat{\rho}}$ (Figures 9.6.3 and 9.6.1, respectively). However, the Normal theory estimator is based on an assumption of bivariate Normality for the data, whereas the bootstrap estimator is completely nonparametric. If simulations had been performed on non-Normal data pairs, we might have found the bootstrap estimator to be superior to $\hat{\sigma}_{\hat{\rho}}$.

The convergence of the bootstrap estimator to Normality can be seen from Figure 9.6.3 to be rather slow. It should also be remembered that because of computational considerations, bootstrapping is mainly a small sample procedure. For large m, one might want to pursue the sectioning ideas discussed earlier in this chapter. Note, too, that this simulation repeats one given in Efron and Gong (1983); however, they studied only the case of $m = 14$. (A larger study is presented in Efron, 1981.)

9.6.2 Improving the Bootstrap

A little thought will make it clear that there must be ways to improve the bootstrap estimate of variability given above. In particular, if the estimator $\tilde{\theta}(m)$ can take on a continuous range of values, why not use a smoothed version of the empirical density or distribution function obtained from $X_1, \ldots, X_j, \ldots, X_m$ to sample from

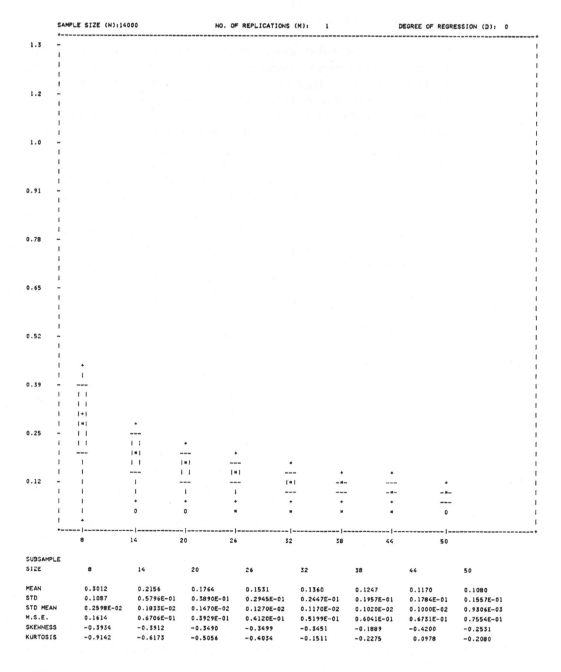

SUBSAMPLE

SIZE	8	14	20	26	32	38	44	50
MEAN	0.3012	0.2156	0.1764	0.1531	0.1360	0.1247	0.1170	0.1080
STD	0.1087	0.5796E-01	0.3890E-01	0.2945E-01	0.2447E-01	0.1957E-01	0.1784E-01	0.1557E-01
STD MEAN	0.2598E-02	0.1833E-02	0.1470E-02	0.1270E-02	0.1170E-02	0.1020E-02	0.1000E-02	0.9306E-03
M.S.E.	0.1614	0.6706E-01	0.3929E-01	0.4120E-01	0.5199E-01	0.6041E-01	0.6731E-01	0.7554E-01
SKEWNESS	-0.3934	-0.3912	-0.3490	-0.3499	-0.3451	-0.1889	-0.4200	-0.2531
KURTOSIS	-0.9142	-0.6173	-0.5056	-0.4034	-0.1511	-0.2275	0.0978	-0.2080

ESTIMATOR: **NORMAL THEORY ESTIMATE OF STD OF CORR. OF BIVARIATE NORMALS
VERTICAL SCALE: YMIN =   0.0179
                 YMAX =   1.3075

Figure 9.6.1   SMTBPC runs to investigate the properties of the Normal theory estimate of the standard deviation of the correlation estimate $\hat{\rho}$ in a bivariate Normal sample. The estimate is $(1 - \hat{\rho}^2)/\{(m - 3)\}^{1/2}$. The true value of ρ is 0.5.

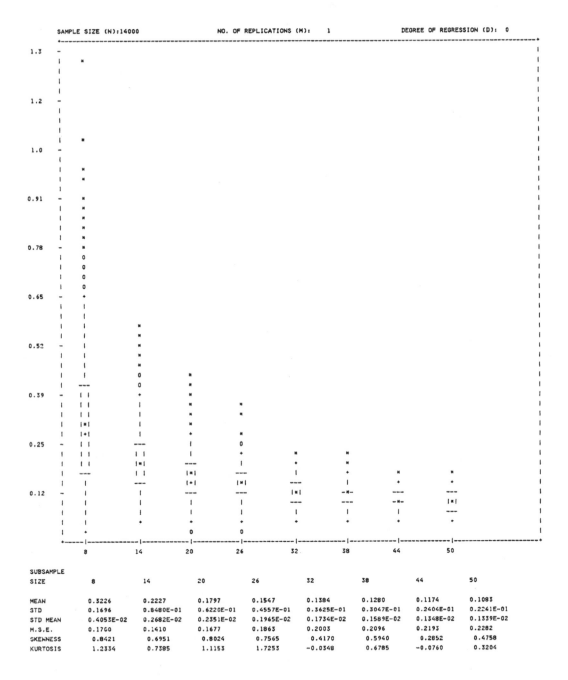

SUBSAMPLE SIZE	8	14	20	26	32	38	44	50
MEAN	0.3226	0.2227	0.1797	0.1547	0.1384	0.1280	0.1174	0.1083
STD	0.1696	0.8480E-01	0.6220E-01	0.4557E-01	0.3625E-01	0.3047E-01	0.2404E-01	0.2241E-01
STD MEAN	0.4053E-02	0.2682E-02	0.2351E-02	0.1965E-02	0.1734E-02	0.1589E-02	0.1348E-02	0.1339E-02
M.S.E.	0.1700	0.1410	0.1677	0.1863	0.2003	0.2096	0.2193	0.2282
SKEWNESS	0.8421	0.6951	0.8024	0.7565	0.4170	0.5940	0.2852	0.4758
KURTOSIS	1.2334	0.7385	1.1153	1.7253	-0.0348	0.6785	-0.0760	0.3204

ESTIMATOR: ***JACKNIFE ESTIMATE OF STD OF CORR. OF BIVARIATE NORMALS
VERTICAL SCALE: YMIN =   0.0179
                 YMAX =   1.3075

Figure 9.6.2 SMTBPC runs to investigate the properties of the jackknife estimate of the standard deviation of an estimate of the correlation coefficient in a bivariate Normal sample ($\rho = 0.5$). The estimate whose standard deviation is being investigated is actually the estimate, $\hat{\rho}$, after complete jackknifing, as at Equation 9.5.1, to remove the bias term of order $1/m$.

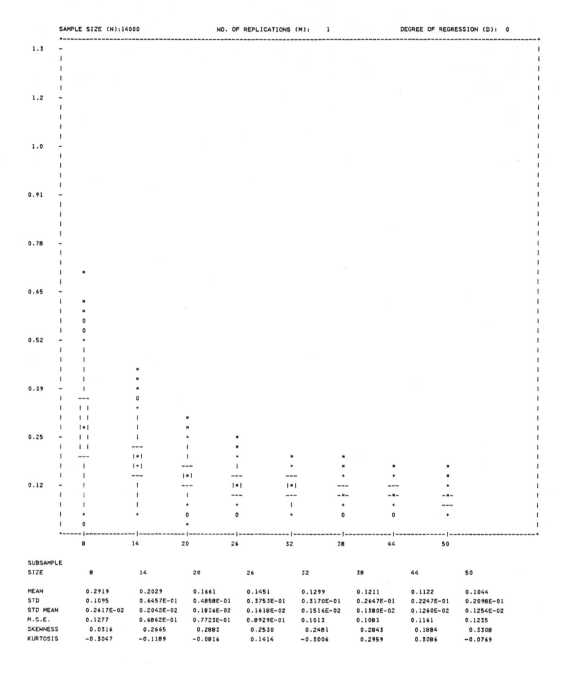

SAMPLE SIZE (N):14000   NO. OF REPLICATIONS (M):   1   DEGREE OF REGRESSION (D): 0

SUBSAMPLE SIZE	8	14	20	26	32	38	44	50
MEAN	0.2919	0.2029	0.1661	0.1451	0.1299	0.1211	0.1122	0.1044
STD	0.1095	0.6457E-01	0.4858E-01	0.3753E-01	0.3170E-01	0.2647E-01	0.2247E-01	0.2098E-01
STD MEAN	0.2617E-02	0.2042E-02	0.1836E-02	0.1618E-02	0.1516E-02	0.1380E-02	0.1260E-02	0.1254E-02
M.S.E.	0.1277	0.6862E-01	0.7723E-01	0.8929E-01	0.1013	0.1083	0.1161	0.1235
SKEWNESS	0.0316	0.2665	0.2883	0.2530	0.2481	0.2843	0.1884	0.3308
KURTOSIS	-0.3047	-0.1189	-0.0816	0.1414	-0.3006	0.2959	0.3086	-0.0769

Figure 9.6.3 SMTBPC run to investigate the properties of the standard boot-strap estimate of the standard deviation of the correlation estimate $\hat{\rho}$ in a bivariate Normal sample ($\rho = 0.5$) for $B = 128$ (number of replications in the bootstrap simulation).

280

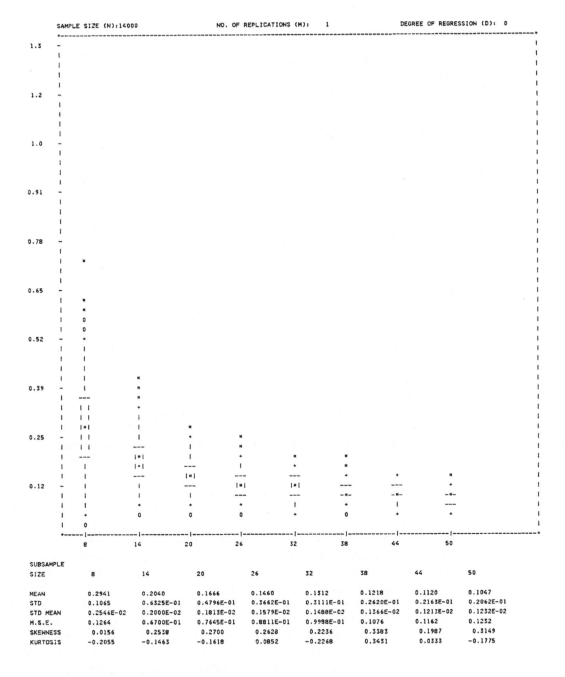

SUBSAMPLE SIZE	8	14	20	26	32	38	44	50
MEAN	0.2941	0.2040	0.1666	0.1460	0.1312	0.1218	0.1120	0.1047
STD	0.1065	0.6325E-01	0.4796E-01	0.3662E-01	0.3111E-01	0.2620E-01	0.2163E-01	0.2062E-01
STD MEAN	0.2546E-02	0.2000E-02	0.1813E-02	0.1579E-02	0.1488E-02	0.1366E-02	0.1213E-02	0.1232E-02
M.S.E.	0.1264	0.6700E-01	0.7645E-01	0.8811E-01	0.9998E-01	0.1076	0.1162	0.1232
SKEWNESS	0.0156	0.2538	0.2700	0.2628	0.2236	0.3383	0.1987	0.3149
KURTOSIS	-0.2055	-0.1463	-0.1618	0.0852	-0.2268	0.3431	0.0333	-0.1775

ESTIMATOR: **BOOTSTRAP ESTIMATE OF STD OF CORR. OF BIVARIATE NORMALS. B=512
VERTICAL SCALE: YMIN =   0.0179
                 YMAX =   1.3075

Figure 9.6.4 SMTBPC run to investigate the properties of the standard bootstrap estimate of the standard deviation of the correlation estimate $\hat{\rho}$ in a bivariate Normal sample ($\rho = 0.5$) for $B = 512$ (number of replications in the bootstrap simulation).

when computing the bootstrap estimate of variance? This, and other ideas for improving the bootstrap may be found in Efron and Tibshirani (1986) and Banks (1988). Here we only consider a simple smoothing of the empirical distribution or density function obtained from $X_1, \ldots, X_j, \ldots, X_m$.

EXAMPLE

9.6.1 The Normal-Smoothed Bootstrap

In Chapter 7 we discussed several methods for obtaining smoothed, kernel density estimates. These involved placing a kernel (e.g., a triangle with area $1/m$) over each data point so that the total area of the smoothed density, which is the sum of these kernels, is 1. In addition, the width of the kernel decreases as m increases in a manner such that the smoothed, kernel density estimate converges, as $m \to \infty$, to the true, unknown density function.

For two-dimensional data, smoothing the data points to obtain a two-dimensional density estimate is similar to that for one-dimensional data. Again, the shape of the two-dimensional kernel that is to be placed at each data point must be chosen.

To obtain a smoothed density estimate for the bootstrapping of the estimate of the correlation coefficient, Efron and Gong (1983) smoothed the two-dimensional empirical density function $\hat{f}_{X,Y}(x, y)$ formed from the points (X_i, Y_i), $i = 1, \ldots, m$. They placed over each point a weighted bivariate Normal density function with covariance matrix $(0.25) \times \hat{\Sigma}$, where $\hat{\Sigma}$ is the estimated covariance matrix of the data sample $Z_i = (X_i, Y_i)$, $i = 1, \ldots, m$. This smoothing is equivalent to generating the bootstrap sample (X_i^*, Y_i^*), $i = 1, \ldots, m$, as an equally weighted mixture of bivariate Normal random variables with means (X_i, Y_i) and covariance $(0.25) \times \hat{\Sigma}$.

The results obtained with this Normal-smoothed bootstrap are illustrated in Figures 9.6.5 and 9.6.6, generated with SMTBPC, for $B = 128$ and $B = 512$, respectively. Close examination and comparison of these figures shows very similar overall behavior of the two estimators; the boxplots for similar sample sizes are almost identical. Closer inspection of the numerical results shows slight improvements with $B = 512$ as compared to $B = 128$. Specifically, in Figure 9.6.5 for $B = 128$, at $m = 8$, the standard deviation of the standard deviation estimate is 0.08215 and the mean square error is 0.1001. The corresponding figures for $B = 512$ from Figure 9.6.6 are 0.07964 and 0.09766. Similar results are found at $m = 50$.

Contrasting Figures 9.6.3 and 9.6.5, however, one can see graphically that the Normal-smoothed bootstrap estimator of the standard deviation of $\hat{\rho}$ is, in most ways, better than the ordinary bootstrap estimator of the standard deviation of $\hat{\rho}$. Not only are there fewer outliers in Figure 9.6.5, but for all m the standard deviations with Normal smoothing are smaller than for the nonsmoothed case. However, there is greater bias with the Normal-smoothed bootstrap than with the ordinary bootstrap.

To conclude this section, we recall that bootstrapping and Normal-smoothed bootstrapping are computer-intensive methods. In many cases it might be easier and cheaper to use larger simulation runs and simpler methods such as sectioning for estimating the precision of the simulation estimates. However, when simulations produce estimates based on too few independent replications to use sectioning, bootstrapping becomes a useful tool. The validity of bootstrapping has been estab-

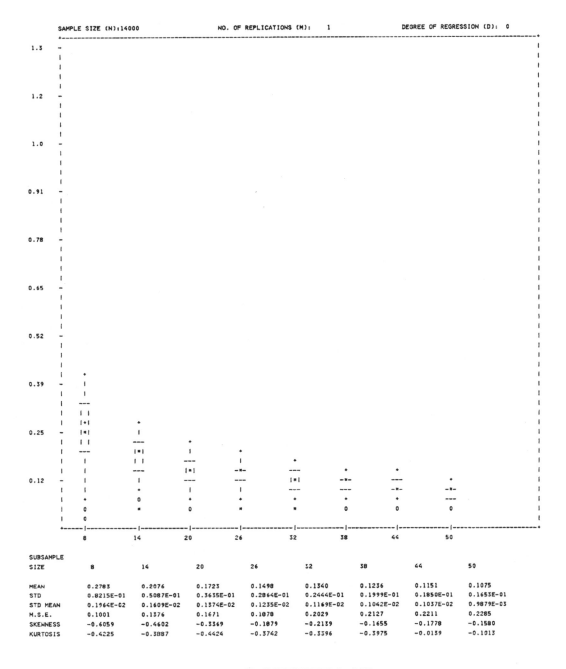

SUBSAMPLE SIZE	8	14	20	26	32	38	44	50
MEAN	0.2783	0.2076	0.1723	0.1498	0.1340	0.1236	0.1151	0.1075
STD	0.8215E-01	0.5087E-01	0.3635E-01	0.2864E-01	0.2444E-01	0.1999E-01	0.1850E-01	0.1653E-01
STD MEAN	0.1964E-02	0.1609E-02	0.1374E-02	0.1235E-02	0.1169E-02	0.1042E-02	0.1037E-02	0.9879E-03
M.S.E.	0.1001	0.1376	0.1671	0.1878	0.2029	0.2127	0.2211	0.2285
SKEWNESS	-0.6059	-0.4602	-0.3369	-0.1879	-0.2139	-0.1655	-0.1778	-0.1580
KURTOSIS	-0.4225	-0.3887	-0.4424	-0.3742	-0.3596	-0.3975	-0.0139	-0.1013

ESTIMATOR: NORMAL-SMOOTH BOOTSTRAP ESTIMATE OF STD OF CORR. OF BIVARIATE NORMALS.   B=128

VERTICAL SCALE: YMIN =   0.0179
                  YMAX =   1.3075

Figure 9.6.5   SMTBPC run to investigate the properties of the Normal-smoothed bootstrap estimate of the standard deviation of the correlation estimate $\hat{\rho}$ in a bivariate Normal sample ($\rho = 0.5$). (Here the number of replications in the bootstrap simulation is $B = 128$.)

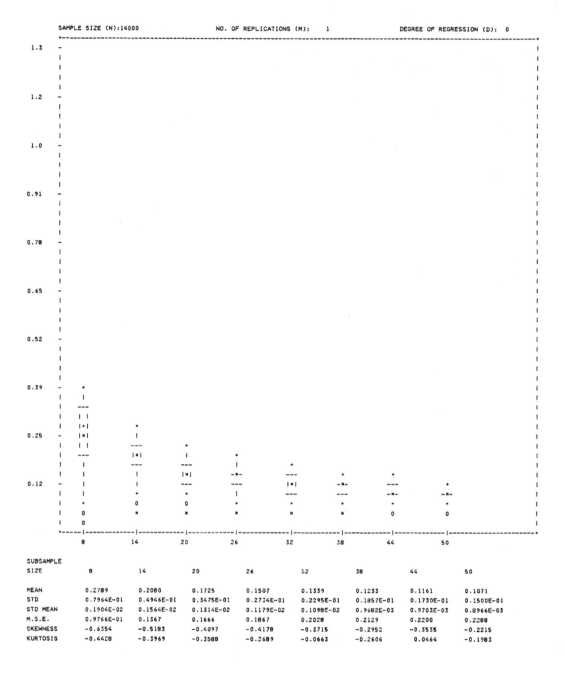

SAMPLE SIZE (N):14000   NO. OF REPLICATIONS (M):   1   DEGREE OF REGRESSION (D):   0

ESTIMATOR: NORMAL-SMOOTH BOOTSTRAP ESTIMATE OF STD OF CORR. OF BIVARIATE NORMALS. B=512
VERTICAL SCALE: YMIN =   0.0179
                 YMAX =   1.3075

SUBSAMPLE SIZE	8	14	20	26	32	38	44	50
MEAN	0.2789	0.2080	0.1725	0.1507	0.1339	0.1233	0.1161	0.1071
STD	0.7964E-01	0.4946E-01	0.3475E-01	0.2734E-01	0.2295E-01	0.1857E-01	0.1730E-01	0.1500E-01
STD MEAN	0.1904E-02	0.1564E-02	0.1314E-02	0.1179E-02	0.1098E-02	0.9682E-03	0.9703E-03	0.8966E-03
M.S.E.	0.9766E-01	0.1367	0.1666	0.1867	0.2028	0.2129	0.2200	0.2288
SKEWNESS	-0.6354	-0.5183	-0.4097	-0.4178	-0.3715	-0.2952	-0.3535	-0.2215
KURTOSIS	-0.4428	-0.3969	-0.3588	-0.2689	-0.0663	-0.2606	0.0464	-0.1983

Figure 9.6.6 SMTBPC run to investigate the properties of the Normal-smoothed bootstrap estimate of the standard deviation of the correlation estimate $\hat{\rho}$ in a bivariate Normal sample ($\rho = 0.5$). (Here the number of replications in the bootstrap simulation is $B = 512$.)

284

lished by simulation techniques; for some of these studies and proposed improvements to bootstrapping see Efron and Tibshirani (1986) and Banks (1988).

9.7 Simulation Studies of Confidence Interval Estimation Schemes

The main thrust in this book has been to assess the variability of a simulation estimate by using an estimated standard deviation of the estimate. The use of confidence intervals—sometimes known as interval estimators—for simulation outputs has been detailed as well, but perhaps in a secondary role. In many books and articles on simulation, the role of interval estimates is primary. The crucial difference between these methods of assessing variability is that confidence intervals usually require assumptions beyond those needed for standard deviation estimates. We have stressed repeatedly the classical case of the variance of $\bar{X}$, which can be estimated without any parametric assumptions. In contrast, confidence intervals using the t statistic require Normality assumptions for either the data or for the sectioned estimators.

A counterargument to the use of a standard deviation estimate is that users consciously or unconsciously convert it to confidence intervals by multiplying the estimated standard deviation by 2 and saying that the "estimate" is $\tilde{\theta}(m)$ plus or minus twice the estimated standard deviation.

From our point of view, the differences between the two approaches are minor; the important point is that some assessment of variability of a simulation output must be made. Either a point estimate and an estimate of its standard deviation, or the width, location, and probability of coverage of an interval estimator, will do.

It is important to note, however, that simulation studies of interval estimation procedures present new problems of output analysis. For example, in the preceding section we studied several procedures for estimating standard deviations of one estimate of ρ, the correlation coefficient in bivariate Normal samples. Usually, different variance assessment procedures are associated with different estimates of ρ, but not always. Since the standard deviation estimate is a point estimate, it is simple enough to use SMTBPC to study these procedures as functions of the sample size. Strictly speaking, though, one should look conjointly at SMTBPC runs of the properties of the point estimate and the estimate of its standard deviation—a very good standard deviation estimate may go along with a very biased point estimate of ρ.

How does one study by simulation the behavior of a confidence interval estimate for a parameter θ? A first question could be whether a procedure for computing the coverage probability for a given scheme works. As an example, the assertion after Equation 9.5.1 says that one should compute an estimate of the standard deviation of the completely jackknifed estimator $\theta^*(m)$ at Equation 9.5.1 as follows:

$$\hat{s.d.}\{\theta^*(m)\} = \left[\frac{1}{m(m-1)} \sum_{i=1}^{m} \{_{(i)}\tilde{\theta}(m) - \theta^*(m)\}^2 \right]^{1/2} \tag{9.7.1}$$

and compute a $100(1 - \alpha)\%$ t confidence interval as follows:

$$(\theta^*(m) - t_{(m-1),(1-\alpha/2)}\hat{s}.d.\{\theta^*(m)\}, \quad \theta^*(m) + t_{(m-1),(1-\alpha/2)}\hat{s}.d.\{\theta^*(m)\}) \qquad (9.7.2)$$

Whether the coverage of this confidence interval is in fact $100(1 - \alpha)\%$ generally can be determined only by a (multifactor) simulation experiment. For example, picking $m = 41$ and $\rho = 0.5$ for a bivariate Normal sample, and $\alpha = 0.05$, one could compute by a simulation with r replications, the probability that this confidence interval covers the value $\rho = 0.5$, using $t_{40,0.975} = 2.021$, from Table 9.3.1, in Equation 9.7.2. How close is this probability to 0.95?

Coverage probabilities are just one aspect of simulation studies of confidence interval estimation; others such as relative widths are discussed later in this section. First we address a weakness in the study of the coverage given above, namely, Why examine only this coverage probability of 0.95?

9.7.1 Schruben's Coverage Function for Studies of Confidence Interval Estimators

The following suggestion by Schruben (1980) gives a convenient graphical method for studying confidence interval coverages.

In the introduction we denoted a confidence interval estimate of θ by (L, U), where L and U are functions of the data sample and the desired coverage probability is $1 - \alpha$. This is called the confidence level. Extending the notation to make this dependence explicit and for convenience, letting $\eta = 1 - \alpha$, we write $(L(\eta, \mathbf{X}), U(\eta, \mathbf{X}))$ for the confidence interval, where $\mathbf{X}$ denotes the sample of size m, and

$$P\{\theta \in (L(\eta, \mathbf{X}), U(\eta, \mathbf{X}))\} = \eta \qquad (9.7.3)$$

Consider now, in Schruben's notation, a random variable

$$\eta^* = \inf\{\eta \in [0, 1]; \theta \in (L(\eta, \mathbf{X}), U(\eta, \mathbf{X}))\} \qquad (9.7.4)$$

Here "inf" may be read as minimum. Now η^* is the smallest confidence coefficient η such that the (known) parameter θ is within the region $(L(\eta, \mathbf{X}), U(\eta, \mathbf{X}))$. Equivalently, the random confidence coefficient $1 - \eta^*$ is the observed level of significance in a hypothesis testing framework (Dempster and Schatzoff, 1965).

Now if the assumptions underlying the computation of η are valid, η^* will be uniformly distributed;

$$P(\eta^* \leqslant \eta) = P\{\theta \in (L(\eta, \mathbf{X}), U(\eta, \mathbf{X}))\} = \eta \qquad (9.7.5)$$

(For a proof see Schruben, 1980.)

The actual distribution of η^*:

$$F_{\eta^*}(\eta) = P(\eta^* \leqslant \eta) = P\{(L(\eta, \mathbf{X}), U(\eta, \mathbf{X})) \text{ actually contains } \theta\} \qquad (9.7.6)$$

is called the *coverage function* of the interval estimation procedure, and departures

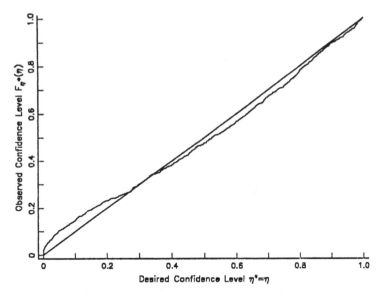

Figure 9.7.1   Confidence interval for the correlation coefficient ρ in a bivariate Normal sample ($\rho = 0.5$) of size m using the completely jackknifed estimator. Case 1: $m = 10$. The figure gives the empirical distribution function $F_{\eta^*}(\eta)$ of η^*. For given η, $F_{\eta^*}(\eta)$ gives the probability that the confidence interval $(L(\eta, \mathbf{X}), U(\eta, \mathbf{X}))$ contains the parameter θ. The straight line represents the ideal case where the desired coverage η equals the actual coverage $F_{\eta^*}(\eta)$. The simulation used 1000 replications.

from a straight line from $(0, 0)$ to $(1, 1)$ in this distribution function show departures from the desired coverage η in the *actual coverage*, namely $F_{\eta^*}(\eta)$.

EXAMPLE   **9.7.1   Confidence Intervals from the Complete Jackknife**

We consider here again estimation of the correlation coefficient in a bivariate Normal population. Our estimate will be the completely jackknifed estimate (Equation 9.4.9), and we will form confidence intervals using the suggestion in Section 9.5. Using Equations 9.7.1 and 9.7.2, the distribution of η^* was determined empirically in a simulation with 1000 replications for $\rho = 0.5$ in a bivariate Normal population. Figure 9.7.1 shows the case of sample size $m = 10$, Figure 9.7.2 the case of $m = 20$, and Figure 9.7.3 the case of $m = 140$.

   The figures give succinct pictures of the departures in the actual coverage probability $F_{\eta^*}(\eta)$ from the desired probability η. Actually, departures are not serious even at $m = 10$, and by $m = 140$ the fit is rather good. Concentrating on the desired value of $\eta = 0.9$, at $m = 10$ $F_{\eta^*}(0.9)$ is actually about 0.89, and about 0.88 at $n = 20$. At $m = 140$ the actual value is approximately the desired value, 0.9.

9.7.2   Other Problems in Studying Confidence Interval Estimators

As discussed above, the robustness of coverage of a confidence interval estimator is not the only desirable criterion of its goodness. Thus, confidence intervals with

Figure 9.7.2   Confidence interval for the correlation coefficient ρ in a bivariate Normal sample ($\rho = 0.5$) of size m using the completely jackknifed estimator. Case 2: $m = 20$. The figure gives the empirical distribution function $F_{\eta^*}(\eta)$, of η^*. For given η, $F_{\eta^*}(\eta)$ gives the probability that the confidence interval, $(L(\eta, \mathbf{X}), U(\eta, \mathbf{X}))$, contains the parameter θ. The straight line represents the ideal case where the desired coverage η equals the actual coverage $F_{\eta^*}(\eta)$. The simulation used 1000 replications.

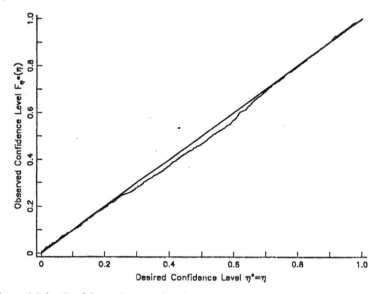

Figure 9.7.3   Confidence interval for the correlation coefficient ρ in a bivariate Normal sample ($\rho = 0.5$) of size m using the completely jackknifed estimator. Case 3: $m = 140$. The figure gives the empirical distribution function, $F_{\eta^*}(\eta)$, of η^*. For given η, $F_{\eta^*}(\eta)$ gives the probability that the confidence interval, $(L(\eta, \mathbf{X}), U(\eta, \mathbf{X}))$, contains the parameter θ. The straight line represents the ideal case where the desired coverage η equals the actual coverage $F_{\eta^*}(\eta)$. The simulation used 1000 replications.

the same coverage may have different widths and, in some sense, bias. For example, the confidence interval based on the t statistic and the t distribution for a Normal population mean is "shorter" than the nonparametric confidence interval based on order statistics using symmetry. (See Chapter 6.)

Clearly one can do simulations and plots (e.g., kernel density estimates or cumulative distribution functions) of the lengths of the confidence interval, and of, say, its "middle." More thought needs to be given to the simulation methodology required for these simulation studies of confidence interval procedures. For some suggestions, see Schruben (1980).

Exercises

9.1. Show that the two expressions for $S_{\hat{p}}$ given at Equation 9.2.5 are equivalent.

9.2. If $\tilde{\theta}(m)$ in Section 9.3.1 is the sample variance S_X^2, will $\bar{\theta}(m)$ defined at Equation 9.3.5 be the same as $\tilde{\theta}(m)$ (as it is for $\bar{X}$)? Consider also the case of the biased estimator of the variance, namely $(n-1)S_X^2/n$.

9.3. Extend the result on bias in sectioned estimators, given at Equation 9.3.11, to terms of order $1/m^2$. Is it possible that when the bias of order $1/m^2$ is included, the sectioned estimator $\bar{x}_\alpha$ or $\bar{\theta}(m)$ may have less bias than the unsectioned estimator when m is small? In other words, is the analysis of the effect of sectioning on the bias in $\tilde{\theta}(m)$ using only terms of order $1/m$ adequate?

T9.4. Extend results in Equations 9.4.3 and 9.4.5 to take into account, respectively, the bias of order $1/m^2$ and the variance of order $1/m^{3/2}$. Is the variance term of order $1/m^{3/2}$ important? That is, would it change one's conclusions about the effect of using a pilot sample to eliminate bias in $\tilde{\theta}(m)$?

T9.5. By dividing a sample size m into three disjoint samples of size m_1, m_2, and m_3, respectively, where $m_1 < m_2 < m_3$ and $m_1 + m_2 + m_3 = m$, it is possible to eliminate bias terms of order $1/m^2$, as well as bias terms of order $1/m$ in an estimate $\tilde{\theta}(m)$. Extend the results based on Equations 9.4.1 and 9.4.2 to obtain a new estimator with this second-order bias elimination. What is the variance term corresponding to Equation 9.4.5 for this new estimator?

9.6. Show that $\delta = m$ in Equation 9.4.8 eliminates bias of order $1/m$ in the resulting estimate $\theta^*(m)$.

9.7. The jackknife scheme for eliminating the bias of order $1/m$ in an estimator $\tilde{\theta}(m)$ can be extended to a jackknife estimator that is bias free up to and including terms of order $1/m^2$. Show how this can be done with a complete (m-fold) jackknife.

C9.8. Perform the following simulations for data sets from samples of size 20, 40, 60, 80, 100, and 120, where the data consist in one case of a Normal$(0,1)$ sample and in the other case of a Gamma$(k = 0.5, \beta = 1)$ sample. We are interested in variance estimates for the sample variance S_X^2. Compare the variance estimates obtained by substituting sample moments and by using a complete jackknife and a bootstrap.

C9.9. Investigate the effects of non-Normal data on estimation of the standard deviation of the correlation coefficient. Do this by running a simulation experiment similar to Example 9.6.1, but this time comparing only the Normal theory estimate (Equation 9.6.6) to the bootstrap estimate (Equation 9.6.2) with $B = 128$. Also, instead of bivariate Normal data, generate pairs of bivariate Gamma data, (X_i, Y_i) by using triplets of exponential data, $(E_{i_1}, E_{i_2}, E_{i_3})$ and defining $X_i = E_{i_1} + E_{i_2}$ and $Y_i = E_{i_2} + E_{i_3}$. Otherwise, use the parameters given in the example.

^c**9.10.** Perform the simple simulation suggested in Section 9.7, comparing the estimated coverage probabilities for confidence intervals from Normal theory, jackknifing, and bootstrapping. The intervals should be for the correlation coefficient for a bivariate Normal sample of size $m = 41$ with true correlation, $\rho = 0.50$. Compare the three types of intervals at $\alpha = 0.05$ and $\alpha = 0.01$.

^c**9.11.** The sample mean $\bar{X}$ and the sample standard deviation S_x are independent in Normal samples and can be shown to be uncorrelated in symmetric samples (Cramér, 1946). Using scatter plots and the sample correlation coefficient given at Equation 9.6.4, investigate the marginal and joint properties of $\bar{X}$ and S_x in Gamma-distributed populations. The investigation should consider various levels of the two factors k and m. The first of these parameters is the shape parameter of the Gamma population, and the second is the sample size.

References

Banks, D. L. (1988). Improving the Bayesian Bootstrap. Forthcoming.

Conover, W. J. (1980). *Practical Nonparametric Statistics*, 2nd ed. Wiley: New York.

Cramér, H. (1946). *Mathematical Methods of Statistics*. Princeton University Press, Princeton, NJ.

Dempster, A. P., and Schatzoff, M. (1965). Expected Significance Level as a Sensitivity Index for Test Statistics. *Journal of the American Statistical Association, 60,* 420–436.

Efron, B. (1981). Nonparametric Estimates of Standard Error: The Jackknife, the Bootstrap, and Other Resampling Methods. *Biometrika, 68,* 589–599.

Efron, B., and Gong, G. (1983). A Leisurely Look at the Bootstrap, the Jackknife and Cross-Validation. *American Statistician, 37,* 36–48.

Efron, B., and Tibshirani, R. (1986). Bootstrap Methods for Standard Errors, Confidence Intervals, and Other Measures of Statistical Accuracy. *Statistical Science, 1*(1), 54–77.

Johnson, N. L., and Kotz, S. (1970). *Distributions in Statistics*, Vol. 2, *Continuous Univariate Distributions*. Wiley: New York.

Miller, R. G. (1974). The Jackknife—A Review. *Biometrika, 61,* 1–17.

Quenouille, M. H. (1949). Approximate Tests of Correlation in Time Series. *Journal of the Royal Statistical Society, B, 11,* 68–84.

Schruben, L. W. (1980). A Coverage Function for Interval Estimators of Simulation Response. *Management Science, 26*(1), 18–27.

Tukey, J. (1958). Bias and Confidence in Not Quite Large Samples. (Abstract). *Annals of Mathematical Statistics, 29,* 614.

10

Bivariate Random Variables: Definitions, Generation, and Graphical Analysis

10.0   Introduction

Up to now we have discussed simulations in which the random variables under consideration, either as the inputs to a model or as the outputs from the simulation, were not only independent but also univariate. Examples include independent replications of a response time at a computer terminal, a waiting time in a queue, or the time to failure of an individual component. However, it may be necessary or desirable to consider joint (bivariate) properties of a system—for example, not only the response time at a computer terminal but also the number of people signed on to the computer at the time of response. Similarly, the times to failure of several components may be related, if for no other reason than that they could fail because of a common shock to the system. An interesting and striking example of this common-mode type of failure was the near crash of an L1011 airliner in Florida in 1983, when all three engines failed because of a mistake in maintenance. Also in Chapter 9 we studied various methods for assessing the variability of estimates of a characteristic θ from a sample $X_1, \ldots, X_j, \ldots, X_m$. In particular, we studied individual (or marginal) properties of the jackknife estimate and the bootstrap estimate of the standard deviation of estimates of θ. However, these two standard deviation estimates probably will be correlated with each other. This correlation may be of interest when comparing jackknife and bootstrap estimates, and it can be assessed using the techniques given in this chapter.

In this chapter we will consider briefly the general situation of many dependent random variables, but to make a large topic viable, we limit attention mainly to the bivariate (pairs) case and primarily to continuous random variables. Discrete bivariate random variables do occur, and are discussed briefly in Section 10.1.5. It is also not hard to think of examples of mixed discrete and continuous random variables such as the computer terminal example above. As another example, consider the waiting time of a customer in a queue and the number of people waiting when the new customer joins the queue. Clearly, the more people there are in the queue when a customer joins it, the longer the new customer will have to wait for service. This observation emphasizes a very important property of many bivariate pairs: they are often *dependent* or *correlated*. In fact, if the two variables were independent of each other, we could analyze each separately, using the methods from previous chapters. Instead, here we assume that the members of the pair are (or at least, may be) dependent.

The first part of the chapter offers a technical discussion of how bivariate random variables are defined through their joint probability distribution function and described through numerical summaries such as the (Pearson product moment) correlation coefficient. The correlation coefficient, which we have used before—particularly in examples in Chapter 9—is presented in some detail, with emphasis on the role it plays in variances of sums of random variables. Finally, ad hoc methods for generating pairs of bivariate random variables are presented. Emphasis is given to the important case of bivariate Normal random variables, often required in empirical sampling studies of many statistical applications (e.g., classification problems, principal components). In Section 10.2 we discuss statistical analysis of independent samples of bivariate pairs, including topics such as scatter plots and bivariate histograms, the former being implemented in the software supplement, *Enhanced Simulation and Statistics Package*, as PLOTPC and the latter as the program BIHSPC. Simulation of statistics, systems, or phenomena with dependent structures is also shown to be simply an extension of the basic crude simulation methodology given in Chapter 4. In Section 10.3 we give the bivariate inverse probability integral transform method both for continuous and discrete random variables. In Section 10.4, methods are discussed for generating bivariate random variables using ad hoc methods and modeling ideas. For illustration, the emphasis is on bivariate exponential and Gamma variates.

Many of the concepts concerning dependence from this chapter will be essential to understanding variance reduction techniques in Chapter 11 and to working with the dependent *sequences* of random variables that occur in the systems simulations of Volume II.

Finally we note that the use of simulation techniques in studying bivariate phenomena is even more important than in the univariate case. This is because analytic techniques, although they may be able to find properties of univariate random variables, face a far more formidable task with bivariate random variables. Thus in looking at a queueing system with congestion at several points, the properties of the waiting line at any single point may be analytically obtainable, but not the joint properties of waiting times at several points in the system. Similarly, the examples in Chapter 9 on techniques of variance assessment are absolutely dependent on simulation techniques for their analysis if similarities or differences in the methods are to be studied.

10.1   Specification and Properties of Bivariate Random Variables

10.1.1   Bivariate Distributions and Densities

We consider here the case of bivariate random variables—that is, a pair (X, Y)—and discuss their probabilistic specification through a bivariate distribution and, for (jointly) continuous distributions, a bivariate probability density function.

Examples of pairs that may be of interest in simulations include the waiting time of the nth customer in a queue, W_n, and the waiting time of the next customer, W_{n+1}. Also, the waiting time of the nth customer and the sum of the first n service times is a pair of bivariate random variables.

Again, if we order two independent random variables X_1 and X_2 and let $X_{(1)}$ be the minimum of the two and $X_{(2)}$ be the maximum, we have a bivariate pair $(X_{(1)}, X_{(2)})$. In this case the random variables are clearly dependent, since $X_{(2)} > X_{(1)}$. Thus if we know that $X_{(1)}$ is greater than x, then we also have that $X_{(2)}$ is greater than x. This is the type of probabilistic dependence that we will try to characterize by some summary numerical measure.

The two random variables $(X_{(1)}, X_{(2)})$ are just *order statistics*, which we have considered in previous chapters (see, in particular, the example in Section 3.4 and the discussion in Section 6.3). In fact, the order statistics $X_{(1)} \leqslant X_{(2)} \leqslant \cdots \leqslant X_{(n)}$ from a sample of n iid random variables is an example of an n-variate random variable.

The probabilistic properties of a pair of random variables (X, Y) are specified by a *joint cumulative distribution function*

$$F_{X,Y}(x, y) = P(X \leqslant x; Y \leqslant y) \qquad -\infty < x, y < \infty \tag{10.1.1}$$

with the properties that $F_{X,Y}(x, y)$:

1. Is a nondecreasing function of each of its arguments.
2. Is continuous from the right in each of its arguments.
3. Satisfies the relationships:

$$\lim_{\substack{x \to \infty \\ y \to \infty}} F_{X,Y}(x, y) = 1 \qquad \lim_{\substack{x \to -\infty \\ y \to -\infty}} F_{X,Y}(x, y) = 0$$

4. Satisfies the restriction that for any $x_1 < x_2$, and $y_1 < y_2$,

$$P\{x_1 < X \leqslant x_2; y_1 < Y \leqslant y_2\}$$
$$= F_{X,Y}(x_2, y_2) - F_{X,Y}(x_1, y_2) - F_{X,Y}(x_2, y_1) + F_{X,Y}(x_1, y_1) \tag{10.1.2}$$

is nonnegative.

The necessity of the last condition is shown by the following example:

$$F_{X,Y}(x, y) = \begin{cases} 0 & \text{if } x < 0 \text{ or } x + y < 1 \text{ or } y < 0 \\ 1 & \text{in the rest of the plane} \end{cases}$$

This function satisfies Properties 1, 2, and 3, (check it), but for example,

$$P(\tfrac{1}{4} < X \leqslant 1; \tfrac{1}{4} < Y \leqslant 1)$$
$$= F_{X,Y}(1, 1) - F_{X,Y}(\tfrac{1}{4}, 1) - F_{X,Y}(1, \tfrac{1}{4}) + F_{X,Y}(\tfrac{1}{4}, \tfrac{1}{4}) = -1$$

which is impossible, since a probability must lie between zero and one. *This example illustrates that definition of joint cumulative distribution functions must be done with great care.*

If there is a nonnegative function $f_{X,Y}(x, y)$ such that

$$F_{X,Y}(x, y) = \int_{-\infty}^{x} \int_{-\infty}^{y} f_{X,Y}(u, v) du\, dv \qquad (10.1.3)$$

for all x, y, then $f_{X,Y}(x, y)$ is called the *bivariate probability density function* of X and Y. In that case, X and Y are said to be *absolutely continuous* random variables and

$$f_{X,Y}(x, y) = \frac{\partial^2}{\partial x\, \partial y} F_{X,Y}(x, y) \qquad (10.1.4)$$

It is a fact that any function of the two variables x and y that is strictly nonnegative and integrates to 1 is a bivariate probability density function.

Similarly in considering discrete random variates, if the possible outcomes of X, Y are the countable set of points (x_k, y_j), with $k = 0, 1, \ldots$, and $j = 0, 1, \ldots$, then $p_{k,j}$ is the *probability function*, where

$$p_{k,j} = P(X = x_k; Y = y_j) \qquad (10.1.5)$$

Clearly the $p_{k,j}$'s must all lie in the interval $[0, 1]$ and the sum over all the $p_{k,j}$'s must be 1. In this case, (X, Y) is a *discrete-valued, bivariate random variable.*

Mixed distributions also occur, and in fact are common rather than exceptional in the bivariate case, as we illustrate in Example 10.1.1, where there are no point probabilities, but there is a concentration of probability along a line in the plane.

EXAMPLE   **10.1.1 The Marshall–Olkin Bivariate Exponential Random Variable**
A bivariate exponential random variable can be constructed in the following way (Marshall and Olkin, 1967a, b).

Let E_1, E_2, and E_3 be univariate, exponentially distributed random variables with common parameter λ. It is easy to show that the minimum of two independent exponentially distributed random variables has an exponential distribution with parameter 2λ:

$$P\{\min(E_1, E_3) > x\} = P\{\text{both } E_1 \text{ and } E_3 > x\} \qquad x > 0$$
$$= P(E_1 > x)P(E_3 > x)$$
$$= e^{-\lambda x}e^{-\lambda x} = e^{-2\lambda x}$$

Thus the bivariate pair X, Y defined as follows:

$$X = \min(E_1, E_3) \qquad \text{and} \qquad Y = \min(E_2, E_3)$$

have exponential marginal distributions and are clearly dependent because of the common random variable E_3 in their constructions. The model also has a simple physical basis in reliability modeling. This basis is the assumption that the two components usually fail independently with identical, exponential times to failure (E_1 and E_2), but both can be put out of commission by a common shock that occurs at random time E_3.

It is not hard to see that the bivariate distribution $F_{X,Y}(x, y)$ for this random variable will have a nonzero concentration of mass along the line $x = y$ even though X and Y are marginally absolutely continuous and exponentially distributed. We will return to this example later and leave it as an exercise (Exercise 10.1) to work out the joint distribution of the random pair. Note that this bivariate exponential pair is easy to simulate—that is, to generate on a computer. A generalization of this model is given by Lawrance and Lewis (1975).

We call $F_X(x) = F_{X,Y}(x, \infty)$ the *marginal distribution* of X, with marginal mean μ_X, variance σ_X^2, and standard deviation σ_X. When X is absolutely continuous, we have

$$\mu_X = E(X) = \int_{-\infty}^{\infty} x f_X(x) dx$$

$$= \int_{-\infty}^{\infty} \int_{-\infty}^{\infty} x f_{X,Y}(x, y) dx\, dy \tag{10.1.6a}$$

$$\sigma_X^2 = \text{var}(X) = E\{(X - \mu_X)^2\} = E(X^2) - \mu_X^2$$

$$= \int_{-\infty}^{\infty} x^2 f_X(x) dx - \mu_X^2$$

$$= \int_{-\infty}^{\infty} \int_{-\infty}^{\infty} x^2 f_{X,Y}(x, y) dx\, dy - \mu_X^2 \tag{10.1.6b}$$

The marginal distributions in Example 10.1.1 are exponential distributions, by construction; the joint cumulative distribution is difficult to construct and integrate (see Exercise 10.1).

In the discrete case we have for the marginal mean and variance

$$\mu_X = E(X) = \sum_{k=0}^{\infty} x_k p_k = \sum_{k=0}^{\infty} \sum_{j=0}^{\infty} x_k p_{k,j} \tag{10.1.7}$$

$$\sigma_X^2 = \text{var}(X) = E\{(X - \mu_X)^2\} = E(X^2) - \mu_X^2$$

$$= \sum_{k=0}^{\infty} \sum_{j=0}^{\infty} x_k^2 p_{k,j} - \mu_X^2 \tag{10.1.8}$$

where p_k is the marginal probability mass function of X: that is, $p_k = P(X = x_k)$, which is the sum of $p_{k,j}$ over all j.

If the joint cdf can be factored into the product of the two marginal cdf's,

$$F_{X,Y}(x, y) = F_X(x) F_Y(y)$$

or, equivalently, if the joint density factors,

$$f_{X,Y}(x, y) = f_X(x)f_Y(y)$$

then X and Y are *independent*. Otherwise, we are dealing with a *dependent* bivariate pair of random variables.

10.1.2   The (Product Moment) Correlation Coefficient

For the purposes of summarizing numerically the properties of a bivariate pair, we concentrate on the second-order *joint* central moment of X and Y. This is called the covariance of X and Y:

$$\text{cov}(X, Y) = E\{(X - \mu_X)(Y - \mu_Y)\} = E(XY) - \mu_X\mu_Y \qquad (10.1.9)$$

If X and Y are independent, then $E(XY) = E(X)E(Y) = \mu_X\mu_Y$, so that $\text{cov}(X, Y) = 0$. For this reason the covariance is used as a measure of dependence, or lack of dependence. Note, however, the following.

1. If X and Y have a bivariate Normal distribution, $\text{cov}(X, Y) = 0$ implies independence (Cramér, 1946) (this distribution is discussed below in Section 10.1.4). Because of the ubiquity of the Normality assumption, a zero covariance is sometimes taken to imply independence in all bivariate cases. But, in fact, it is easy to construct bivariate random variables that are highly dependent, but for which $\text{cov}(X, Y) = 0$. For example, if X is a Normal$(0, 1)$ random variable and $Y = X^2$, these random variables are clearly dependent. Thus, if $X = 5$, then $Y = 25$; and if $Y = 4$, then $X = -2$ or $+2$, each with a probability of one-half. However, since odd moments of a Normal$(0, 1)$ random variable are, by symmetry, zero,

$$\begin{aligned}
\text{cov}(X, Y) &= E(X \cdot X^2) - E(X)E(X^2) \\
&= E(X^3) - 0 = 0
\end{aligned}$$

Thus X and Y have covariance 0 but are clearly highly dependent. The argument in fact goes through if X is a random variable with *any* distribution that is symmetric about zero.

2. The covariance is *shift* invariant but not scale invariant, since for constants a, b, α, and β, with neither α nor β equal to zero,

$$\begin{aligned}
\text{cov}\left(\frac{X - a}{\alpha}, \frac{Y - b}{\beta}\right) &= E\left[\left\{\frac{(X - a)}{\alpha} - \frac{(\mu_X - a)}{\alpha}\right\}\left\{\frac{(Y - b)}{\beta} - \frac{(\mu_Y - b)}{\beta}\right\}\right] \\
&= \frac{\text{cov}(X, Y)}{\alpha\beta}
\end{aligned}$$

The covariance therefore measures the scale (magnitude and units) of X and Y, as well as their dependence.

It is for the second reason that we define a normalized covariance, called the *Pearson product moment correlation coefficient*, $\rho(X, Y)$:

$$\rho(X, Y) = \text{corr}(X, Y) = \frac{\text{cov}(X, Y)}{\sigma_X \sigma_Y}$$

$$= E\left\{\frac{(X - \mu_X)}{\sigma_X} \frac{(Y - \mu_Y)}{\sigma_Y}\right\} = \text{cov}(X', Y') \qquad (10.1.10)$$

In words, it is the covariance between the random variables X' and Y', which are, respectively, the random variables X and Y after they have been standardized to have mean 0 and standard deviation 1.

10.1.3   Properties of the Correlation Function

There are several important properties of the correlation coefficient. By definition, the correlation coefficient is commutative: that is, $\rho(X, Y) = \rho(Y, X)$. In addition, one can show, using Schwartz's inequality (Cramér, 1946, p. 88), that

$$|\rho(X, Y)| \leqslant 1 \qquad (10.1.11)$$

If X is a linear function of Y, so that $X = aY + c$, then $\rho(X, Y)$ is 1 if a is positive, or -1 if a is negative. Thus, we have a measure of dependence and association independent of location and scale, which is 0 if the random variables X and Y are independent, which ranges between -1 and 1 if X and Y are dependent, and whose sign reflects the direction of the association.

The correlation coefficient also arises when one considers the problem of the linear function of Y which is, in the sense of minimum variance, the best predictor of X. (See Cramér, 1946, or Mood, Graybill, and Boes, 1974.) This in turn leads to the rough idea that for positive correlation, X is on the average large when Y is large and small when Y is small. Similarly for negative correlation, X is on the average small when Y is large and large when Y is small.

The correlation coefficient is not the only useful measure of association or dependence (see, e.g., the discussion in Gibbons, 1971, Chap. 12). Two nonparametric measures, Kendall's tau, which we discussed in Example 4.4.1, and Spearman's rank correlation coefficient, share many of the correlation coefficient's properties and have the advantage of being invariant under all transformations of X and Y in which order of magnitude is preserved. This is not true for $\rho(X, Y)$.

On the other hand, the correlation coefficient comes in quite naturally when the variance is used as a measure of scale or dispersion for sums of dependent random variables. In simulation, we do this frequently when working with estimators based on random variables from replications in which we have induced correlation to achieve variance reduction (see Chapter 11). Again, in systems simulation (Volume II), we average over successive waiting times W_n in a single realization of a stationary queueing system to estimate $E(W_n) = E(W)$. The W_n are, by their very nature, correlated (see the queueing examples in Chapter 3).

Thus we have, for the case of two random variables,

$$
\begin{aligned}
\text{var}(X + Y) &= E(\{X + Y - \mu_X - \mu_Y\}^2) \\
&= E[\{(X - \mu_X) + (Y - \mu_Y)\}^2] \\
&= E\{(X - \mu_X)^2 + (Y - \mu_Y)^2 + 2(X - \mu_X)(Y - \mu_Y)\} \\
&= E\{(X - \mu_X)^2\} + E\{(Y - \mu_Y)^2\} + 2E\{(X - \mu_X)(Y - \mu_Y)\} \\
&= \text{var}(X) + \text{var}(Y) + 2\,\text{cov}(X, Y) \\
&= \sigma_X^2 + \sigma_Y^2 + 2\sigma_X\sigma_Y\rho(X, Y) \qquad\qquad (10.1.12)
\end{aligned}
$$

For simplicity, consider the case of $\sigma_X = \sigma_Y = \sigma$. Then

$$
\text{var}(X + Y) = 2\sigma^2\{1 + \rho(X, Y)\} \qquad\qquad (10.1.13)
$$

Since the variance of a random variable is greater than or equal to zero, and $|\rho(X, Y)| \leqslant 1$, we have:

$$
0 \leqslant \text{var}(X + Y) = 2\sigma^2\{1 + \rho(X, Y)\} \leqslant 4\sigma^2
$$

Thus the variance of the sum of two random variables, which is $2\sigma^2$ if the random variables are uncorrelated, can be inflated by positive correlation, but not by more than a factor of 2. This is, of course, the case of complete dependence when $X = Y$ so that $X + Y = 2X$. And, again, if the correlation is negative, the variance of the sum is reduced in comparison to the independence case. This fact will be exploited in variance reduction techniques in Chapter 11.

Note too that for the *average* of the two random variables,

$$
0 \leqslant \text{var}\left(\frac{X + Y}{2}\right) = \frac{\sigma^2}{2}\left\{1 + \rho(X, Y)\right\} \leqslant \sigma^2 \qquad\qquad (10.1.14)
$$

These results on the variance of the sum of two random variables can be extended to the sum of more than two random variables, $X_1 + X_2 + \cdots + X_m$, where the X_i's are dependent. For simplicity, let the variances be equal, $\text{var}(X_i) = \text{var}(X)$, and define $\rho_{ij} = \text{corr}(X_i, X_j)$. Then we have the following important result:

$$
\text{var}(X_1 + \cdots + X_m) = \text{var}(X)\left\{m + 2\sum_{i=1}^{m-1}\sum_{j=i+1}^{m}\rho_{ij}\right\} \qquad\qquad (10.1.15)
$$

For $m = 3$ this is

$$
\text{var}(X_1 + X_2 + X_3) = \text{var}(X)\{3 + 2(\rho_{12} + \rho_{13} + \rho_{23})\} \qquad\qquad (10.1.16)
$$

Another important relationship that will be useful subsequently, and whose proof is left as an exercise, is that if we have sample averages $\bar{X}$ and $\bar{Y}$ from a bivariate random sample $(X_1, Y_1) \cdots (X_m, Y_m)$, then

$$\text{cov}(\bar{X}, \bar{Y}) = \frac{1}{m} \text{cov}(X, Y) \tag{10.1.17}$$

and

$$\text{corr}(\bar{X}, \bar{Y}) = \text{corr}(X, Y) \tag{10.1.18}$$

Note that "bivariate random sample" means that *within* pairs (X_j, Y_j) we have dependence, but *between* pairs: that is, between (X_j, Y_j) and (X_k, Y_k), $k \neq j$, we have independence.

10.1.4   The Bivariate Normal Distribution

The best known and most widely used bivariate distribution is the *bivariate Normal distribution*. This distribution arises quite naturally when one considers the limit of averages of bivariate pairs $(\bar{X}_n, \bar{Y}_n) = (\sum X_i/n, \sum Y_i/n)$. The bivariate Normal distribution is used, for example, to model the distribution of heights and weights of people, or the horizontal and vertical displacement of shots from the center of a target. Again, X and Y may be the north–south and east–west components of wind speed at some location. (Whether bivariate Normal is an appropriate model in these cases would have to be checked against empirical data.)

The bivariate Normal distribution is absolutely continuous with *joint* probability density function

$$f_{X,Y}(x, y) = \frac{1}{2\pi\sigma_X\sigma_Y(1 - \rho^2)^{1/2}} \exp[-(\tfrac{1}{2})Q^{-1}(x - \mu_X, y - \mu_Y)] \tag{10.1.19}$$

where the bivariate function $Q^{-1}(x, y)$ is given by

$$Q^{-1}(x, y) = \frac{1}{1 - \rho^2} \left(\frac{x^2}{\sigma_X^2} - \frac{2\rho xy}{\sigma_X\sigma_Y} + \frac{y^2}{\sigma_Y^2} \right) \tag{10.1.20}$$

This bivariate probability density has five parameters $\mu_X, \sigma_X, \mu_Y, \sigma_Y$, and ρ. Of these, the first four were introduced previously as the marginal means and standard deviations of X and Y, while $\rho = \rho(X, Y)$ is the *correlation* between X and Y. This parameter must have value $|\rho| < 1$ for the density to be defined. The cumulative distribution function cannot be obtained in closed form but is tabulated for the standardized variables

$$X' = \frac{X - \mu_X}{\sigma_X} \quad \text{and} \quad Y' = \frac{Y - \mu_Y}{\sigma_Y}$$

with the single parameter ρ. See Johnson and Kotz (1972, Chap. 36) for a complete discussion of the bivariate Normal distribution and tables of probabilities. The density and its contours are shown in Figure 10.1.1 for $\rho = 0$ and $\rho = 0.75$.

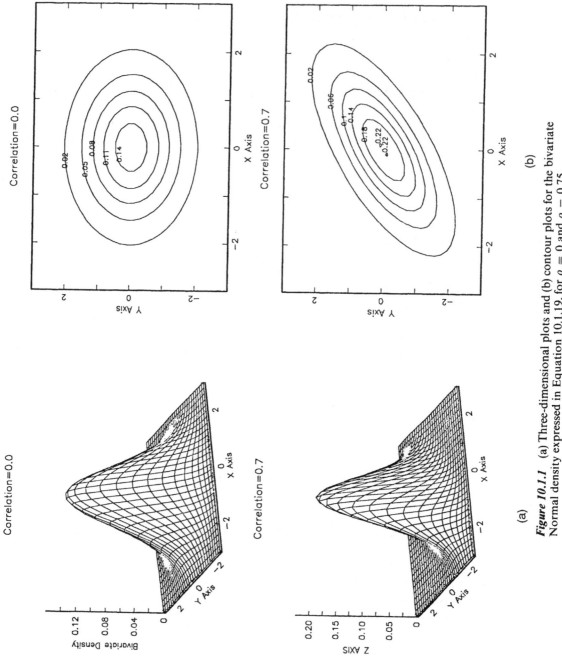

Figure 10.1.1   (a) Three-dimensional plots and (b) contour plots for the bivariate Normal density expressed in Equation 10.1.19, for $\rho = 0$ and $\rho = 0.75$.

Several properties of the bivariate Normal distribution are particularly useful in simulation:

1. X and Y have univariate marginal Normal distributions with shift and scale parameters μ_X, σ_X and μ_Y, σ_Y, respectively.

2. If we have two *independent* Normal random variables X_1 and X_2 with zero means and unit variances and consider the two linear transformations

$$X' = aX_1 + bX_2 \tag{10.1.21}$$
$$Y' = cX_1 + dX_2 \tag{10.1.22}$$

then X' and Y' are both Normally distributed with zero means and variances $\sigma_{X'}^2 = a^2 + b^2$ and $\sigma_{Y'}^2 = c^2 + d^2$, respectively. It can be shown that (X', Y') is a bivariate Normal pair with the joint density in Equation 10.1.19 and

$$\rho = \text{corr}(X', Y') = \frac{ac + bd}{[(a^2 + b^2)(c^2 + d^2)]^{1/2}} \tag{10.1.23}$$

It is important to note that while Equations 10.1.21 and 10.1.22 provide a simple way to generate a pair (X', Y') with the joint Normal distribution whose density is given at Equation 10.1.19, the transformation is not unique. That is, many values of a, b, c, and d can produce a pair (X', Y') with fixed variances $\sigma_{X'}^2$, $\sigma_{Y'}^2$, and fixed correlation ρ (see Exercise 10.4).

We can also consider the opposite question: for specified σ_X^2, σ_Y^2, and ρ, can we find any a, b, c, and d to compute the desired (X', Y') pair? The answer is yes, and proving the existence of such a transformation and finding the coefficients is simple. We assume, without loss of generality (see Exercise 10.4) that $\sigma_X^2 = \sigma_Y^2 = 1$, so that $(a^2 + b^2) = (c^2 + d^2) = 1$. If $\rho > 0$, set $c = 0$, so that $d = 1$. Then Equation 10.1.23 gives $\rho = b = (1 - a^2)^{1/2}$, which is valid for all $b = (0, 1)$. Similarly, if $\rho < 0$, set $c = -1$, so that $d = 0$. Then, from Equation 10.1.23, $\rho = -a$, so that $b = (1 - \rho^2)^{1/2}$, which is valid for all a in $(0, 1)$. Thus we have the following *algorithm to generate a standardized bivariate Normal pair (X, Y) with correlation* $\rho \in (-1, 1)$:

- Generate independent $N(0, 1)$ variables X_1 and X_2.
- If $\rho = 0$ set $X = X_1$ and $Y = X_2$.
- Else, set $X = X_1$ and $Y = \rho X_1 + (1 - \rho^2)^{1/2} X_2$.

3. Continuing with the properties of the bivariate Normal distribution, the conditional density of X, given $Y = y$, is Normal with conditional mean and variance

$$E(X|y) = \mu_X + \frac{\rho \sigma_X}{\sigma_Y}(y - \mu_y) \tag{10.1.24}$$

$$\text{var}(X|y) = \sigma_X(1 - \rho^2) \tag{10.1.25}$$

while the conditional density of Y, given $X = x$, is also Normal with analogous moments.

Thus, while it is generally difficult in simulation to generate the conditional distributions associated with a given bivariate distribution, it is simple in the case of the bivariate Normal distribution. Note, too, that the conditional variances are independent of the fixed x and y values. From a modeling point of view, this can be an unrealistic restriction. If X represents the score of an individual on the mathematics part of an SAT test, and Y the score on the verbal part, should one expect the same spread in the verbal scores for all given math scores?

10.1.5   Discrete Dependent Bivariate Random Variables and 2 × 2 Contingency Tables

We digress here briefly to illustrate the extension of the preceding ideas concerning dependence and correlation to discrete random variables. A complete treatment of measures of association and agreement is left to outside reading—for example, in Bishop, Fienberg, and Holland (1975)—while the generation of sequences of dependent discrete variables is discussed in McKenzie (1985).

The simplest example of a discrete bivariate pair is a pair of Bernoulli random variables (X, Y), where we define:

$$P\{(X, Y) = (1, 1)\} = p_{11}$$
$$P\{(X, Y) = (1, 0)\} = p_{10}$$
$$P\{(X, Y) = (0, 1)\} = p_{01}$$
$$P\{(X, Y) = (0, 0)\} = p_{00}$$

The marginal distributions of the binary X and Y variables are given by:

$$P(X = 1) = p_{1+} = p_{10} + p_{11}$$
$$P(Y = 1) = p_{+1} = p_{01} + p_{11}$$

These marginal distributions are obtained by the formula given after Equation 10.1.8. In this bivariate Bernoulli pair, once the marginal distributions of X and Y have been determined, there is only one parameter (one degree of freedom) left to specify the dependency structure of the pair. Although this degree of freedom can be expressed through the correlation coefficient given in Equation 10.1.26 below, we note two things. First, the correlation coefficient will not have the complete range of -1 to 1 as it does for the bivariate Normal distribution. Second, $\rho = 0$ does not generally imply independence, but it is equivalent to independence for this pair. The random variables X and Y are independent if and only if (Exercise 10.6):

$$p_{11} = p_{1+}p_{+1} = (p_{10} + p_{11})(p_{01} + p_{11})$$

Such bivariate Bernoulli pairs occur if we are observing two characteristics of one item. In one typical situation, for example, X is an indicator of lung cancer in a person while Y indicates whether that person smokes. In this case, we might be interested in measuring and testing the association (dependence) between smoking and lung cancer. As another example, X may represent the true state of an experimental unit, say defective or not defective, while Y represents the results of a diagnostic test meant to assess the condition of the experimental unit. If Y is a good test, then Y will agree with X (i.e., be highly correlated) for most or all of the items inspected.

If we observe n independent pairs of such variates, (X_i, Y_i), $i = 1, 2, \ldots, n$, then the data are classically displayed in a 2×2 table with the X variable representing the row and the Y variable representing the column:

$$
\begin{array}{c}
 & & Y & \\
 & 1 & 0 & \\
\end{array}
$$

		1	0	
	1	n_{11}	n_{10}	n_{1+}
X	0	n_{01}	n_{00}	n_{0+}
		n_{+1}	n_{+0}	

If we let $I'\{(X_i, Y_i) = (k, l)\}$ be 1 when $(X_i, Y_i) = (k, l)$, and 0 otherwise, then

$$
n_{kl} = \sum_{i=1}^{n} I'\{(X_i, Y_i) = (k, l)\} \qquad \text{for } k = 0, 1; l = 0, 1
$$

with $n_{11} + n_{10} + n_{01} + n_{00} = n$, and with n_{1+}, n_{0+}, n_{+1}, and n_{+0} the row and column marginal totals (i.e., $n_{1+} = n_{11} + n_{10}$, etc.). This simple display of the outcomes is possible because there are only four possible values for the (X, Y) pairs.

i. Testing for Independence in 2×2 Tables

Specifying and testing for independence between the X and Y variables in the 2×2 tables defined above is a very important topic in applied statistics and can be approached in several ways, all of which involve simulation for their evaluation.

If we proceed by analogy to the development of the correlation coefficient in the previous sections, it is not hard to show (see Exercise 10.7) that for the bivariate Bernoulli pair (X, Y)

$$
\text{corr}(X, Y) = \frac{p_{11} - p_{1+}p_{+1}}{\{p_{1+}(1 - p_{1+})p_{+1}(1 - p_{+1})\}^{1/2}} \tag{10.1.26}
$$

which is commonly estimated by

$$
r = \frac{n n_{11} - n_{1+}n_{+1}}{(n_{1+}n_{0+}n_{+1}n_{+0})^{1/2}} \tag{10.1.27}
$$

It is worthwhile to note that this correlation coefficient is related to the Pearson chi-square test for independence in a 2×2 table (Bishop, Fienberg, and Holland, 1975), which is itself equivalent to a t test for H_0: $P(X = 1 | Y = 1) = P(X = 1 | Y = 0)$ comparing n_{11}/n_{+1} to n_{10}/n_{+0} (Fienberg, 1977). The chi-square test for independence, based on the correlation coefficient, requires the calculation of

$$\chi^2 = \frac{n(n_{11}n_{00} - n_{10}n_{01})^2}{n_{1+}n_{0+}n_{+1}n_{+0}} \tag{10.1.28}$$

which is then compared to the quantiles of a chi-square distribution with one degree of freedom. One might ask whether this distributional assumption is valid for all n, p_{11}, p_{10}, p_{01}, and p_{00}. An answer requires the kind of multifactor simulation discussed in Chapter 8.

An alternative measure of association, widely used in genetic and medical contexts, is the odds ratio or cross-product ratio. The odds ratio is defined as

$$\alpha = p_{11}p_{00}/p_{10}p_{01} \tag{10.1.29}$$

which may be easier to interpret in its alternative form:

$$\alpha = \frac{P(X = 1 \mid Y = 1)}{P(X = 0 \mid Y = 1)} \Big/ \frac{P(X = 1 \mid Y = 0)}{P(X = 0 \mid Y = 0)}$$

The usual estimate of the odds ratio is given by

$$\hat{\alpha} = n_{11}n_{00}/n_{10}n_{01} \tag{10.1.30}$$

Under the hypothesis of independence of X and Y, $\alpha = 1$. We can test $\hat{\alpha}$ through use of the fact (Fienberg, 1977) that

$$\frac{(\log \hat{\alpha})^2}{(1/n_{11}) + (1/n_{10}) + (1/n_{01}) + (1/n_{00})} \tag{10.1.31}$$

has, asymptotically, a chi-square distribution with one degree of freedom.

Exercise 10.8 asks you to perform a simulation experiment to compare the power of the tests based on χ^2 versus the power of the tests based on the statistic $\hat{\alpha}$. Similarly, Exercise 10.9 asks for a simulation exploration of competing methods for obtaining confidence intervals for the odds ratio.

ii. Generating 2 × 2 Tables

In performing simulations involving the 2×2 tables discussed above for fixed n, we can approach the computer generation of simulated data in a number of ways, of which we will consider two. First, there is the straightforward algorithm for generation of single (X, Y) pairs using a uniform$(0, 1)$ variate U:

If $U \leqslant p_{11}$, let $(X, Y) = (1, 1)$
Else if $p_{11} < U \leqslant (p_{10} + p_{11})$, let $(X, Y) = (1, 0)$
Else if $(p_{11} + p_{10}) < U \leqslant (p_{11} + p_{10} + p_{01})$, let $(X, Y) = (0, 1)$
Else let $(X, Y) = (0, 0)$

Repeating this algorithm n times with n independent uniform variates will produce the desired 2×2 table. Alternatively, we can generate the values $n_{11}, n_{10}, n_{01}, n_{00}$ in the table as outcomes from a multinomial distribution. This can be done by sophisticated generation schemes discussed in Volume II. One such scheme is implemented in the IMSL Stat/Library as subroutine RNMTN (IMSL Library, 1987).

Another and different model for a 2×2 table also occurs. In this case, we fix the column marginal totals at some prespecified values n_{+1} and n_{+0}, which sum to n. Many experimental situations fit this model. For example, we may wish to compare smoking habits between $n_{+1} = 500$ lung cancer patients and $n_{+0} = 500$ matched controls. The statistical results for estimation and testing are the same for the two contingency table models (Bishop, Fienberg, and Holland, 1975).

To generate this type of 2×2 table, generate independently:

$$Z_1 \sim \text{binomial}[n_{+1}, \{p_{11}/(p_{11} + p_{01})\}]$$
$$Z_2 \sim \text{binomial}[n_{+0}, \{p_{10}/(p_{10} + p_{00})\}]$$

(see Knuth, 1981) and set

$$n_{11} = Z_1; \qquad n_{01} = n_{+1} - n_{11}$$
$$n_{10} = Z_2; \qquad n_{00} = n_{+0} - n_{10}$$

The X's will have the desired conditional distribution: that is,

$$P(X = 1 | Y = 1) = \frac{p_{11}}{p_{11} + p_{01}}$$

and

$$P(X = 1 | Y = 0) = \frac{p_{10}}{p_{10} + p_{00}}$$

but the Y's are dependent since $\sum_{i=1}^{n} Y_i = n_{+1}$ is fixed. Because both n_{+1} and n_{+0} are fixed, rather than just n, this sampling scheme is different from the first formulation.

We close this discussion by considering a third model for a 2×2 table, and a method for testing for dependence between X and Y: Fisher's exact test (Cox, 1970; Bishop, Fienberg, and Holland, 1975). In this case, we treat all the row and column marginal totals as fixed (i.e., $n, n_{+1}, n_{+0}, n_{1+}$, and n_{0+} are fixed). Thus the probability of any 2×2 table is given by the hypergeometric probability: $\binom{n_{1+}}{n_{11}}\binom{n_{0+}}{n_{01}} / \binom{n}{n_{+1}}$.

To test for significant dependence in an observed table of this type, we simply add the probabilities of all tables that show equal or more association than the observed table (i.e., we adjust the n_{ij} to put more emphasis on the main diagonal while maintaining the marginal totals). If the marginal totals are large, Fisher's exact test is computationally intensive. This fact is even more pronounced when dealing with tables larger than 2×2, or with exact distributions for other testing situations. In these cases, we can use simulation methodology to choose a random subset of all possible data configurations, and estimate a test's significance from that subset. Enhancements of such Monte Carlo exact tests are illustrated in Mehta (1986) and relate to the material on importance sampling discussed in Chapter 11.

10.2   Numerical and Graphical Analyses for Bivariate Data

10.2.1   General Considerations

We consider now three techniques for the display and estimation of the properties of a bivariate pair (X, Y). These properties are to be inferred from a random sample $(X_1, Y_1), \ldots, (X_n, Y_n)$ from n independent replications of the pair. The first technique is the estimation of the basic numerical summary of dependence, the correlation coefficient ρ defined at Equation 10.1.10. The second is the scatter plot, which graphically displays the dependence of X and Y and, to some extent, the "density" of occurrences of pairs. This "density," essentially an estimate of the joint pdf $f_{X,Y}(x, y)$, is more directly presented through the third technique we examine, the bivariate histogram. In the software supplement, a program called BIHSPC provides a bivariate histogram and estimates of correlation and other numerical summaries. A scatter plot is provided by the program PLOTPC.

To fix ideas, we will focus in Section 10.2.5 on a specific simulation problem: the marginal *and* the joint probability properties of estimates of the coefficient of variation $C(E)$, the coefficient of skewness γ_1, and the coefficient of kurtosis γ_2, from an iid sample of n exponentially distributed random variables. Thus, if the sample of exponentials is $E_1, \ldots, E_n$ and we define $\tilde{C}(E)$, $\hat{\gamma}_1$, and $\hat{\gamma}_2$ as at Equations 6.1.19, 6.1.13, and 6.1.17, respectively, then $\{\tilde{C}(E), \hat{\gamma}_1, \hat{\gamma}_2\}$ is a random triple with a trivariate distribution. However, we will not try to investigate the trivariate distribution but will instead successively consider the three possible bivariate pairs: $(\tilde{C}(E), \hat{\gamma}_1)$, $(\tilde{C}(E), \hat{\gamma}_2)$, and $(\hat{\gamma}_1, \hat{\gamma}_2)$. For an underlying sample that is Normal instead of exponential, the bivariate distribution of quantities related to $\hat{\gamma}_1$ and $\hat{\gamma}_2$ is known and tabulated (Pearson and Hartley, 1966, Tables 34B and 34C). In that case, tests of Normality can be based on the hypothesis that $\gamma_1 = \gamma_2 = 0$ within reasonable limits set by sample size n. For an exponential sample, we recommended, in Chapter 6, an analogous test for exponentiality that requires us to ascertain whether $C(E) = 1$, $\gamma_1 = 2$, and $\gamma_2 = 6$. To do this we need to know the sampling distributions of $\tilde{C}(E)$, $\hat{\gamma}_1$, and $\hat{\gamma}_2$. In particular, since the estimates of these quantities may be tightly correlated, there may not be much more information in all three estimates than, say, in two of them. It is this issue we will look at in Section 10.2.5, using the bivariate tools we learn in this section.

10.2.2   Estimating the Correlation Coefficient

The covariance between two random variables X and Y, $\text{cov}(X, Y)$, can be estimated from a sample of n pairs $(X_1, Y_1), (X_2, Y_2), \ldots, (X_n, Y_n)$ by the sample product moment

$$\text{côv} = \frac{1}{n} \sum_{i=1}^{n} (X_i - \bar{X})(Y_i - \bar{Y}) \tag{10.2.1}$$

Assuming all second-order moments to be finite, this is a consistent estimator of $\text{cov}(X, Y)$ with the following properties (Cramér, 1946, Chap. 27):

1. $E(\text{côv}) = \left(\dfrac{n-1}{n}\right) \text{cov}(X, Y)$

2. $\text{var}(\text{côv}) = \dfrac{E[\{X - E(X)\}^2 \{Y - E(Y)\}^2] - \{\text{cov}(X, Y)\}^2}{n} + O\left(\dfrac{1}{n^2}\right)$

To estimate the correlation coefficient $\rho(X, Y)$ we divide Equation 10.2.1 by the sample standard deviations of the X and Y samples, S_X and S_Y, to obtain

$$r = \frac{\dfrac{1}{n} \sum_{i=1}^{n} (X_i - \bar{X})(Y_i - \bar{Y})}{S_X S_Y} \tag{10.2.2}$$

(This estimator was used in the examples in Chapter 9.)

As we noted previously, the ratio of two unbiased estimators is not necessarily unbiased, and this fact holds in general for r. However, the following properties are known:

1. If X and Y are *independent* and from continuous distributions, then

$$E(r) = 0 = \text{corr}(X, Y)$$

and

$$\text{var}(r) = \frac{1}{n}$$

(Kendall and Stuart, 1977, p. 425).

2. If X and Y are from a continuous bivariate distribution and subject to some regularity conditions, then

$$E(r) = \text{corr}(X, Y) + O\left(\frac{1}{n}\right)$$

so that the bias decreases as $1/n$. The variance of r also decreases as $1/n$, and a complicated expression for it is available (Cramér, 1946, Chap. 27).

For bivariate Normal random variables X and Y, the properties of this correlation estimate are well known and usable (see, e.g., Kendall and Stuart, 1977, Chap. 26). We have

$$E(r) = \rho \left\{ 1 - \left(\frac{1 + \rho^2}{2n} \right) + O(n^{-2}) \right\} \tag{10.2.3}$$

$$\text{var}(r) = \frac{(1 - \rho^2)^2}{n - 1} \left(1 + \frac{11\rho^2}{2n} \right) + O(n^{-3}) \tag{10.2.4}$$

$$\gamma_1(r) = \frac{-6\rho}{n^{1/2}} + o(n^{-1/2}) \tag{10.2.5}$$

$$\gamma_2(r) = \frac{6(12\rho^2 - 1)}{n} + o(n^{-1}) \tag{10.2.6}$$

where $\rho = \rho(X, Y)$. The notation $o(n^{-1})$ means that $no(n^{-1})$ goes to zero as n goes to infinity; similarly, $n^{1/2}o(n^{-1/2})$ goes to zero as n goes to infinity. It is also known that the distribution of r converges to Normality very slowly.

We will use the estimate r of the correlation coefficient in Chapter 11 in discussing variance reduced estimators of quantities appearing in simulations. When necessary in non-Normal situations, the sampling variance of r may be estimated by sectioning, with due regard for the problem of bias, or by bootstrapping or jackknifing. In fact, the use of bootstrapping and jackknifing for this problem was investigated using SMTBPC in Chapter 9.

The estimated correlation coefficient is also the basis of tests for independence of two random variables, although because of the slow convergence to Normality, it is generally recommended that Fisher's z transform be used; other alternative nonparametric tests such as those based on Kendall's tau (Example 4.4.1) or the Spearman rank correlation coefficient may also be preferable (see Gibbons, 1971, Chap. 13, or Kendall and Stuart, 1974).

10.2.3   Scatter Plots

The most basic statistical tool for looking at bivariate data (or successive pairs of variables from higher dimensional data) is the scatter plot: every pair (X_i, Y_i) in a sample of n pairs is used as a pair of coordinates in a two-dimensional plot. We have already used this idea in Chapter 4 to demonstrate what we mean by "really" random pseudo-random pairs: if the joint distribution is uniform in the plane, we should see approximately equal numbers of points in similar-sized little squares and no "pattern" in the scatter plot. Moreover, the idea of no "pattern" can be quantified by chi-square tests of fit (Conover, 1980, Chap. 4).

The scatter plot and its interpretation are closely tied up with three quantities. The first is the correlation coefficient, $\rho(X, Y)$, which, if high enough in magnitude, will be expressed in the scatter plot as data falling close to a straight line. The slope of the line will have the same sign as the correlation coefficient. Unlike the numerical summary, however, the scatter plot will let us see *all* the data and look for nonlinear relationships, clusters, and outliers.

The second set of numerical summaries is the regression of X on Y and the regression of Y on X, respectively, $E(X|y)$ and $E(Y|x)$. These were discussed in Section 10.1.4 for bivariate Normal random variables; more generally, for continuous bivariate random variables they are given by:

$$E(X|y) = \int_{-\infty}^{\infty} x \frac{f_{X,Y}(x, y)}{f_Y(y)} \, dx \tag{10.2.7}$$

and

$$E(Y|x) = \int_{-\infty}^{\infty} y \frac{f_{X,Y}(x, y)}{f_X(x)} \, dy \tag{10.2.8}$$

Graphically, one may think of Equation 10.2.7 as follows: divide the y-axis into thin strips and find the mean of the x values of the observations that fall in each strip. Then, join the means in the successive strips to form a curve; this is approximately $E(X|y)$. Your eye will be performing this procedure when you look for pattern in a scatter plot. More formal methods for doing this approximation that are also available include the LOWESS procedure discussed in Chambers et al. (1983).

The third quantity involved in the interpretation of a scatter plot is the joint probability density function $f_{X,Y}(x, y)$ and, in the discrete case, the joint probability function $p_{k,j}$. The eye usually integrates over areas of a scatter plot and relates "numbers of points (pairs) in a given area" with probability mass in that area (height of the joint probability density function).

EXAMPLE   **10.2.1 Scatter Plot for a Bivariate Exponential Distribution**
An example of a scatter plot is given in Figure 10.2.1, where 1000 pairs from the bivariate exponential distribution defined in Example 10.1.1 are plotted. The shape

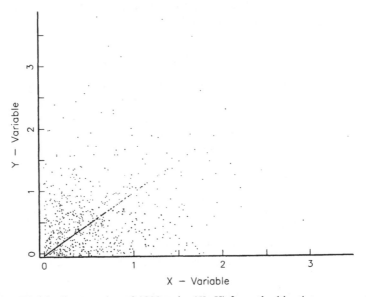

Figure 10.2.1 Scatter plot of 1000 pairs (X_i, Y_i) from the bivariate exponential distribution defined in Example 10.1.1. Note the concentration of mass along the line $x = y$. The plot was made using a very high resolution graphics monitor and printer. The correlation $\rho(X, Y) = 0.6$.

of the joint distribution can be seen roughly in the scatter plot, and most particularly the concentration of probability along the line $x = y$.

What one reads in a scatter plot is highly dependent on the resolution of the graphical device used. The scatter plot in Figure 10.2.1 was made with a very high resolution device and details are clear. If the resolution is not high, as when one uses a printer plot, one point on the graph may represent many pairs, especially if the data are discrete. There are several ways around this (see, e.g., Chambers et al., 1983), one being the bivariate histogram described in the next section.

A final point relates to the scaling of scatter plots. In Normal bivariate data, the regression of X on $Y = y$, is linear (Equation 10.1.24) in y. Often, however, *incorrect* extrapolations are made from this. In particular, people mistakenly interpret the following situations:

- If $r = 0$, then the points form a "round" scatter.
- If r is close to 1, the points lie close to a line of slope 1.
- If r is close to -1, the points lie close to a line of slope -1.

In fact, the slope of the scatter is controlled as much by σ_X and σ_Y as by ρ, since the slope of the regression X on $Y = y$ is, in Normal data (Equation 10.1.24), actually $\rho\sigma_X/\sigma_Y$, and the lengths of the major and minor axes of the elliptical scatter are determined by σ_X and σ_Y. For this reason, it is usually better to plot in a scatter diagram the standardized data, $(X_i - \bar{X})/S_X$ and $(Y_i - \bar{Y})/S_Y$. Then, the interpretation of slope is at least correct in Normal data, with the proviso that very different regressions are possible in data lacking a bivariate Normal distribution. An example of this is obtained by considering the dependent, but uncorrelated pairs (X, Y), where X has a Normal$(0, 1)$ distribution and $Y = X^2$. We recall that in this case $\rho = 0$. What would the scatter plot look like?

10.2.4 Bivariate Histograms: BIHSPC

A bivariate histogram, like a univariate histogram, is a rather crude attempt to enhance (smooth) a scatter plot and/or to obtain a picture of the bivariate probability density function $f_{X,Y}(x, y)$. It is a crude attempt at estimating $f_{X,Y}(x, y)$, because data are grouped and counted in equal-sized rectangles in the plane with no attempt to use the data to guide the amount of grouping. (Recall the problem of window size for the kernel density estimates given in Chapter 7 for univariate data.) It is also usually crude in the sense that a three-dimensional plot has to be rendered in two dimensions, unless three-dimensional graphics are available.

One such attempt, the output from the program BIHSPC described in the software supplement, can be seen in Figure 10.2.8. The details of the random variables with which the bivariate histogram is generated will be discussed later. However, the relative heights of the estimated frequencies (counts in each cell) are indicated by symbols, which fill up the area more and more for 16 different increasing frequency levels. The numerical values corresponding to these levels are indicated below Figure 10.2.8 and, in this example, the darkest symbol corresponds to 15 or more observations per cell. Printer plot resolution limits the accuracy of this bivariate histogram, since bin sizes are determined by the number of characters per line and per page.

In addition, the estimated covariance and correlation coefficients, are given, as well as Spearman's rank correlation coefficient, which is similar in concept to Kendall's tau statistic, which was discussed in Chapter 4. If the pairs (X_i, Y_i) are replaced by their ranks in their respective X and Y samples, then Spearman's rank correlation has the same formula as r given at Equation 10.2.2 (see Conover, 1980, pp. 252–256). It is a distribution-free statistic.

The bivariate histogram also produces univariate statistics—an abbreviated form of those given in HISTPC—and statistics that test for the equidistribution of the X and Y samples, assuming that they are independent. These tests are discussed in Conover (1980).

10.2.5   An Extended Example: Estimated Moment Summaries in Exponential Samples

i. The Problem

In Chapter 6 we discussed three moment summaries, the coefficient of variation $C(E)$, the coefficient of skewness γ_1, and the coefficient of kurtosis γ_2. These can be used to characterize a data sample, say $E_1, \ldots, E_n$, and to guide determination of a suitable parametric distribution function $F_E(x, \theta)$ for the data. This function will presumably be used to generate inputs to a simulation model.

Estimators $\tilde{C}(E)$, $\hat{\gamma}_1$, and $\hat{\gamma}_2$ for these coefficients were given at Equations 6.1.19, 6.1.13, and 6.1.17, respectively, and to use them for identification purposes, it was emphasized that some idea of their sampling variability should be obtained. This can be done either analytically, using results in Chapter 6, or from the sample using the resampling methods such as the bootstrap given in Chapter 9.

For Normal samples, it is known that $\text{var}(\hat{\gamma}_1) = 6/n$ and $\text{var}(\hat{\gamma}_2) = 24/n$. In fact, the bivariate distribution of $\hat{\gamma}_1$ and $(\hat{\gamma}_2 + 3)^{1/2}$ is known and tabulated (Pearson and Hartley, 1966, Table 34), the purpose being to use $\hat{\gamma}_1$ and $\hat{\gamma}_2$ to test jointly for Normality in a sample.

Another very commonly occurring case is that of exponentially distributed samples for which we should get $C(E) = 1, \gamma_1 = 2$, and $\gamma_2 = 6$. To have convenient methods for judging exponentiality without going through the sectioning methods of Chapter 9, approximations to the joint and marginal distributions of $\tilde{C}(E)$, $\hat{\gamma}_1$, and $\hat{\gamma}_2$ in exponential samples of size n must be found. In this example, we proceed to do this by simulation.

Several questions can be asked.

1. What are the marginal distributions of the three statistics $\tilde{C}(E)$, $\hat{\gamma}_1$, and $\hat{\gamma}_2$, and can we tabulate selected quantiles for them?

2. Are the statistics seriously biased for small n?

3. Are the statistics approximately Normally distributed from some n on, and what are good approximations to their variances?

4. What do the joint bivariate distributions of $\tilde{C}(E)$ and $\hat{\gamma}_1$, $\tilde{C}(E)$ and $\hat{\gamma}_2$, and $\hat{\gamma}_1$ and $\hat{\gamma}_2$ look like? In particular, how correlated (dependent) are the estimates of the coefficients? If the statistics are too dependent, looking at all three statistics to see how close they are to 1, 2, and 6, respectively, may be misleading.

ii. SUPER-SMTBPC

To investigate these questions, simulations of all three statistics, based on the same exponential samples, were run in an extension of the program SMTBPC described in Chapter 8 and in the software supplement. This SUPER-SMTBPC program allows for super-replications of the type of output obtained in Chapter 8, as well as examination of (joint) bivariate distributions, as a function of sample size, for the statistics under consideration. We discuss the idea of super-replications first.

Thus, Figure 10.2.2 shows SMTBPC output for the statistic $\tilde{C}(E)$ evaluated at subsample sizes $n_i = 10, 20, 30, 40, 50, 60, 70, 80$ with a total of $M \times N = 25 \times 10,000 = 250,000$ independent exponential random variables used at each n_i. Thus, the boxplot at $n_i = 10$ is a boxplot of $250,000/10 = 25,000$ independent replications of $\tilde{C}(E)$.

In SUPER-SMTBPC, an additional parameter (NSR = number of super-replications) specifies how many times the procedure summarized in Figure 10.2.2 is (independently) replicated. All the statistics normally computed in each replication of the graphical output are averaged across replications and printed, with

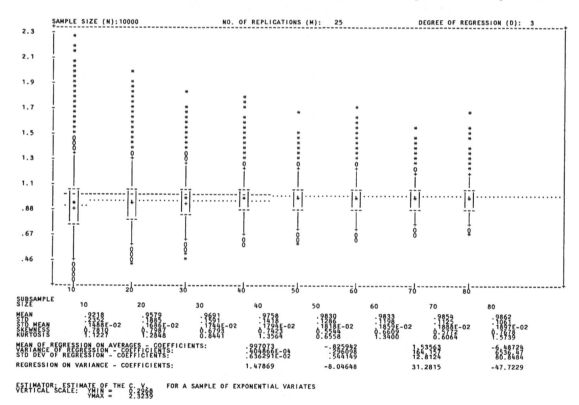

Figure 10.2.2 SMTBPC display of the first super-replication of the standard estimate of the coefficient of variation, $\tilde{C}(E) = S/\bar{E}$, in an exponential population. Note that the bias is small and has become insignificant at sample size $n = 50$. However, positive skewness in the distribution of the estimated coefficient of variation persists.

SUMMARY STATISTICS (MEAN/STD)   20 REPETITIONS

SUBSAMPLE SIZE	10	20	30	40	50	60	70	80
MEAN	.9217/.4481E-03	.9563/.5302E-03	.9687/.5666E-03	.9767/.5668E-03	.9809/.4008E-03	.9840/.5302E-03	.9863/.5302E-03	.9873/.4908E-03
STD	.2360/.2964E-03	.1861/.5025E-03	.1594/.3407E-03	.1420/.5459E-03	.1297/.4176E-03	.1198/.3459E-03	.1119/.4293E-03	.1053/.4070E-03
SKEWNESS	.7606/.6648E-02	.7901/.6905E-02	.7636/.1332E-01	.7395/.1338E-01	.6978/.1968E-01	.6539/.1312E-01	.6136/.1358E-01	.6089/.1487E-01
KURTOSIS	1.070/.2859E-01	1.076/.2751E-01	1.031/.7358E-01	1.008/.8300E-01	1.220/.1209	1.036/.6500E-01	.9167/.7873E-01	.9249/.6634E-01
SER.COR.	.1869E-03/.1320E-02	-.2189E-02/.1895E-02	.1505E-02/.2005E-02	-.4274E-02/.3250E-02	-.2744E-03/.2933E-02	.3537E-03/.3296E-02	.8227E-03/.3350E-02	-.7423E-02/.4243E-02
QUANTILES								
0.010	.4815/.7984E-03	.6071/.1162E-02	.6476/.1166E-02	.7047/.1071E-02	.7397/.1324E-02	.7596/.1093E-02	.7669/.9359E-03	.7789/.1087E-02
0.025	.5379/.8257E-03	.6517/.9200E-03	.7062/.7980E-03	.7396/.9292E-03	.7640/.9558E-03	.7829/.6887E-03	.7958/.7190E-03	.8085/.7948E-03
0.050	.5877/.6934E-03	.6926/.5864E-03	.7404/.5465E-03	.7716/.7716E-03	.7927/.6923E-03	.8105/.8031E-03	.8212/.5734E-03	.8317/.8332E-03
0.100	.6485/.4805E-03	.7412/.7191E-03	.7839/.5999E-03	.8192/.9297E-03	.8279/.8600E-03	.8422/.8378E-03	.8534/.8053E-03	.8617/.5949E-03
0.250	.7583/.3381E-03	.8296/.4699E-03	.8589/.8093E-03	.8786/.5488E-03	.8912/.8712E-03	.9093/.8013E-03	.9084/.5668E-03	.9140/.6646E-03
0.500	.8264/.4481E-03	.9367/.6337E-03	.9537/.6337E-03	.9637/.6012E-03	.9687/.5982E-03	.9728/.6337E-03	.9769/.6337E-03	.9785/.6646E-03
0.750	1.059/.6012E-03	1.063/.6942E-03	1.061/.7761E-03	1.060/.6942E-03	1.057/.6337E-03	1.055/.6337E-03	1.053/.6646E-03	1.050/.6337E-03
0.900	1.233/.7761E-03	1.192/.1186E-02	1.177/.8735E-03	1.161/.9399E-03	1.159/.1022E-02	1.149/.1151E-02	1.132/.1112E-02	1.123/.9817E-03
0.950	1.352/.9817E-03	1.294/.1770E-02	1.255/.1060E-02	1.230/.1591E-02	1.212/.9610E-03	1.197/.1329E-02	1.186/.1513E-02	1.174/.1459E-02
0.975	1.466/.1339E-02	1.385/.1912E-02	1.332/.1339E-02	1.298/.2186E-02	1.272/.1792E-02	1.250/.1972E-02	1.235/.1735E-02	1.222/.1314E-02
0.990	1.608/.2287E-02	1.499/.2925E-02	1.428/.2462E-02	1.386/.3091E-02	1.348/.2543E-02	1.323/.2841E-02	1.296/.2972E-02	1.282/.2693E-02

MEAN OF REGRESSION ON AVERAGES – COEFFICIENTS: .983345/.107911E-02   -.960955/.900819E-01   2.89084/2.21084   -8.42002/14.2002

STD DEV OF REGRESSION – COEFFICIENTS: .556122E-02/.119059E-03   .493846/.114958E-01   11.6548/1.276020   73.1589/7.71589

REGRESSION ON VARIANCE – COEFFICIENTS: 1.04502/.907823E-01   -.952052/1.41486   -5.04988/6.97808   18.8501/18.8695

ESTIMATOR: ESTIMATE OF THE C. V.   FOR A SAMPLE OF EXPONENTIAL VARIATES

Figure 10.2.3   Summary of 20 super-replications of the estimated coefficient of variation, giving estimated standard deviations of all estimated properties. Thus, the kurtosis of $\tilde{C}(E)$ for $n = 10$ is estimated to be 1.070, and this estimate has estimated standard deviation of 0.02659. The kurtosis of $\tilde{C}(E)$ is thus clearly nonzero. In fact, even at $n = 80$ it is significantly different from zero (0.9249 with estimated s.d. of 0.06634).

estimated standard deviations, using the basic technique of Chapter 4. The results for 20 super-replications of Figure 10.2.2 are shown in Figure 10.2.3. Notice that while the procedure is repeated 20 times and estimates are based on 20 super-replications, only the first three super-replications produce graphical output—all the rest are suppressed. The use of super-replications to estimate standard deviations (of estimated moments, quantiles, etc.) of the statistic being considered is really the sectioning technique introduced in Chapter 9.

Thus the quantity 1.608 at the bottom of the $n = 10$ column of Figure 10.2.3 is the average of 20 estimates of the 0.990 quantile of $\tilde{C}(E)$, each estimate having been computed from 25,000 replications of $\tilde{C}(E)$. The quantity below 1.608, namely 0.00227, is the estimated standard error of the average quantile estimate (1.608), computed as the sample standard deviation of the 20 quantile estimates, divided by $(20)^{1/2}$. The reasons for the super-replications are as follows:

1. As we have stressed, it is necessary to assess variability of estimates that come out of a simulation. The sectioning technique of Chapter 9 has been adopted here, so that when $n_i = 10$, instead of using $20 \times 25,000$ realizations of $\tilde{C}(E)$ for one estimate of the 0.99 quantile, we have sectioned these into 20 sets, with each set

providing an estimate of the 0.99 quantile. There is, of course, some danger of bias, but hopefully not with such large numbers of replications.

2. It is difficult to use much more than $M \times N = 250{,}000$ inputs to the program in any one replication, because of computer storage limitations. Also, a boxplot is too coarse a display to show the greater accuracy in estimating the distribution of the statistic that is obtainable with 20 times the data.

3. The program optionally plots figures like Figure 10.2.2 for the first three super-replications. This way, the stability of the regression curves can be assessed. Typically with $M \times N = 10{,}000$ they will vary considerably from figure to figure; with $M \times N = 250{,}000$ they will stabilize. This 25-fold increase in the number of replications increases the precision of the estimates of the coefficients in the regression by a factor of 5.

iii. Marginal Distributions of the Estimated Coefficients

Figure 10.2.2 shows that the distribution of $\tilde{C}(E)$ is quite positively skewed, even out at $n_i = 80$, but the bias is small by $n_i = 50$. Note that the estimated skewness and kurtosis of this statistic fluctuate up and down with positive values with no apparent convergence to zero, although the standard deviation is decreasing. The greater precision allowed from 20 super-replications in Figure 10.2.3 shows that the skewness of $\tilde{C}(E)$ increases slightly from $n_i = 10$ to $n_i = 20$ and then decreases slowly as n_i increases. Similarly, the kurtosis of $\tilde{C}(E)$ increases up to $n_i = 30$. Convergence to Normality is clearly very slow.

In fact, from the last column of Figure 10.2.3, i.e., when $n_i = 80$, we have the following normalized 0.025 and 0.975 quantiles:

$$\frac{\hat{x}_{0.025} - \bar{x}}{s} = \frac{0.8085 - 0.9873}{0.1053} = -1.698$$

$$\frac{\hat{x}_{0.975} - \bar{x}}{s} = \frac{1.2220 - 0.9873}{0.1053} = +1.97$$

Therefore, although the 0.975 quantile is close to the standardized Normal theory value of 1.96, the 0.025 quantile of -1.698 is quite far from the standardized Normal theory value of -1.96.

Figures 10.2.4 through 10.2.7 show that there is serious bias in both the skewness and the kurtosis estimates $\hat{\gamma}_1$ and $\hat{\gamma}_2$ in exponential populations. In fact, for the estimated kurtosis the bias is on the order of the standard deviation. From the column in Figure 10.2.7 headed $n_i = 80$, we see that the bias is $(6 - 3.951) = 2.049$, while the standard deviation of $\hat{\gamma}_2$ is 3.803. Thus the root-mean-square error is $(3.803^2 + 2.049^2)^{1/2} = 4.32$. This suggests that jackknifing might be in order to reduce the bias, hopefully without inflating the variance of the estimated kurtosis too much.

The tabulations of the quantiles of the marginal distributions of $\tilde{C}(E)$, $\hat{\gamma}_1$, and $\hat{\gamma}_2$ in Figures 10.2.3, 10.2.5, and 10.2.7 can, with possible interpolation, be used to decide with given confidence if an observed estimate of these quantities is consistent

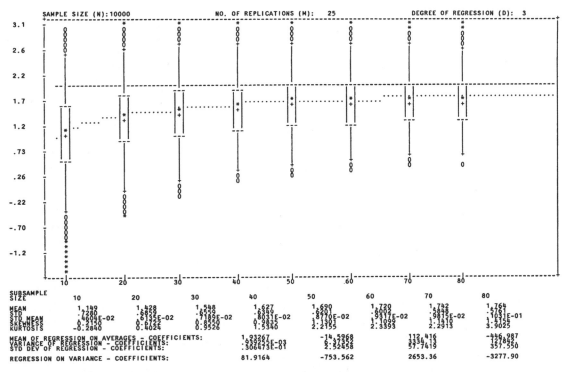

SAMPLE SIZE (N):10000   NO. OF REPLICATIONS (M):   25   DEGREE OF REGRESSION (D): 3

```
SUBSAMPLE
SIZE           10        20        30        40        50        60        70        80

MEAN          1.149     1.428     1.548     1.627     1.690     1.720     1.742     1.764
STD            .7280     .6450     .6550     .6969     .6969     .6002     .4691     .4421
STD.MEAN      .4505E-02 .6155E-02 .71895E-02 .8031E-02 .8710E-02 .93917E-02 .5848E-02 .1031E-01
SKEWNESS      0.6155    0.6820    0.9835    1.1301    2.1155    2.3393    1.1415    3.4154
KURTOSIS     -0.2840    0.6024    0.9520    1.5340    2.2155    2.3393    2.2913    3.9025

MEAN OF REGRESSION ON AVERAGES - COEFFICIENTS:     1.93267   -14.5968   112.416   -446.987
VARIANCE OF REGRESSION - COEFFICIENTS:            .930255E-03   6.37352  3334.13   127892.
STD DEV OF REGRESSION - COEFFICIENTS:             .306473E-01   2.52458   57.7419   357.550

REGRESSION ON VARIANCE - COEFFICIENTS:             81.9164    -753.562  2653.36   -3277.90
```

ESTIMATOR: ESTIMATE OF THE SKEWNESS FOR A SAMPLE OF EXPONENTIAL VARIATES
VERTICAL SCALE: YMIN = -1.5550
                 YMAX =   3.1150

Figure 10.2.4   SMTBPC display of the first super-replication of the standard estimate (Equation 6.1.13) of the coefficient of skewness $\hat{\gamma}_1$ in an exponential population. True value is $\gamma_1 = 2$. There is still bias at $n = 80$ and clearly non-Normality. In fact, kurtosis and skewness of the distribution of $\hat{\gamma}_1$ appear to be increasing with sample size. For validation of this result see Figure 10.2.5.

with the hypothesis of an exponential distribution. Of course, in general, this brings up questions of the joint distributions of $\tilde{C}(E)$, $\hat{\gamma}_1$, and $\hat{\gamma}_2$, which we consider now.

iv. Joint Bivariate Distributions from SUPER-SMTBPC

In Figure 10.2.8 we show a bivariate histogram, computed in BIHSPC, of $m = 1000$ replications of $\tilde{C}(E)$ and $\hat{\gamma}_1$ for sample size $n = 10$. The case $n = 50$ is shown in Figure 10.2.10. These are to be considered in conjunction with the much more precise simulations of the marginal distributions in previous figures.

It is clear from the two figures that these two variables, $\tilde{C}(E)$ and $\hat{\gamma}_1$, are fairly independent. The numerical summary given by the estimated correlation coefficient r confirms this: it is -0.0123 at $n = 10$ and -0.0276 at $n = 50$. Given the number of replications, $m = 1000$, these are <u>not</u> significantly different from zero, since for independent samples, s.d.(r) is $1/\sqrt{m-1} = 0.032$ (from Equation 10.2.4).

The bivariate histograms for $\hat{\gamma}_1$ and $\hat{\gamma}_2$ at sample sizes $n = 10$ and 50 (Figures 10.2.9 and 10.2.11) show a completely different picture. The correlation between the estimated coefficients is very strong, with $r = 0.919$ at $n = 10$ and $r = 0.964$ at

SUMMARY STATISTICS (MEAN/STD)   20 REPETITIONS

SUBSAMPLE SIZE	10	20	30	40	50	60	70	80
MEAN	1.150 / .6942E-03	1.423 / .1486E-02	1.551 / .1847E-02	1.630 / .2968E-02	1.685 / .2204E-02	1.722 / .2240E-02	1.751 / .2337E-02	1.775 / .2044E-02
STD	.7298 / .6032E-03	.6795 / .1219E-02	.6573 / .1723E-02	.6401 / .1976E-02	.6259 / .2166E-02	.6072 / .2325E-02	.5838 / .2830E-02	.5791 / .2156E-02
SKEWNESS	.2088 / .2361E-02	.6772 / .4475E-02	.8803 / .7903E-02	1.032 / .8206E-02	1.140 / .1583E-01	1.181 / .1577E-01	1.235 / .1856E-01	1.312 / .1306E-01
KURTOSIS	-.2925 / .4313E-02	.4657 / .1095E-01	1.126 / .3195E-01	1.433 / .4300E-01	2.195 / .8412E-01	2.440 / .9147E-01	2.740 / .1122	3.224 / .1100
SER. COR.	-.5670E-03 / .1499E-02	.1898E-03 / .2030E-02	.4646E-03 / .2025E-02	-.2641E-02 / .3092E-02	-.4393E-02 / .3167E-02	.6153E-03 / .3243E-02	.8793E-02 / .3773E-02	-.4755E-02 / .3554E-02
QUANTILES								
0.010	-.3918 / .2989E-02	.1635 / .3769E-02	.3280 / .3608E-02	.5401 / .3847E-02	.6387 / .3923E-02	.7185 / .4775E-02	.7848 / .3878E-02	.8325 / .4381E-02
0.025	-.1658 / .1933E-02	.3225 / .2566E-02	.5393 / .2296E-02	.6692 / .2836E-02	.7624 / .3155E-02	.8354 / .3096E-02	.8924 / .3060E-02	.9411 / .3329E-02
0.050	.2625E-01 / .1669E-02	.4635 / .1593E-02	.6637 / .2081E-02	.7892 / .2452E-02	.8799 / .2692E-02	.9376 / .2718E-02	.9919 / .2282E-02	1.037 / .2493E-02
0.100	.2486 / .1313E-02	.6342 / .1432E-02	.8093 / .1799E-02	.9296 / .2576E-02	1.007 / .2337E-02	1.064 / .2454E-02	1.114 / .2421E-02	1.152 / .2204E-02
0.250	.6327 / .8740E-03	.9419 / .1431E-02	1.088 / .1994E-02	1.183 / .1943E-02	1.252 / .1901E-02	1.302 / .1796E-02	1.345 / .1539E-02	1.378 / .2024E-02
0.500	1.101 / .9817E-03	1.237 / .1473E-02	1.454 / .2258E-02	1.531 / .2258E-02	1.581 / .2111E-02	1.622 / .2102E-02	1.651 / .2231E-02	1.674 / .2495E-02
0.750	1.631 / .1816E-02	1.817 / .2222E-02	1.911 / .2805E-02	1.967 / .2590E-02	2.003 / .3026E-02	2.030 / .3813E-02	2.045 / .4058E-02	2.059 / .3039E-02
0.900	2.173 / .1953E-02	2.345 / .3749E-02	2.430 / .4399E-02	2.472 / .5065E-02	2.499 / .5351E-02	2.508 / .6283E-02	2.511 / .5678E-02	2.510 / .4855E-02
0.950	2.465 / .1901E-02	2.706 / .5058E-02	2.793 / .6157E-02	2.842 / .6406E-02	2.872 / .8705E-02	2.871 / .9927E-02	2.858 / .7580E-02	2.860 / .7848E-02
0.975	2.668 / .2328E-02	3.017 / .5222E-02	3.137 / .8707E-02	3.181 / .8251E-02	3.232 / .1144E-01	3.222 / .1113E-01	3.215 / .1292E-01	3.216 / .1128E-01
0.990	2.840 / .2328E-02	3.352 / .6065E-02	3.551 / .1144E-01	3.638 / .1519E-01	3.629 / .1847E-01	3.674 / .1482E-01	3.661 / .1881E-01	3.687 / .1989E-01

MEAN OF REGRESSION ON AVERAGES - COEFFICIENTS: 1.95903 / .611753E-02   -15.8715 / .520660   112.9405 / 12.3587   -540.801 / 74.0241

STD DEV OF REGRESSION - COEFFICIENTS: .273027E-01 / .895046E-03   2.31608 / .892446E-01   23.1378 / 2.16028   328.388 / 13.4883

REGRESSION ON VARIANCE - COEFFICIENTS: 79.8642 / 1.60586   -707.236 / 24.5081   2372.19 / 118.241   -2789.49 / 179.947

ESTIMATOR: ESTIMATE OF THE SKEWNESS FOR A SAMPLE OF EXPONENTIAL VARIATES

Figure 10.2.5 Summary of 20 super-replications of the estimated coefficient of skewness $\hat{\gamma}_1$ in an exponential population. It is clear that both the estimated skewness and the kurtosis of the skewness estimate $\hat{\gamma}_1$ are increasing with sample size n, so that convergence to Normality ($\gamma_1 = \gamma_2 = 0$) is slow.

$m = 50$. In fact, in Figure 10.2.11 the dependence is strong enough to suggest that $\hat{\gamma}_2$ gives no more information about exponentially than $\tilde{C}(E)$ and $\hat{\gamma}_1$ alone! An interesting question is whether this holds true for all n.

The situation is very different in the case of Normal random variables. In Figure 10.2.12, we show a bivariate histogram computed with BIHSPC of $\hat{\gamma}_1$ and $\hat{\gamma}_2$ from a Normal sample of size $n = 10$. The true values are $\gamma_1 = \gamma_2 = 0$. Since $r = 0.095$, the dependence is small; the horn in the histogram suggests dependence that is slightly stronger than is suggested by the estimated correlation. The horn reflects the fact that $Y = \hat{\gamma}_2$ has a marginal distribution with positive skewness and that large values of $\hat{\gamma}_2$ are accompanied by either very small or very large values of $X = \hat{\gamma}_1$.

10.3   The Bivariate Inverse Probability Integral Transform

We have introduced several simple bivariate random variables, all generated by simple operations on univariate random variables. Thus, the bivariate Normal pair (X, Y) in Section 10.1.4 was generated by simple additions of random variables,

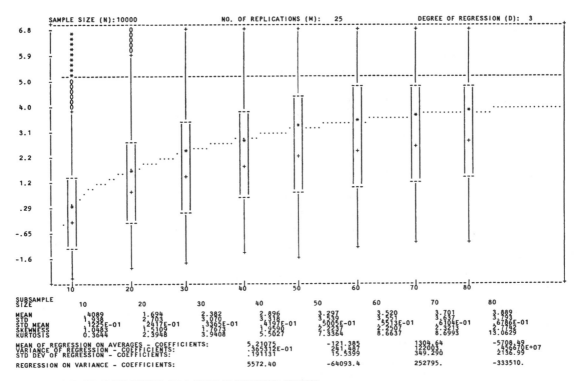

Figure 10.2.6 SMTBPC display of the first super-replication of the standard estimate $\hat{\gamma}_2$ (Equation 6.1.17) of the coefficient of kurtosis in an exponential population. True value is $\gamma_2 = 6$. The bias is severe enough to warrant consideration of jackknifing. Since the estimate is a ratio, this may not entail much variance inflation.

using the fact that linear combinations of Normal random variables are Normally distributed. Bivariate exponential pairs were generated in Example 10.1.1 using the idea that the minimum of two independent exponentially distributed random variables has an exponential distribution. Again, in Exercise 9.9 a bivariate Gamma pair was generated using the additivity of Gamma variates with the same rate parameter, β.

Several of these ad hoc schemes will be examined in Section 10.4, together with more specialized schemes to achieve effects such as negative correlation. Before doing this, we discuss the general approach to generating bivariate random variables using an analogue of the inverse probability integral transform for univariate random variables given in Chapter 4.

The schemes that will be given for generating bivariate random variables are meant to be illustrative, not comprehensive. The literature on bivariate random variables, and on generating bivariate random variables, is vast. For further references and details, see either N. L. Johnson and Kotz (1972), DeVroye (1986), or M. E. Johnson (1987).

SUMMARY STATISTICS (MEAN/STD)   20 REPETITIONS

SUBSAMPLE SIZE	10	20	30	40	50	60	70	80
MEAN	.4144 / .1960E-02	1.666 / .5893E-02	2.329 / .7963E-02	2.891E-01 / .1031E-01	3.281E-01 / .1321E-01	3.546E-01 / .1329E-01	3.769E-01 / .1502E-01	3.951E-01 / .1374E-01
STD	1.939 / .1747E-02	2.670 / .6561E-02	3.093 / .1138E-01	3.368 / .1408E-01	3.565 / .2011E-01	3.653 / .2954E-01	3.719 / .2481E-01	3.893 / .1972E-01
SKEWNESS	1.244 / .2543E-02	1.533 / .4835E-02	1.822 / .1259E-01	2.050 / .1897E-01	2.212 / .2687E-01	2.303 / .2882E-01	2.423 / .3519E-01	2.562 / .3466E-01
KURTOSIS	3.400 / .3700E-02	2.543 / .2777E-01	4.309 / .7665E-01	6.138 / .1313	7.327 / .2321	8.232 / .2838	9.337 / .3822	10.431 / .4439
SER.COR.	-.3759E-03 / .7437E-02	.1630E-02 / .1919E-02	.8770E-03 / .2293E-02	-.8948E-02 / .3168E-02	-.2665E-02 / .2985E-02	-.1010E-02 / .3200E-02	.6793E-03 / .5733E-02	.8835E-03 / .3835E-02
QUANTILES								
0.010	-2.017E-02 / .1552E-02	-1.478E-02 / .3505E-02	-1.455E-02 / .4973E-02	-.9175E-02 / .5287E-02	-.7242E-02 / .6900E-02	-.5561E-02 / .6104E-02	-.3907E-02 / .8699E-02	-.2546E-01 / .1159E-01
0.025	-1.889E-02 / .1986E-02	-1.329E-02 / .2985E-02	-.9648E-02 / .3937E-02	-.6948E-02 / .5988E-02	-.4724E-02 / .6240E-02	-.2938E-02 / .6598E-02	-.1154E-02 / .9314E-02	.1833E-01 / .8233E-01
0.050	-1.749E-02 / .1459E-02	-1.144E-02 / .3162E-02	-.7504E-02 / .3749E-02	-.4642E-02 / .5595E-02	-.2139E-02 / .5661E-02	-.2969E-02 / .2929E-02	.1548 / .8883E-02	.3078 / .6038E-02
0.100	-1.555E-02 / .1251E-02	-.8777E-02 / .8455E-02	-.4440E-02 / .3547E-02	-.1376 / .3268E-02	.1430 / .5619E-02	.3404 / .6915E-02	.5393 / .6423E-02	.6842 / .9008E-02
0.250	-1.084E-02 / .1816E-02	-.2853E-02 / .3038E-02	.2705 / .4908E-02	.6194 / .8162E-02	.9442 / .7338E-02	1.197 / .5275E-02	1.374 / .6032E-02	1.539 / .8599E-02
0.500	-.1541E-02 / .2555E-02	.2012 / .4578E-02	1.872E-02 / .8150E-02	1.958 / .9999E-02	2.285E-01 / .1125E-01	2.531 / .1002E-01	2.732 / .1156E-01	2.916 / .1356E-01
0.750	1.467 / .4763E-02	2.779 / .9689E-02	3.552 / .1159E-01	4.982 / .1946E-01	4.256 / .1698E-01	4.734 / .2133E-01	4.928 / .2576E-01	5.119 / .2257E-01
0.900	3.542 / .5225E-02	5.368 / .1885E-01	6.431 / .2491E-01	7.324 / .3241E-01	7.631 / .3438E-01	7.921 / .4751E-01	8.133 / .4350E-01	8.346 / .4408E-01
0.950	4.501 / .5250E-02	7.314 / .2358E-01	8.739 / .3981E-01	9.497 / .4219E-01	10.28 / .8843E-01	10.26 / .9219E-01	10.90 / .9555E-01	11.019 / .7033E-01
0.975	5.236 / .7860E-02	8.999 / .2997E-01	11.02 / .6088E-01	12.22 / .2855E-01	13.13 / .9010	13.59 / .9547E-01	14.04 / .9834	14.35 / .9814E-01
0.990	5.848 / .8700E-02	10.183 / .3159E-01	13.72 / .7904E-01	15.52 / .1154	17.02 / .1490	17.51 / .1532	18.18 / .1820	18.83 / .1887

MEAN OF REGRESSION ON AVERAGES - COEFFICIENTS:			5.34507 / .3278E-01		-129.528 / 3.02522		1439.40 / 68.2228		-6370.589 / 477.589

STD DEV OF REGRESSION - COEFFICIENTS:			.363741 / .397590E-02		13.3868 / .464050		298.827 / 18.9824		1818.129 / 181.6123

REGRESSION ON VARIANCE - COEFFICIENTS:			5595.84 / 90.6821		-63593.7 / 1355.13		247651.55 / 6415.55		-322952.70 / 9807.70

ESTIMATOR: ESTIMATE OF THE KURTOSIS FOR A SAMPLE OF EXPONENTIAL VARIATES
*** WIDEST Y VALUES FOUND: YMIN=-2.442   YMAX= 48.46

Figure 10.2.7 Summary of 20 super-replications of the estimated coefficient of kurtosis $\hat{\gamma}_2$ in an exponential population. Even at $n = 80$, the bias $(6.00 - 3.951)$ in the estimate is comparable to the standard deviation (3.803) of the estimate, confirming that jackknifing is in order. The convergence to Normality, measured by the estimated skewness and kurtosis of the estimate, is slower than the convergence of the estimates of $\tilde{C}(E)$ and γ_1. This is typical behavior of estimates of higher order moments. Note that the median of the estimate $(2.916$ at $n = 80)$ is even further from the true value of 6.0 than is the mean (3.951).

10.3.1   Generating Continuous Bivariate Random Variables: Conditional Approach

In Equation 10.2.7 for the conditional expectation of X given y, the quantity in the integral is the conditional density of X given by:

$$f_{X|Y}(x, y) = \frac{f_{X,Y}(x, y)}{f_Y(y)} \qquad f_Y(y) > 0 \tag{10.3.1}$$

and the conditional distribution of X given y is

$$F_{X|Y}(x, y) = \int_{-\infty}^{x} \frac{f_{X,Y}(u, y)}{f_Y(y)} \, du = \frac{\dfrac{\partial}{\partial y} F_{X,Y}(x, y)}{\displaystyle\int_{-\infty}^{\infty} \dfrac{\partial^2}{\partial x \, \partial y} F_{X,Y}(x, y) \, dx} \tag{10.3.2}$$

BIVARIATE SAMPLE SUMMARY   SAMPLE SIZE =   1000

MEASURES OF ASSOCIATION

COVARIANCE	-2.108469E-03
CORRELATION COEFFICIENT	-.012349
SPEARMAN'S RANK CORRELATION COEF.	-.004913

TESTS FOR EQUIDISTRIBUTION

	TEST STATISTIC	NORMALIZED STATISTIC
KOLMOGOROV-SMIRNOV TEST	.353000	
WALD-WOLFOWITZ RUNS TEST	655	-15.47746
MANN-WHITNEY TEST	-12713	-39.70465
WILCOXSON TEST	-3733	-77.76811
SIEGEL-TUKEY TEST	5497	-77.05334

UNIVARIATE STATISTICS

	X SAMPLE	Y SAMPLE
MEAN	1.168705E+00	9.279585E-01
MEDIAN	1.134827E+00	8.992730E-01
VARIANCE	5.160483E-01	5.649039E-02
STD DEV	7.183650E-01	2.376771E-01
RANGE	4.148465E+00	1.834233E+00
SKEWNESS	1.948493E-01	8.670235E-01
KURTOSIS	-2.606908E-01	1.470041E+00
MAXIMUM	3.095154E+00	2.258827E+00
MINIMUM	-1.053311E+00	4.245944E-01

Y-axis labels: 2.202E+00, 1.972E+00, 1.743E+00, 1.514E+00, 1.284E+00, 1.055E+00, 8.258E-01, 5.966E-01

X-axis labels:
-9.237E-01   1.134E-01   1.151E+00   2.188E+00
   -4.051E-01   6.320E-01   1.669E+00   2.706E+00

KEY

SYMBOL PRINTED	-	=	₸	₮	‡	♦	♦	I	‡	‡	▌	▌	▌	▌	▌
NO. OBSERVATIONS	0	1	2	3	4	5	6	7	8	9	10	11	12	13	14	15

BIV SKEWNESS VS C.V.   EXPON. N,SEED=   10   1872.

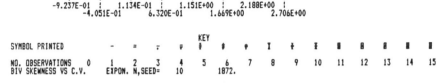

Figure 10.2.8 Bivariate histogram produced by BIHSPC. The (X, Y) pair are the estimated coefficient of variation $\tilde{C}(E)$ and the estimated coefficient of kurtosis $\hat{\gamma}_1$, computed from the sample of $n = 10$ iid exponential(1) random variables. Sample size (number of replications) is $m = 1000$. Both the general scatter and estimated correlation coefficient, $r = -0.0123$, indicate low correlation and dependence between the estimated coefficients. Marginal properties of these coefficients are simulated in more detail in Figures 10.2.2 through 10.2.5 at $n = 10$.

319

Figure 10.2.9 Bivariate histogram for a simulation similar to that in Figure 10.2.8: the (X, Y) pair are the estimated coefficients of skewness $\hat{\gamma}_1$ and kurtosis $\hat{\gamma}_2$, from an exponential sample of size $n = 10$. Sample size (number of replications) is $m = 1000$. It is clear from the scatter and the estimated correlation of $r = 0.919$ that $\hat{\gamma}_1$ and $\hat{\gamma}_2$ are highly dependent. Note that the x and y scales are different, a consequence of automatically scaling the plot. There is a definite ridge in the bivariate density and the regression of X on Y is probably quadratic.

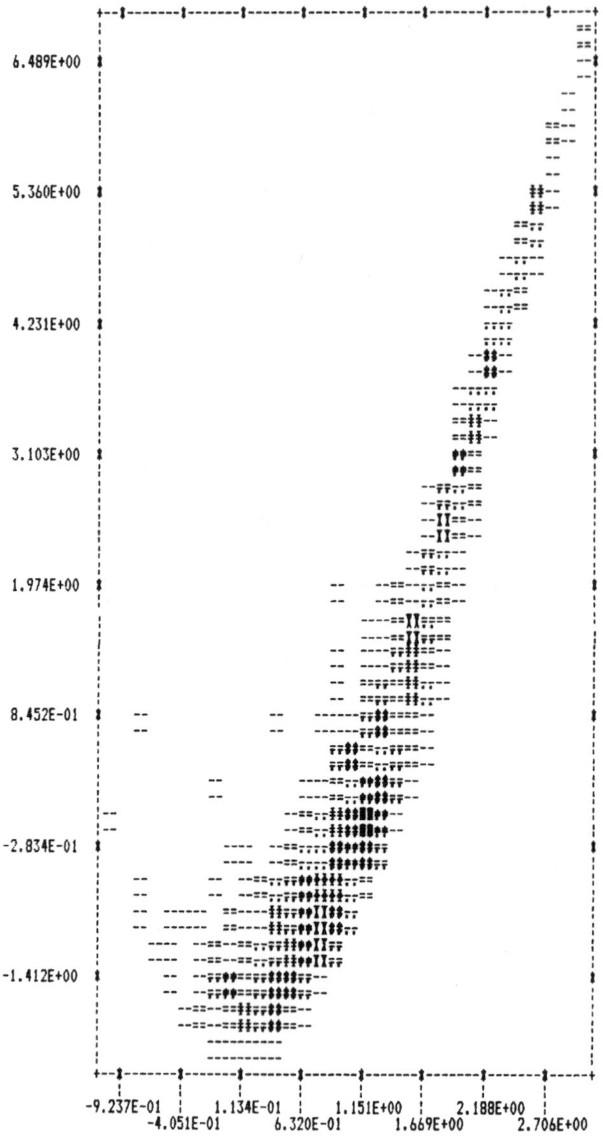

Figure 10.2.10 Bivariate histogram similar to that in Figure 10.2.8, except that the underlying (exponential) sample size has been raised to $n = 50$. The marginal standard deviations have decreased and the correlation is still low.

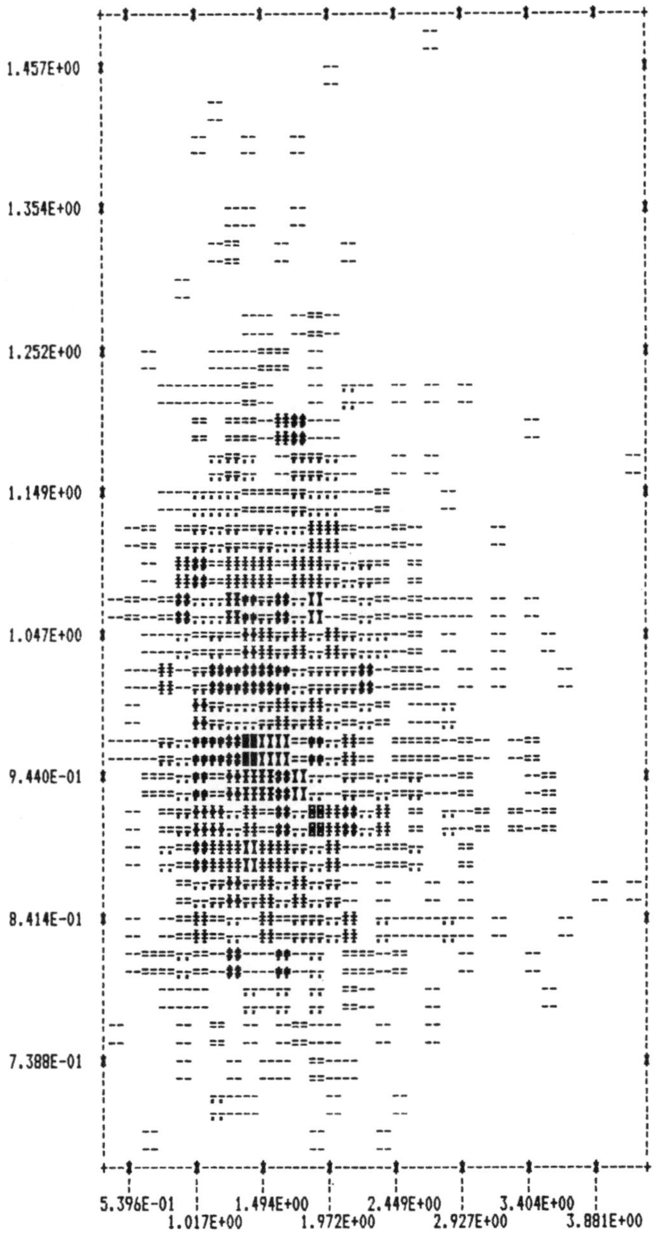

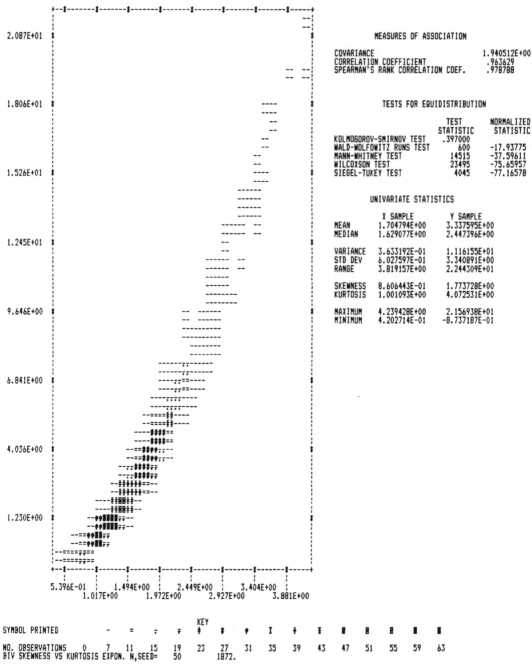

Figure 10.2.11 Bivariate histogram for $(\hat{\gamma}_1, \hat{\gamma}_2)$ similar to that in Figure 10.2.9, but with the exponential sample size increased to $n = 50$. The joint density is even tighter than for the $n = 10$ case, and the estimated correlation has *increased* to 0.964. This calls into question the correctness of any assumption that the correlation decreases to zero as sample size $n \to \infty$.

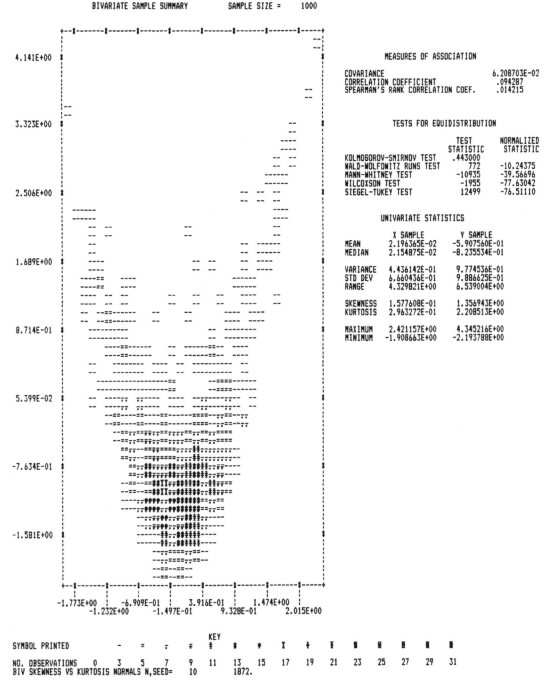

Figure 10.2.12 Bivariate histogram of $(\hat{\gamma}_1, \hat{\gamma}_2)$ similar to Figure 10.2.9, with which it should be compared. In this case, the sample from which the kurtosis and skewness has been estimated is 10 iid Normal$(0, 1)$ random variables. Note the very small estimated correlation ($r = 0.0943$) in contrast to that in the exponential case ($r = 0.919$).

The second equality in Equation 10.3.2 follows only upon the (strong) assumption of absolute continuity of $(\partial/\partial y)F_{X,Y}(x, y)$.

It is clear that one can generate the pair (X, Y) by the following conditional approach.

1. Generate Y from the (marginal) distribution $F_Y(y)$. This could be done by using the inverse probability integral transform, so that $Y = F_Y^{-1}(U_1)$, or by any of the more sophisticated means to be considered, for example, in Volume 2 or Kennedy and Gentle (1980).

2. Generate X from the conditional distribution of X, given the value y of Y generated in Step 1. This could be done as $X = F_{X|Y}^{-1}(U_2, y)$ if no other methods are available, and if the inverse distribution function is not too hard to compute. Note one big problem with this method; each value of y obtained in step (1) requires computation of a new inverse conditional distribution function for X.

This method could be used when the joint distribution of X and Y has a known functional form and the marginal and conditional distributions are derived from this using Equations 10.3.1 and 10.3.2. More often, though, the marginal and conditional distributions and densities are known, or given. In practice, one can then calculate the joint distribution of X and Y. However, if the joint distribution is not absolutely continuous, it may not be easily derivable from the conditional and marginal distributions $F_{X|Y}(x, y)$ and $F_Y(y)$, which will also not be absolutely continuous. The algorithm may still hold, if care is taken with discontinuities, as shown in Example 10.3.1.

EXAMPLE   **10.3.1 The Lawrance–Lewis Bivariate Exponential Pair**
This bivariate pair is constructed as follows. Let E_1 and E_2 be independent, exponential(λ) random variables. Let $Y = E_2$, so that Y has marginal distribution $F_Y(y)$ that is exponential(λ). Now let

$$
\begin{aligned}
X &= \rho Y & \text{with probability } \rho & \qquad (0 \leqslant \rho < 1) \\
&= \rho Y + E_1 & \text{with probability } 1 - \rho & \qquad (10.3.3)
\end{aligned}
$$

It is left as an exercise to show that the random linear combination (Equation 10.3.3) of two independent exponential random variables gives an exponentially distributed random variable X (Exercise 10.12).

Now, given $Y = y$, one sees that the conditional random variable X has discrete value ρy with probability ρ. Otherwise, with probability $1 - \rho$, it is a shifted exponential random variable, $\rho y + E_1$. The correlation between X and Y is ρ.

Elaborations on this method of generating bivariate exponentials are given in Lawrance and Lewis (1983). It is not hard to show that the joint distribution of this bivariate exponential pair has a discontinuity along the line $x = \rho y$. In fact, X can never be less than ρY. This pair is closely related to the Marshall–Olkin bivariate exponential pair discussed in Example 10.1.1.

10.3.2   Generating Discrete Bivariate Random Variables: Conditional Approach

Bivariate, discrete-valued random variables are functionally much simpler to deal with than continuous or mixed continuous–discrete random variables. Thus, as at Equation 10.1.5, the joint probability function is given by $p_{k,j}$, where the $p_{k,j}$ sum to one and satisfy $0 \leqslant p_{k,j} \leqslant 1$. Then

$$p_{k|j} = \frac{p_{k,j}}{p_j} = \frac{p_{k,j}}{\sum\limits_{k=0}^{\infty} p_{k,j}} \qquad (p_j > 0) \tag{10.3.4}$$

is the conditional probability that $X = k$, given that $Y = j$, where $p_j = P(Y = j)$. The conditional algorithm for generating the pair X and Y is as follows:

1. Generate Y with probability function $p_j, j = 0, 1, \ldots$. This could be done using the discrete version of the inverse probability integral transform method given in Chapter 4.
2. Given the value $Y = j$, generate X conditional on $Y = j$ from the conditional probability function given in Equation 10.3.4.

A bivariate geometric random pair is discussed in Exercise 10.2. Further discussion of this example is given in Exercise 10.13.

10.4   Ad Hoc and Model-Based Methods for Bivariate Random Variable Generation

We conclude this chapter with a brief discussion of bivariate pairs X and Y that are useful in modeling situations that arise in operations research and engineering. Thus, the discussion is mostly of exponential and Gamma marginal distributions. Moreover, the emphasis is on generation schemes that are motivated by physical models. Specialized schemes also become necessary if one is to generate pairs with negative correlation, since most of the simple schemes generate positively correlated pairs.

10.4.1   Generation Methods Based on Additivity and Antithetics

There are other random variables besides Normal variables that can be additively combined to generate bivariate pairs with fixed marginal distributions. One of these is the Gamma(k, β) random variable. However, not every linear combination of Gamma variates is Gamma: the scale parameters in the added Gamma variables must have the same value. That is, a Gamma(k_1, β_1) variable added to a Gamma(k_2, β_2) variable is Gamma distributed if and only if $\beta_1 = \beta_2$. See Exercise 9.9 for details.

Another case of interest is the discrete-valued Poisson variate. If E_1 is Poisson(λ_1) and E_2 is independently Poisson(λ_2), then $E_1 + E_2$ is Poisson$(\lambda_1 + \lambda_2)$. Thus, bivariate Poisson pairs such as $(E_1, E_1 + E_2)$ can be easily generated. This example is explained further in Exercise 10.15.

As another example, a random variable with an l-Laplace distribution is a long-tailed symmetric alternative to Normally distributed random variables. The characteristic function of this random variable X is

$$\phi_X(\omega) = E(e^{i\omega X}) = \left(\frac{\lambda^2}{\lambda^2 + \omega^2}\right)^l \qquad l > 0; \lambda > 0; -\infty < \omega < \infty \qquad (10.4.1)$$

and when $l = 1$, this is a double-exponential (Laplace)-distributed random variable with

$$f_X(x) = \frac{1}{2\lambda} e^{-|x|/\lambda} \qquad -\infty < x < \infty \qquad (10.4.2)$$

This Laplace density is actually the probability density function of the difference of two independent exponential random variables. When l gets large, the l-Laplace random variable approximates a Normal random variable.

An l-Laplace random variable with parameters λ and l_1, when added to an independent l-Laplace random variable with parameters λ and l_2, is l-Laplace$(l_1 + l_2)$. This follows from the form of the characteristic function (Equation 10.4.1). Consequently, it is simple to generate bivariate l-Laplace pairs.

Schemes other than fixed-coefficient additive combinations for generating random variables are also useful. A particularly simple and useful one uses the Beta–Gamma transformation. Thus, if B is a Beta$(\rho k, (1 - \rho)k)$ random variable, independent of X, a Gamma(k, β) random variable, then BX is Gamma$(\rho k, \beta)$. Using this idea, Gamma variates can be combined just like Normal random variables to obtain Gamma-distributed random pairs. These have particularly smooth bivariate distributions (Lewis, McKenzie, and Hugus, 1988).

Most of the schemes above give positively correlated random pairs (X, Y) when the marginal random variables are positive valued. One should note that a degenerate *negatively* correlated pair with continuous marginal distributions can always be obtained by the antithetic transformation:

$$Y = F_X^{-1}\{1 - F_X(X)\} \qquad (10.4.3)$$

In this case, the value of X completely determines Y.

This scheme was discussed for exponential pairs X and Y in Section 4.2.5, where it was shown that the correlation between X and Y in that case is -0.6649. For symmetric random variables it is easily shown that Equation 10.4.3 gives $Y = -X$ and $\rho(X, Y) = -1$; moreover, in this case of symmetric random variables, any nondegenerate scheme for combining random variables to give positively correlated bivariate pairs can be converted into a scheme for negatively correlated pairs by a change of sign. The problem of generating nondegenerate negatively correlated pairs is not so simple, however, for positive-valued and nonsymmetric variables.

10.4.2 Bivariate Exponential Random Variables

The exponential distribution plays a key role in modeling for operations research, and it is not surprising that much effort has gone into deriving bivariate exponential pairs. The special properties of the exponential distribution make this particularly easy. Several recent schemes are discussed here; for general references see N. L. Johnson and Kotz (1972) and M. E. Johnson (1987).

i. Positive Correlation

Of the schemes discussed previously for generating bivariate exponential pairs, only the antithetic scheme (Equation 10.4.3) gives negative correlation. The Marshall–Olkin bivariate exponential based on minima in Example 10.1.1 gives the full range of positive correlations, as do the pairs based on Equation 10.3.3 and elaborated in Lawrance and Lewis (1983). Again, application of the Beta–Gamma transformation to give linear random coefficient, additive combinations of Gamma variables gives a calculus similar to that for Normal random variables in Equations 10.1.21 and 10.1.22; see Lewis (1985). The correlations are, however, all positive.

ii. Gaver's Negatively Correlated Exponential Model

Generating negatively correlated bivariate exponential random pairs is difficult, particularly if the full range of attainable negative correlations $(0, -0.6649)$ is to be achieved in a fairly "smooth" bivariate exponential distribution. In particular, one would like to control the degree of negative correlation by changing a single parameter.

The following ingenious scheme of Gaver (1972) for a negatively correlated bivariate exponential pair (X, Y) has the additional advantage that it is based on modeling considerations from an operational situation, not simply on arbitrary combinations of random variables.

Suppose that a system, when put into operation, has a random number N of defective items. The times at which each of these components is used are assumed to be independent and identically distributed random variables, $Y_1, \ldots, Y_N$, where N takes values $1, 2, \ldots$. Then, the time to failure of the system is the time when an attempt is first made to use one of these defective items, or

$$Y = \min(Y_1, Y_2, \ldots, Y_N) \tag{10.4.4}$$

We let $F_Y(y)$ denote the common distribution of the Y_i variables. Now it is assumed that after failure all N defective items are found and repaired one by one (i.e., serially). If individual repair times $X_1, \ldots, X_N$ are iid with distribution $F_X(x)$, then, given $N = n$, the time to complete repair has distribution $F_X^{n*}(x)$. Here $F_X^{n*}(x)$ denotes the n-fold convolution of $F_X(x)$.

Thus, if p_k, $k = 1, 2, \ldots$, is the discrete probability function for N, the joint distribution function of $X = X_1 + \cdots + X_N$ and Y can be calculated from:

$$P(X \leqslant x; Y > y) = \sum_{k=1}^{\infty} \{1 - F_Y(y)\}^k p_k F_X^{k*}(x) \tag{10.4.5}$$

The more faults there are in the system, the shorter will be the time Y to failure and the longer will be the time to complete repair, X. Hence, one can expect the pair (X, Y) to be negatively correlated.

General results can be obtained for this model, but we specialize to the case where N has a "geometric plus one" distribution:

$$p_k = P(N = k) = (1 - p)p^{k-1} \qquad 0 \leqslant p < 1; k = 1, 2, \ldots \tag{10.4.6}$$

Then it can be shown that Y is exponentially distributed if the Y_i's are exponentially distributed.

We then need to know whether a distribution $F_X(x)$ for the X_i exists so that $X = X_1 + \cdots + X_N$ is exponential(λ). For this to hold, we must have, in general:

$$\sum_{k=1}^{\infty} \{1 - F_X(x)\}^k p^{k-1}(1 - p) = e^{-\lambda t} \tag{10.4.7}$$

Gaver (1972) showed that for this to hold, the X_i must have a positive logistic distribution

$$F_X(x) = 1 - \frac{1}{p + (1 - p)e^{\lambda x}} \qquad x \geqslant 0; 0 \leqslant p < 1; \lambda > 0 \tag{10.4.8}$$

The inverse probability transform method of Chapter 4 shows that this can be generated from a uniform random variable U as follows:

$$X = \frac{1}{\lambda} \ln\left[\frac{1}{1 - p}\left(\frac{1}{1 - U} - p\right)\right] \tag{10.4.9}$$

It can then be shown from Equation 10.4.5 that

$$\text{corr}(X, Y) = \rho(X, Y) = -\frac{1}{2}\left(1 - \frac{1}{E(N)}\right) = -\frac{p}{2} \tag{10.4.10}$$

Thus the model will generate bivariate exponential pairs with correlation approaching $-\frac{1}{2}$ as p gets close to 1. Of course, the generation of the pair becomes computationally expensive as p gets close to 1.

This scheme can also be iterated to produce even more negatively correlated Gamma pairs. This is done by making the pairs (Y_i, X_i), $i = 1, \ldots, N$, negatively correlated, but independent from pair to pair.

10.4.3   Bivariate Gamma Random Variables

Gamma-distributed variables, including the special exponential case when the shape parameter k is 1, are also important in operational analysis and have been extensively studied. We study bivariate Gamma variables briefly here, concentrating on the case where the marginal distributions of X and Y are both Gamma(k, β).

i. Positive Correlation

In Exercise 9.9 we discussed the use of the additivity property of Gamma variates to generate bivariate Gamma pairs (X, Y). Another scheme is obtained from results of Gaver and Lewis (1980) and Lewis (1985). The result is that if E_1 is Gamma(k, β), and if $Y = E_1$, then

$$X = \rho Y + E_2 \qquad 0 \leqslant \rho < 1 \tag{10.4.11}$$

is Gamma(k, β) if E_2 has Laplace–Stietjes transform:

$$L_{E_2}(s) = E(e^{-sE_2}) = \left(\frac{\rho + \beta s}{1 + \beta s}\right)^k \qquad k > 0 \tag{10.4.12}$$

Moreover, the correlation $\rho(X, Y)$ is ρ. Unfortunately E_2 is not simple to generate unless k is an integer. For $k = 1$, the result reduces to Equation 10.3.3. Recently Lawrance (1982) and McKenzie (1987) have given methods for generating E_2 for general k, so that the method is quite generally useful. Note that this bivariate pair does not have an absolutely continuous bivariate distribution.

Much "smoother" bivariate Gamma bivariate random variables can be generated using the Beta–Gamma transformation. For such details, see Lewis (1983).

ii. Lewis's Negatively Correlated Gamma Model

Generating negatively correlated bivariate Gamma pairs is difficult, but the task has received considerable attention recently. Even the antithetic pair is difficult to generate because the inverse Gamma probability transform is not easy to compute. Schmeiser and Lal (1982) have investigated this problem and given the value of the minimum correlation attainable for a given value of k, the shape parameter in the Gamma distribution.

Other schemes are given in Schmeiser and Lal (1982) and it is perhaps possible to extend Gaver's scheme for negatively correlated exponentially distributed random variables to Gamma(k, β) marginal distributions. However, we give here only a very simple scheme due to Lewis (1983). The scheme requires merely a source of independent Gamma(k, β) variables to generate a negatively correlated pair (X, Y) with Gamma(k, β) marginal distributions.

Thus, let $G_1(k)$, $G_2(k)$, $G_3(k)$, and $G_4(k)$ denote independent Gamma(k, β) random variables. Now let

$$X = G_1(k) \tag{10.4.13}$$

and

$$Y = \frac{G_2(k)}{G_1(k) + G_2(k)}\{G_3(k) + G_4(k)\} = \frac{G_2(k)}{X + G_2(k)}\{G_3(k) + G_4(k)\} \tag{10.4.14}$$

The variable X is Gamma(k, β) by construction. The ratio $G_2(k)/\{G_1(k) + G_2(k)\}$ is a Beta(k, k) random variable multiplying the independent variable $\{G_3(k) + G_4(k)\}$, which is Gamma$(2k, \beta)$. The product Y is thus, by the Beta–Gamma transform, a Gamma(k, β) random variable.

Negative correlation is induced because when the X variable in the denominator of the ratio in Equation 10.4.14 is large, Y will tend to be small. In fact, it can be shown that

$$\rho(X, Y) = \frac{-k}{1 + 2k}$$

which is, for example, $-\frac{1}{3}$ if $k = 1$. This compares to a minimum value of $\rho(X, Y) = -\frac{1}{2}$ for the Gaver scheme given in the preceding section, which is attained when p goes to one.

This scheme can be iterated (Lewis, 1983), but the minimum attainable correlation at $k = 1$ is $\rho(X, Y) = -\frac{1}{2}$. The virtue of the scheme is its extreme simplicity from a computational viewpoint. However, properties of the pair beyond the correlation (e.g., joint distributions) are hard to obtain.

Exercises

10.1. Generalize the bivariate exponential distribution given in Example 10.1.1 to the case of E_1, E_2, and E_3 independently exponentially distributed with parameters λ_1, λ_2, and λ_3. In particular, find:
   a. $\rho(X, Y) = \text{corr}(X, Y)$.
   b. $F_{X,Y}(x, y)$, the joint distribution of X, Y. In particular, is it an absolutely continuous bivariate distribution function? If not, describe the discontinuities or concentrations.

T10.2. Is the minimum of two independent geometrically distributed random variables, each with parameter p (Table 6.1.3), geometrically distributed? If so, define a bivariate geometric distribution analogous to the bivariate exponential distribution in Exercise 10.1 and repeat that exercise for the bivariate geometric pair.

10.3. Prove the relationship of Equation 10.1.23 for the correlation coefficient of two linear transformations of independent Normal random variables.

T10.4. Using Equations 10.1.21, 10.1.22, and 10.1.23, find the locus of points a that give a bivariate Normal distribution with unit variances and correlation ρ. Without assuming $\sigma_X = \sigma_Y = 1$, find a, b, c, and d that give a bivariate Normal pair with parameters σ_X^2, σ_Y^2, and ρ.

10.5. Let $\bar{X} = (X_1 + \cdots + X_m)/m$ and $\bar{Y} = (Y_1 + \cdots + Y_m)/m$. Show that if the bivariate pairs (X_i, Y_i) are independent over i and have the same means, variances, and correlation, then the following are true.
   a. $\text{cov}(\bar{X}, \bar{Y}) = \dfrac{1}{m}\text{cov}(X, Y)$
   b. $\text{corr}(\bar{X}, \bar{Y}) = \text{corr}(X, Y)$

T10.6. Verify the result that the Bernoulli random variables defined in Section 10.1.5 are independent if and only if $p_{11} = p_{1+}p_{+1} = (p_{10} + p_{11})(p_{01} + p_{p11})$. Hint: Start by writing out conditions for independence.

T10.7. Assume that X and Y are bivariate Bernoulli random variables with $P\{(X, Y) = (k, l)\} = p_{kl}$, for $k = 0, 1; l = 0, 1$, and $P(X = 1) = p_{1+}$ and $P(Y = 1) =$

p_{+1}. Verify the following:

a. $\text{corr}(X, Y) = \dfrac{p_{11} - p_{1+}p_{+1}}{\{p_{1+}(1 - p_{1+})p_{+1}(1 - p_{+1})\}^{1/2}}$

b. $\text{corr}(X, Y) = \dfrac{p_{11}p_{22} - p_{21}p_{22}}{\{p_{1+}(1 - p_{1+})p_{+1}(1 - p_{+1})\}^{1/2}}$

c. $r = \dfrac{nn_{11} - n_{1+}n_{+1}}{(n_{1+}n_{0+}n_{+1}n_{+0})^{1/2}}$

satisfies the definition of Equation 10.2.2 when applied to bivariate Bernoulli data as arranged in a 2×2 table.

^{C}**10.8.** Design and perform a simulation to compare the power of the chi-square test for association in a 2×2 table based on Equation 10.1.28 to the odds ratio test based on Equation 10.1.31. One factor in your experiment should be the true amount of association, say $\alpha = 1, 2, 5$; another factor could be the marginal probability $p_{+1} = 0.1, 0.5$; a final factor should be the total sample size $n = 10, 25, 50$. Comment on both the validity ($\alpha = 1$) and comparative power of the tests. By "validity," we mean here that we want to check whether the null distributions of the statistics are in fact those of chi-square random variables with one degree of freedom.

^{C}**10.9.** Design and perform a simulation to compare the coverage probabilities of the following confidence intervals for α:

a. $[(\ln \hat{\alpha}) \pm z_\gamma \{(1/n_{11}) + (1/n_{10}) + (1/n_{01}) + (1/n_{00})\}^{1/2}]$

where z_γ is the Normal γ quantile

b. $\left[\dfrac{a_L(n_{+0} - n_{1+} + a_L)}{\{(n_{1+} - a_L)(n_{+1} - a_L)\}}, \dfrac{a_U(n_{+0} - n_1 + a_U)}{\{(n_{1+} - a_U)(n_{+1} - a_U)\}} \right]$

where a_L and a_U are found iteratively through the formula:

$$a_0 = n_{11}$$
$$a_i = n_{11} \pm 0.5 \pm z_\gamma [(1/a_{i-1}) + \{1/(n_{1+} - a_{i-1})\}$$
$$+ \{1/(n_{+1} - a_{i-1})\} + \{1/(n_{+0} - n_{1+} + a_{i-1})\}]^{-1/2}$$

where i is incremented until convergence. We reach a_U by using the plus whenever $\pm$ appears, and we converge to a_L by using the minus. Again, z_γ is the Normal γ quantile.

The second method, is called Cornfield's method (Schlesselman, 1982), while the first is based on Equation 10.1.31. Use the idea of coverage functions from Section 9.7.1 for this study.

10.10. Using the Normal theory results that $\text{var}(\hat{\gamma}_1) = 6/n$ and $\text{var}(\hat{\gamma}_2) = 24/n$ and Figures 10.2.5 and 10.2.7, discuss the validity of these approximations in the exponential case.

^{C}**10.11.** Use results in Table 6.1.4 and the known moments of the exponential distribution to obtain $\text{var}(\hat{\gamma}_1)$ and $\text{var}(\hat{\gamma}_2)$ up to order $1/n^2$. Compute these for several values of n (say $n = 10, 100$) and compare them with the simulated values in Figures 10.2.5 and 10.2.7 to validate the simulation.

^{C}**10.12.** Show that the random linear combination of two independent exponential random variables given in Equation 10.3.3 yields an exponential random variable X. Generate a sample of size $m = 1000$ from this bivariate exponential pair and produce a scatter plot of the pairs. With this scatter plot in mind, calculate the joint distribution of X and Y. Show that the correlation between X and Y is ρ.

10.13. From the bivariate geometric random variable discussed in Exercise 10.2, calculate the conditional distributions of X and Y. Also generate $m = 1000$ pairs from this bivariate distribution and study the scatter plot generated by these points.

[T] **10.14.** Is there an analogue for geometric random variables of the random combination of two exponential random variables used in Exercise 10.12 to generate bivariate exponential random variables? *Hint:* Replace multiplication of Y in Equation 10.3.3 with "thinning" of Y ones. That is, each of the Y ones is set to zero with probability $1 - \rho$ (McKenzie, 1985).

10.15. Using the fact that if E_1 is Poisson(γ_1) and E_2 is Poisson(γ_2) then E_1 plus E_2 is Poisson$(\gamma_1 + \gamma_2)$, explore the possibility of generating bivariate Poisson pairs (X, Y) with identical Poisson(γ) marginal distributions, by combining three independent Poisson variables. What is the bivariate probability function for the pair? What is the correlation between X and Y, and what range of correlations can be achieved?

[T] **10.16.** Show that the l-Laplace random variable whose characteristic function is given at Equation 10.4.1 can be generated as the difference of two Gamma(l, γ) random variables. Give schemes for generating bivariate l-Laplace random variables.

[C] **10.17.** For the negatively correlated exponential pair (X, Y) described in subsection ii of Section 10.4.2, prove the results of Equations 10.4.5, 10.4.7, 10.4.8, 10.4.9, and 10.4.10. Generate $m = 1000$ pairs from this distribution and examine the joint distribution of X and Y via a scatter plot. Additional investigation should be made, if possible, by examining Equation 10.4.5 for this special case.

References

Bishop, Y. M. M., Fienberg, S. E., and Holland, P. W. (1975). *Discrete Multivariate Analysis: Theory and Practice.* MIT Press: Cambridge, MA.

Chambers, J. M., Cleveland, W. S., Kleiner, B., and Tukey, P. A. (1983). *Graphical Methods for Data Analysis.* Wadsworth: Belmont, CA.

Conover, W. J. (1980). *Practical Nonparametric Statistics,* 2nd ed. Wiley: New York.

Cox, D. R. (1970). *The Analysis of Binary Data.* Chapman & Hall: London.

Cramér, H. (1946). *Mathematical Methods of Statistics.* Princeton University Press: Princeton, NJ.

DeVroye, L. (1986). *Non-Uniform Random Variate Generation.* Springer: Berlin.

Fienberg, S. E. (1977). *The Analysis of Cross-Classified Categorical Data.* MIT Press: Cambridge, MA.

Gaver, D. P. (1972). Point Processes in Reliability. In *Stochastic Point Processes: Statistical Analysis, Theory and Applications,* P. A W. Lewis, Ed. Wiley: New York.

Gaver, D. P., and Lewis, P. A. W. (1980). First-Order Autoregressive Gamma Sequences and Point Processes. *Advances in Applied Probability, 12,* 727–745.

Gibbons, J. D. (1971). *Nonparametric Statistical Inference.* McGraw-Hill: New York.

IMSL Library (1987). IMSL Stationary: User Manual, Version 1.0. IMSL, Inc.: Houston.

Johnson, M. E. (1987). *Multivariate Statistical Simulation.* Wiley: New York.

Johnson, N. L., and Kotz, S. (1972). *Distributions in Statistics: Continuous Multivariate Distributions.* Wiley: New York.

Kendall, M. G., and Stuart, A. (1974). *The Advanced Theory of Statistics,* Vol. 2, 3rd ed. Griffin: London.

Kendall, M. G., and Stuart, A. (1977). *The Advanced Theory of Statistics,* Vol. 1, 4th ed. Griffin: London.

Kennedy, W. J., and Gentle, J. E. (1980). *Statistical Computing.* Dekker: New York.

Knuth, D. E. (1981). *The Art of Computer Programming*, Vol. 2, *Seminumerical Algorithms.* Addison-Wesley: Reading, MA.

Lawrance, A. J. (1982). The Innovation Distribution of a Gamma-Distributed Autoregressive Process. *Scandinavian Journal of Statistics, 9*, 234–236.

Lawrance, A. J., and Lewis, P. A. W. (1975). Properties of the Bivariate Delayed Poisson Process. *Journal of Applied Probability, 12*, 257–268.

Lawrance, A. J., and Lewis, P. A. W. (1983). Simple Dependent Pairs of Exponential and Uniform Random Variables. *Operations Research, 31*, 1179–1197.

Lewis, P. A. W. (1983). Generating Negatively Correlated Gamma Variates Using the Beta–Gamma Transformation. In *Proceedings of the 1983 Winter Simulation Conference.* S. Roberts, J. Banks, and B. Schmeiser, Eds. IEEE Press: New York, pp. 175–176.

Lewis, P. A. W. (1985). Some Simple Models for Continuous Variate Time Series. *Water Resources Bulletin, 21*(4), 635–644.

Lewis, P. A. W., McKenzie, E., and Hugus, D. (1988). Gamma Processes. *Stochastic Models, 4*, to appear.

Marshall, A. W., and Olkin, I. (1967a). A Multivariate Exponential Distribution. *Journal of the American Statistical Association, 62*, 30–44.

Marshall, A. W., and Olkin, I. (1967b). A Generalized Bivariate Exponential Distribution. *Journal of Applied Probability, 4*, 291–302.

McKenzie, E. (1985). Some Simple Models for Discrete Variate Time Series. *Water Resources Bulletin, 21*(4), 645–650.

McKenzie, E. (1987). Innovation Distributions for Gamma and Negative Binomial Autoregressions. *Scandinavian Journal of Statistics, 14*, 645–650.

Mehta, C. (1986). Algorithm 643 FEXACT: A FORTRAN Subroutine for Fisher's Exact Test on Unordered $r \times c$ Contingency Tables. *ACM Transactions on Mathematical Software, 12*, 154–161.

Mood, A. M., Graybill, F. A., and Boes, D. C. (1974). *Introduction to the Theory of Statistics*, 3rd ed. McGraw-Hill: New York.

Pearson, E. S., and Hartley, H. O. (1966). *Biometrika Tables for Statisticians*, Vol. 1. The University Press: Cambridge.

Schlesselman, J. J. (1982). *Case-Control Studies, Design, Conduct, Analysis.* Oxford University Press: New York.

Schmeiser, B. W., and Lal, R. (1982). Bivariate Gamma Random Vectors. *Operations Research, 30*, 355–374.

Chapter 11

Variance Reduction

11.0   Introduction

We have pointed out that in using straightforward sampling to obtain an estimate of a response or quantification of a stochastic system, the variance of the estimate generally decreases as a/m, where a is a finite constant, often the variance of a single response, and m is the number of *independent* replications of the quantity being examined. Thus, to halve the (absolute) precision of the estimate, as measured by its standard deviation, or the relative dispersion of the estimate, as measured by its standard deviation divided by its expectation, we have to generate a sample four times as large (i.e., $4m$). Consequently, to reduce the standard deviation of the estimate by a factor of $1/c$, we need, in general, a sample that is c^2 as big as before. This "frustration factor" may make it difficult to obtain results precise enough to answer questions being posed by the simulator, such as, Which of two queueing disciplines has a waiting time with the smaller 0.95 quantile?

As a first step in decreasing the standard deviation of an estimate in a given amount of computing time, one can look to more efficient computation of the functions involved or to faster generation of uniform (Chapter 5) or nonuniform (Volume II) random variables. If the simulation has been initially written in a high-level language, say FORTRAN, or an even higher-level simulation language like SIMSCRIPT, one might even redo the whole simulation program in C or in assembly language. These approaches can produce large time savings and always should be looked at first. The programming problems that arise beyond the first cleanup, however, can be formidable and errors may well be introduced. An alternative is to simplify the model!

The techniques for increasing precision (reducing variance) discussed in this chapter are based on the ideas that simulation is a *controlled statistical experiment* and that we can reduce the variance obtained for a given number m of replications by using probabilistic ideas. Thus, in the examples in Chapter 3, the variance of an estimate (e.g., the mean waiting time W_n of the nth customer in a queue or the trimmed t statistic $T_{n,1}$ or the maximum M_n, of n random variables) may be reduced using variance reduction methods from say, $10/m$ to $1/m$. This would be equivalent to the effect of a 10-fold increase in sample size.

These *variance reductions* may be obtained by one or more of the following:

- Generating realizations that are not independent, usually by using *common random numbers* or transformations of common inputs
- Using information about similar output and input random variables (responses from similar systems) that have known probabilistic properties
- Using partial information, known from theory, about the output and input random variables
- Using more sophisticated sampling methods than the straightforward random sampling discussed in Chapter 4

Reducing variance in paired comparisons by using differences and *common random numbers* was introduced in Chapter 8. The present chapter can be considered an extension of that simple and broadly applicable technique to more sophisticated and complex methodology for variance reduction.

The following limitations of variance reduction techniques, however, should also be noted:

1. The variance (or the standard deviation) is used as the measure of precision, but variance reduction techniques may induce non-Normality and very wild observations (outliers). *Thus, the smaller variance of the new estimate may not tell the whole story of what is being done by a variance reduction scheme.*

2. The requirements of variance reduction may have an impact on the methods of generating nonuniform random numbers or stochastic processes; in particular, the probability integral transform method may be preferable to the more efficient methods discussed in Volume II.

3. We are considering here only single outputs: that is, one random variable such as the waiting time W_n of the nth customer in a queue. In some cases when, for example, W_n is to be compared for different queue disciplines, creating correlated outputs with (individually) smaller variances may make standard two-sample statistical techniques based on independent samples (e.g., analysis of variance) nonapplicable and may increase the variance of the differences. In other words, in multifactor situations, the techniques discussed here may be even more difficult to apply. In the two-sample case, this problem is handled advantageously by using differences, as discussed in Chapter 8. For multiple factors and comparisons, the gain from using correlated outputs is not always clear; again, see Chapter 8 for a discussion.

4. Implementation of the techniques may require more *computing time*. This increase should not be greater than the time it would have taken to reduce the variance the same amount, by simply generating more replications of the original simulation.

5. Finally, the decision to use a particular variance reduction technique, or a combination of variance reduction techniques, should reflect consideration of the extra time required to devise and program the variance reduction methods. Once you go beyond the basic ideas, variance reduction is an art that can require great skill, experience, and time to implement. In fact, Handscomb (1969) calls a technique

"variance reducing" if it reduces the variance proportionally more than it increases the work involved. For simple to moderate problems with limited numbers of changing factors, crude simulation will often be the most efficient way to proceed, perhaps with longer computer runs than were originally envisioned.

J. R. Wilson (1984) discusses formal measures of variance reduction which take into account both the additional computing time and the analyst's time; we do not discuss them further here.

In this chapter, a technique is called variance reducing if it produces, for a given estimate, a smaller variance than would be produced, for given number of replications m, by the straightforward technique described in Chapters 1 through 4 on crude simulation. In general, in any large simulation involving investigation of attributes of stochastic systems or systems that have several parameters (concomitant variables, queue disciplines) that can vary deterministically, variance reduction techniques are essential if sufficiently precise answers are to be obtained.

There are roughly seven main categories of variance reduction techniques. Common random numbers and controlled dependence are used in:

Antithetic variates
Control variables
Control variables with regression
Systematic sampling

whereas conditioning arguments and known properties of subpopulations are used in:

Conditional sampling
Importance sampling
Stratification

The categories are not airtight, and the techniques can and should be used together. Above all, their intelligent use depends on delineating as far as possible, before the simulation, what questions are to be asked and answered. Thus a variance reduction technique that works well for estimating the mean value of a statistic may not work well for estimating an extreme quantile of that statistic.

For an early exposition of variance reduction techniques, see Kahn (1956). The categories in that paper do not exactly correspond to those given above and, in fact, the reader should be aware that many of the techniques discussed assume different names in the literature. A short summary of variance reduction techniques with applications in operations research is given in Gaver and Thompson (1973, Chap. 12). For a survey of actual applications of variance reduction, see Kleijnen (1974, Chap. III), and for a recent, rather technical but excellent survey see J. R. Wilson (1984), who also presents a taxonomy for the categories of variance reduction techniques similar to the one given above. An attempt to systematize and provide an analytical framework for variance reduction is given by Nelson and Schmeiser (1986); for a summary, see J. R. Wilson (1984).

This chapter proceeds from the variance reduction techniques that are conceptually the simplest—antithetic variates and then control variables—to those that are conceptually quite difficult. Thus, the reader with less background might consider reading only through to Section 11.2, "Control Variables." Control variables are likely to be the most useful of the variance reduction techniques because they can be applied to simulations of *all* aspects of the distribution of the response, and because controls are simpler to find than the characteristics required for other variance reduction techniques. Moreover, simulating a control variable, which must have a *known* or approximately known expectation, is invaluable for the all-important problem of validating the computer program that generates the simulation.

Finally, the examples in the chapter tend to be oriented toward the use of variance reduction in statistical problems and in problems in physics. More detailed applications of the techniques developed here to problems in systems simulation will be given in Volume II.

11.1   Antithetic Variates: Induced Negative Correlation

11.1.1   Simple Antithetic Variates

This simple variance reduction technique is based on the idea that the variance of two random variables, say X_1 and X_2, when they are averaged is, from Equation 10.1.14,

$$\operatorname{var}\left(\frac{X_1 + X_2}{2}\right) = \frac{1}{4}\{\operatorname{var}(X_1) + \operatorname{var}(X_2) + 2\operatorname{cov}(X_1, X_2)\} \qquad (11.1.1)$$

and, if $\operatorname{var}(X_1) = \operatorname{var}(X_2) = \sigma^2$, then

$$\operatorname{var}\left(\frac{X_1 + X_2}{2}\right) = \frac{1}{2}\sigma^2(1 + \rho) \qquad (11.1.2)$$

Thus if ρ, the correlation between X_1 and X_2, is very negative, the average of the negatively correlated pair has a variance that is much smaller than the variance of the average of an independent pair.

Note that this is different from the variance reduction scheme discussed in Chapter 8. There the *difference* of two positively correlated responses was taken to reduce the variance of their difference, since it was their difference, not their actual values, which was of interest.

To illustrate the idea of antithetic variates, consider the simple situation of a *single* random variable X, where X has a known distribution $F_X(x)$, and assume we wish to estimate some attribute of a function $g(X)$ of this random variable X. Here $g(X)$ could be $\ln(X)$, or $(X - E(X))^2$ if the variance of X is of interest; it is assumed that the complete distribution of $g(X)$ is difficult to handle and that moments or percentiles are difficult to compute. In the real examples involving multiple random variables considered later, it will certainly be true that analytic results are difficult to obtain.

Instead of the m realizations of X, say

$$X_1 = F_X^{-1}(U_1),\ X_2 = F_X^{-1}(U_2),\ \ldots,\ X_m = F_X^{-1}(U_m) \tag{11.1.3}$$

which would be used in the crude simulation scheme given in Chapter 4, we generate $m/2$ *antithetic pairs* (assuming m is even):

$$
\begin{aligned}
X_1 &= F_X^{-1}(U_1), & X_1^a &= F_X^{-1}(1 - U_1) \\
X_2 &= F_X^{-1}(U_2), & X_2^a &= F_X^{-1}(1 - U_2) \\
&\ \ \vdots & &\ \ \vdots \\
X_{m/2} &= F_X^{-1}(U_{m/2}), & X_{m/2}^a &= F_X^{-1}(1 - U_{m/2})
\end{aligned}
$$

(The savings in time for using only $m/2$ uniform pseudo-random numbers may be significant, although it does not appear explicitly in what follows.)

Note that since $1 - U_1$ is uniformly distributed, X_1^a has the marginal distribution $F_X(x)$, just as X_1 has distribution $F_X(x)$. There is an assumption here that $F_X^{-1}(\cdot)$ is relatively simple to compute, as would be the case when X is exponential(λ), where $F_X^{-1}(\alpha) = -(1/\lambda)\ln(1 - \alpha)$. Again if X is a symmetric random variable, then $X_1^a = -X_1$.

Now, if U_i is large, $1 - U_i$ is small, and since $F_X^{-1}(\cdot)$ is a monotonic transformation, the pair X_i and X_i^a, will be *negatively correlated* for each $i = 1, 2, \ldots, m/2$. Also, $g(X_i)$ and $g(X_i^a)$ *may* be negatively correlated if, for example, $g(\cdot)$ is a monotonic transformation. Denote this correlation by ρ_A.

Thus we form the average

$$\bar{g}(X) = \frac{g(X_1) + g(X_2) + \cdots + g(X_{m/2})}{m/2}$$

where, for simplicity, X is used to denote the vector of observations $X_1, \ldots, X_{m/2}$. Here $\bar{g}(X)$ is the sum of $m/2$ independent random variables.

Again we form the average

$$\bar{g}(X^a) = \frac{g(X_1^a) + g(X_2^a) + \cdots + g(X_{m/2}^a)}{m/2}$$

where X^a denotes the vector of observations $X_1^a, \ldots, X_{m/2}^a$ and this is again the average of $m/2$ independent random variables. Finally the *antithetic* estimate of the mean of $g(X)$ is

$$\bar{g}^A(X) = \frac{\bar{g}(X) + \bar{g}(X^a)}{2} = \frac{1}{m/2}\sum_{j=1}^{m/2}\frac{g(X_j) + g(X_j^a)}{2} \tag{11.1.4}$$

$$= \frac{1}{m/2}\sum_{j=1}^{m/2} g_j^A(X) \tag{11.1.5}$$

where

$$g_j^A(X) = \frac{\{g(X_j) + g(X_j^a)\}}{2} \qquad (11.1.6)$$

Note that $\bar{g}^A(X)$ is still the average of the function $g(\cdot)$ of m variables having marginal distribution $F_X(x)$, but the variables are not independent. They are, in fact, pairwise dependent.

Now by Equation 10.1.18

$$\text{corr}\{\bar{g}(X), \bar{g}(X^a)\} = \text{corr}\{g(X_j), g(X_j^a)\} = \rho_A \qquad (11.1.7)$$

and

$$\text{var}\{\bar{g}(X)\} = \text{var}\{\bar{g}(X^a)\} = 2\sigma_m^2$$

where σ_m^2 is the variance of the average of m *independent* realizations of $g(X)$. Thus we have, from Equation 11.1.2, that the variance of the antithetic estimate is

$$\text{var}\{\bar{g}^A(X)\} = \sigma_m^2(1 + \rho_A) = \frac{\text{var}\{g(X)\}}{m}(1 + \rho_A) \qquad (11.1.8)$$

where $\text{var}\{g(X)\}/m$ would be the variance obtained through crude simulation using m independent $g(X_j)$'s. Again note that instead of generating m *independent* variables, we have generated m variables that are pairwise dependent.

This equation shows that a considerable variance reduction can be obtained if ρ_A is very negative. For example, if $\rho_A = -\frac{3}{4}$ then, by substituting this value in Equation 11.1.8, we find that we have achieved the same result we would have achieved by quadrupling the number m of replications; $\text{var}\{\bar{g}^A(X)\} = \frac{1}{4}\text{var}\{\hat{g}(X)\}$, where $\hat{g}(X)$ is the straightforward average over m independent $g(X_j)$'s. Note, however, that the standard deviation of $\hat{g}(X)$ is reduced only by a factor of $\frac{1}{2}$.

To *estimate* the variance of the antithetic estimate $\text{var}\{\bar{g}^A(X)\}$, we take the sample of size $m/2$ independent quantities $g_j^A(X)$, $j = 1, 2, \ldots, m/2$, defined at Equation 11.1.6, and since $\bar{g}^A(X)$ is, from Equation 11.1.5, the sample average of the $g_j^A(X)$'s we have, as at Equation 4.1.2,

$$\hat{\text{var}}\{\bar{g}^A(X)\} = \frac{1}{m/2} S_{\bar{g}^A(X)}^2 = \frac{1}{m/2} \sum_{j=1}^{m/2} \frac{\{g_j^A(X) - \bar{g}^A(X)\}^2}{\{(m/2) - 1\}} \qquad (11.1.9)$$

Confidence intervals can be formed, if required, by methods given in Chapter 6; in particular, if tests on the sample of $g_j^A(X)$'s show them to be approximately Normally distributed, then confidence intervals based on the t statistic are appropriate.

There are a number of problems involved in this method of antithetic variates.

1. It generally requires that we use the inverse transform method to generate the random variables, and the increased (computer) time needed to do this as

opposed to using more sophisticated variate generating techniques (Volume II) may mitigate against the overall variance reduction obtained. Actually with decomposition methods (Volume II), it should be simple to generate antithetic pairs by using U and $1 - U$ to select suitably ordered parts of the decomposed density, but we know of no package that does this. Also, in some cases such as exponential random variables, the inverse probability integral transform method is quite fast. For *symmetric random variables* such as the Normal, however, any generation method can be used, since $X^a = -X$. In other words, no special considerations are required to generate X^a. Symmetry of random variables is extensively exploited in variance reduction methodology.

2. We are assuming that $g\{F_X^{-1}(U)\}$ and $g\{F_X^{-1}(1 - U)\}$ will be negatively correlated. However, assume that $g(\cdot)$ is monotonic and takes negative and positive numbers. Then $g^2\{F_X^{-1}(U)\}$ and $g^2\{F_X^{-1}(1 - U)\}$ will be positively correlated. Thus the method may give a better estimate than straightforward independent sampling for the mean of $g(X)$, but a worse estimate for the variance of $g(X)$. Thus the antithetic technique may be *very particular to the property of $g(X)$ we are examining.*

3. Note that it is not always possible to achieve correlations close to -1 for every random variable. For example, for exponentially distributed random variables with mean 1, the antithetic pair is $X = -\ln(1 - U)$ and $X^a = -\ln(U)$ and we obtain the maximum possible negative correlation $1 - \pi^2/6 = -0.64$. In general for antithetic pairs, we have

$$X = F_X^{-1}(U)$$
$$X^a = F_X^{-1}(1 - U)$$

and thus

$$X^a = F_X^{-1}\{1 - F_X(X)\}$$

so that fixing X completely determines X^a. However, this does not imply that the correlation between X and X^a has the value of -1! (See Moran, 1967, for details.)

To go on from this rather simple illustration of antithetic variables, we note that when the random variable to be examined is a complex function of *not one, but many random variables*, it is not always easy to see how to obtain an antithetic realization of the random variable. This reduces the utility of the technique. Also, if we are interested in estimates that are not sample averages (e.g., quantile estimates) then estimating the variance of the antithetic estimate may be difficult.

The classic situation for which it is easy to see how to obtain antithetic pairs is the waiting time W_n in the GI|GI|1 queue (Section 3.2).

EXAMPLE   **11.1.1 Antithetic Variates for the GI|GI|1 Queue**
Let $U_i, U_i', i = 1, \ldots, n$ be $2n$ independent uniform variables and let

$$S_{i-1} = F_S^{-1}(U_i) \qquad \text{and} \quad X_i = F_X^{-1}(U_i') \qquad\qquad i = 1, \ldots, n$$
$$\phantom{S_{i-1}} = (i - 1)\text{st service time} \qquad = i\text{th interarrival time}$$

and, as in Section 3.2:

$$W_n = [W_{n-1} + S_{n-1} - X_n]^+ \qquad n > 0$$
$$W_0 = 0 \tag{11.1.10}$$

where the function $[x]^+$ is x when x is zero or greater and zero otherwise. Recall, too, that the waiting time W_n is defined to be the time from the nth customer's arrival to the *start* of his service.

Now define the *antithetic realization* of the GI|GI|1 queue by letting

$$S_{i-1}^a = F_S^{-1}(1 - U_i) \qquad \text{and} \qquad X_i^a = F_X^{-1}(1 - U_i') \qquad i = 1, \ldots, n$$

and

$$W_n^a = [W_{n-1}^a + S_{n-1}^a - X_n^a]^+ \qquad n > 0$$

with

$$W_0^a = 0 \tag{11.1.11}$$

Note that in a realization of this queue in which, for $i = 1, \ldots, n$, the S_i's are all large (slow service) and the X_i's all short (speedy arrivals), then W_n, the waiting time of the nth customer, will be large. In the antithetic realization, however, the S_i^a's are then short and the X_i^a's are long, so that W_n^a is short (probably zero). Thus we should have negatively correlated pairs W_n^a and W_n. The negative correlation achieved will depend on factors like the mean service and interarrival times, and other distributional properties, as we will illustrate below.

Thus, to apply the technique to find an estimate of, for example, the expected waiting time, $E(W_n)$, of the nth customer, generate m independent sets of uniform random numbers, each of size $2n$. For each of the sets, use n of the uniforms from each of m sets to determine the X_i's and n uniforms to determine the S_i's, and then generate from each a single realization of W_n and its antithetic W_n^a. Repeating this m times will give $W_n(j)$ and $W_n^a(j), j = 1, \ldots, m$. The average in each pair is:

$$W_n^A(j) = \frac{W_n(j) + W_n^a(j)}{2} \qquad j = 1, \ldots, m$$

To estimate $E(W_n)$, proceed as before with the m independent replications $W_n^A(j)$ to get an overall average:

$$\overline{W}_n^A = \frac{1}{m} \sum_{j=1}^{m} W_n^A(j)$$

The variance of $\overline{W}_n^A$ can be estimated from the sample variance of the m independent $W_n^A(j)$'s. Note, though, that $W_n^A(j)$ does not have the same distribution as $W_n(j)$. Clearly we must have $E\{W_n(j)\} = E\{W_n^A(j)\}$, but var$\{W_n(j)\}$ must be greater than

$2 \operatorname{var}\{W_n^A(j)\}$ if the variance reduction is to work. Thus to proceed further in investigating the distribution of W_n—for example, to simulate $\operatorname{var}(W_n)$—one must start from the beginning or possibly accept a degradation in precision in the simulation of $\operatorname{var}(W_n)$.

In particular, to estimate $E(W_n^2)$, as well as $E(W_n)$, form

$$\frac{\{W_n(j)\}^2 + \{W_n^a(j)\}^2}{2} \qquad j = 1, 2, \ldots, m \tag{11.1.12}$$

and average over these m quantities. *Again note that even if $W_n(j)$ and $W_n^a(j)$ are negatively correlated, $\{W_n(j)\}^2$ and $\{W_n^a(j)\}^2$ may not be negatively correlated.* The negative correlation will often be preserved if the simulated quantities are positive random variables, which is the case for the $W_n(j)$'s; in other areas of application this positivity may not hold, and in such cases the squared variables may be *positively* correlated.

EXAMPLE   **11.1.2 Antithetics in a Specific GI|GI|1 Queue**
A simulation of a GI|GI|1 queue in which the service times and interarrival times are assumed to have Weibull distributions (see Table 6.1.2) with shape parameters k_S and k_X, respectively, is presented here to fix ideas on the use of antithetic variates. Analytical results for the waiting times in a GI|GI|1 queue, for example, are known when $k_S = k_X = 1$, the MI|MI|1 case, but not in the general case. Since the inverse distribution function for the Weibull distribution is, from Table 6.1.2,

$$F^{-1}(\alpha) = \frac{\{-\ln(1 - \alpha)\}^{1/k}}{\beta}$$

where β is the scale parameter in the Weibull distribution, it is possible to generate antithetic pairs X and X^a quite simply.

The other parameter in the simulation is the traffic intensity

$$t = E(S)/E(X) = \frac{\Gamma\left(\dfrac{1}{k_S} + 1\right)}{\beta_S} \times \frac{\beta_X}{\Gamma\left(\dfrac{1}{k_X} + 1\right)}$$

where $\Gamma(\cdot)$ is the (complete) Gamma function. If the shape parameters k_S and k_X are chosen, and $E(X)$ is set to 1, then choosing the traffic intensity t sets β_X and β_S, the scale parameters of the Weibull distributions of the interarrival and service times, respectively.

Results of a simulation of the waiting times of the $n = 1000$th customer when $t = 0.90$ are shown below (Figure 11.1.2). The queue is started empty, with the server idle, and the number of replications is $m = 300$. Figure 11.1.1 shows the estimated joint distribution of W_{1000} and W_{1000}^a, generated with the subroutine BIHSPC in the software supplement, while Figures 11.1.2 and 11.1.3 give histograms of W_{1000}

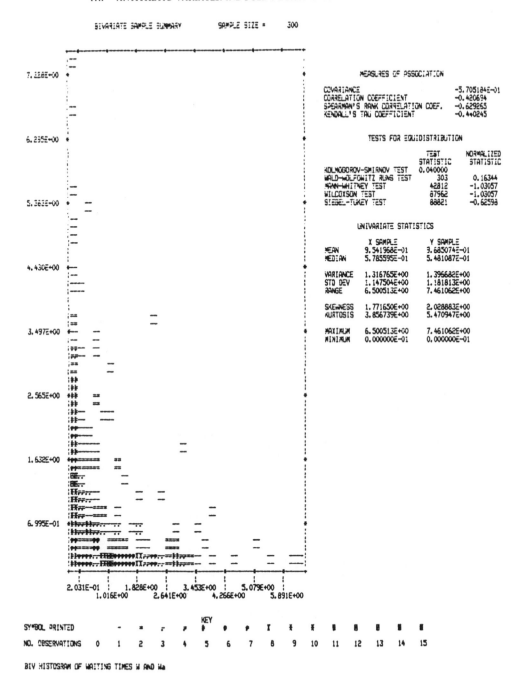

BIV HISTOGRAM OF WAITING TIMES W AND Wa

Figure 11.1.1 Bivariate histogram generated from $m = 300$ independent pairs of waiting times $\{W_{1000}(j), W^a_{1000}(j)\}$ in a GI|GI|1 queue. The horizontal axis (X-sample) is the W variable and the vertical axis (Y-sample) is the W^a variable. The service and waiting times both have Weibull distributions with parameters k_s and $k_x = 3$. Thus, these distributions are *tightly dispersed*, in the sense that their coefficients of variation σ/μ are both 0.363, which is small compared to the value of 1 for an exponential distribution. The traffic intensity is $t = 0.90$ and $E(X) = 1$.

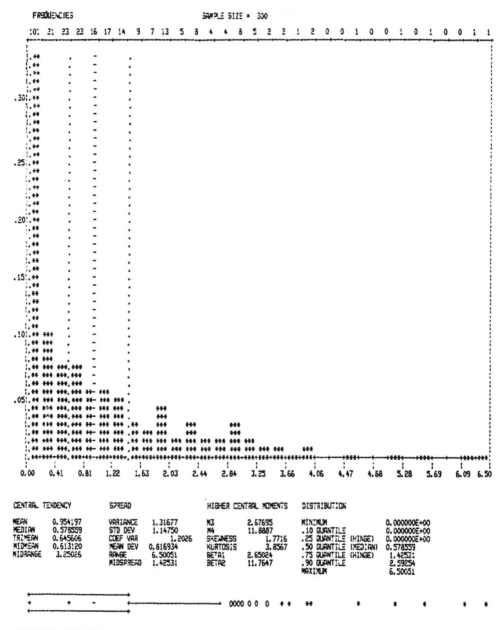

Figure 11.1.2 Marginal histogram generated from $m = 300$ independent waiting times $\{W_{1000}(j)\}$ in a GI|GI|1 queue. The details of this simulation are the same as those in Figure 11.1.1.

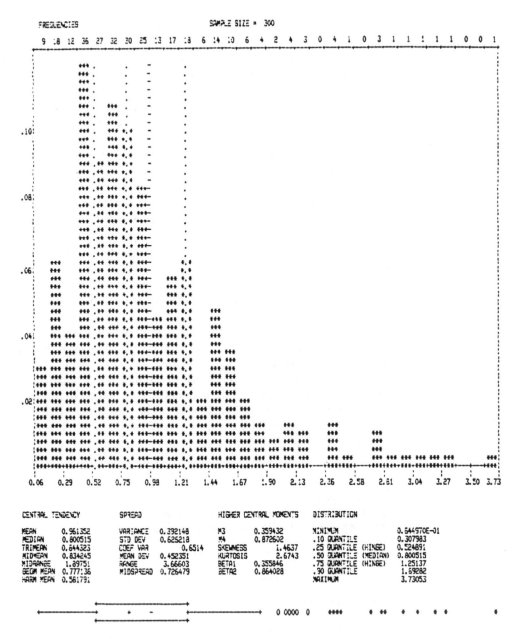

Figure 11.1.3 Marginal histogram from $m = 300$ independent waiting time averages $W_{1000}^A(j) = \{W_{1000}(j) + W_{1000}^a(j)\}/2$. Note that the distribution of W_{1000}^A is not the same as the distribution of W_{1000}. In fact, the probability density function has a mode at about 0.60 and there are no zero values.

and $W_{1000}^A = (W_{1000} + W_{1000}^a)/2$, respectively. (Recall that W_{1000} and W_{1000}^a have identical distributions.)

The correlation between W_{1000} and W_{1000}^a is estimated in Figure 11.1.1 to be -0.421, which should reduce the variance of W_{1000}^A to $1 - 0.421 = 0.579$ of the variance of W_{1000}. The variance of W_{1000} is estimated in Figure 11.1.2 to be 1.317, and the variance of W_{1000}^A is estimated in Figure 11.1.3 to be 0.392, where in both cases the sample size is 300. However, since $\overline{W}_{1000}^A$ is actually estimated from a sample of size $2m = 600$, namely, 300 $W_{1000}(j)$'s and 300 $W_{1000}^a(j)$'s, the comparison should be to $\overline{W}_{1000}$ estimated from a sample of size 600. Thus the comparison of $(1.317)/(2 \times 300) = 0.00220$ with $0.392/300 = 0.00131$ yields a ratio of 0.56 compared to the predicted value of 0.579 from the estimated correlation.

Note, too, that the waiting times have a mixed distribution with a positive probability that the waiting time will be zero. Thus, although the correlation comes into the formula for variance reduction, association between the variables W_{1000}^a and W_{1000} may be greater than indicated by the correlation coefficient. This is suggested by the value of -0.629 attained by the Spearman rank correlation coefficient in Figure 11.1.1. (See Conover, 1980, for a discussion of the Spearman rank correlation test; it is a permutation test closely related to the Kendall's tau statistic discussed in Example 4.4.1.)

It must be emphasized that the results above are "best cases" and that the gains achieved by variance reduction can be fragile, depending quite critically on some of the parameters of the problem being investigated. Thus in Table 11.1.1, we show the results of a simulation that was performed to estimate the correlation between W_{1000}^a and W_{1000} for various values of the parameters t, k_S, and k_X. (It is not possible to calculate any of these correlations analytically.) The standard deviation of these correlation estimates is approximately $m^{-1/2} = (300)^{-1/2} = 0.059$. The indication from this *simulation* is that the antithetic idea works best for high traffic intensities and quite regular queues ($k_S = k_X = 3.0$). For more skewed queueing schemes (i.e.,

Table 11.1.1 Simulation estimates of the correlation between the waiting time W_{1000} and its antithetic W_{1000}^a in a GI|GI|1 queue with Weibull-distributed service and interarrival times.

Here k_S is the Weibull shape parameter of the service time distribution and k_X is the Weibull shape parameter of the interarrival time. The traffic intensity is t, and $E(X) = 1$. Simulation is based on $m = 300$ replications. Thus the standard deviation of the correlation estimates is approximately 0.059.

	Shape parameters		
t	$k_S = 3.0$ $k_X = 3.0$	$k_S = 1.0$ $k_X = 1.0$	$k_S = 0.5$ $k_X = 0.5$
0.9	-0.353	-0.277	-0.101
0.75	-0.282	-0.235	-0.056
0.5	-0.086	-0.235	-0.046

$k_S = k_X = 0.5$), the correlation is apparently not large enough to warrant the use of antithetic variables. However, our simulation estimates of the correlations are very imprecise and the sample size should be increased from $m = 300$ to obtain more precise results and firmer conclusions.

Another possible improvement in the simulation of the waiting times in the GI|GI|1 queue would be to simulate delay time, $D_n = W_n + S_n$, and from these, obtain $E(W_n)$, since $E(S_n)$ is known. The correlation between D_n and D_n^a could be greater than that between W_n and W_n^a.

11.1.2 Exchanges and Systematic Sampling

Negatively correlated realizations of the GI|GI|1 queue can also be created by a technique called *exchanges* (Page, 1965). Thus, let U_i and U_i' be the iid uniform variates generated in Example 11.1.1, from which we obtained

$$S_{i-1} = F_S^{-1}(U_i) \qquad \text{and} \qquad X_i = F_X^{-1}(U_i')$$

Now let

$$S_i^e = F_S^{-1}(U_i') \qquad \text{and} \qquad X_i^e = F_X^{-1}(U_i) \tag{11.1.13}$$

As compared to the scheme in Example 11.1.1, we have simply *exchanged* the uniform variates that are used to generate the service and waiting times. This creates a realization W_n^e that will be negatively correlated with W_n, as defined at Equation 11.1.10, and perhaps with W_n^a. It is left as an exercise to explain why W_n^e and W_n^a will be negatively correlated.

The exchange estimate of the waiting time is then defined to be

$$\overline{W_n}^E = \frac{\overline{W_n} + \overline{W_n}^e}{2} \tag{11.1.14}$$

where $\overline{W_n}$ is the average of the W_n's defined at Equation 11.1.10 and $\overline{W_n}^e$ is the analogous average over the exchange realizations. For the GI|GI|1 queue considered in Example 11.1.2, Figure 11.1.4 shows the bivariate histogram of W_n and W_n^e.

Table 11.1.2 shows the correlation between W_n and W_n^e for six combinations of traffic intensities and Weibull shape parameters. If anything, the exchange idea creates greater negative correlation than the original antithetic idea of replacing U in all cases by $1 - U$. The effectiveness of the scheme is shown in Table 11.1.2 to depend on the traffic intensity parameter t. The negative correlation for queues with highly positively skewed service and interarrival time distributions ($k_X = k_S = 0.5$) is greater than the corresponding correlations achieved between W_{1000} and W_{1000}^a (Table 11.1.1).

Exchanges is just one of several ideas to extend antithetic variates beyond *pairs* of negatively correlated random variables (see Gaver and Thompson, 1973, and Kleijnen, 1974, for further details; see also Cheng, 1981, 1982, 1984). Antithetics are, in turn, closely related to *systematic sampling*. To see this, note that in Equation

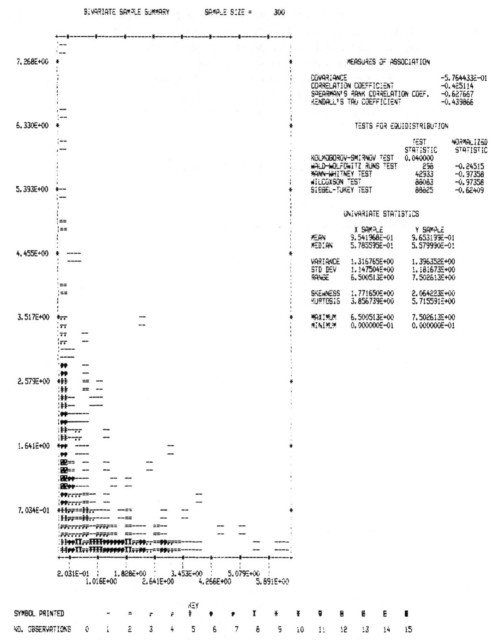

Figure 11.1.4 Bivariate histogram generated from $m = 300$ independent pairs of waiting times $\{W_{1000}(j), W_{1000}^e(j)\}$ in a GI|GI|1 queue. The service and waiting times both have Weibull distributions with parameters k_s and $k_x = 3$. Thus, these distributions are *tightly dispersed*, in the sense that their coefficients of variation σ/μ are both 0.363, which is small compared to the value of 1 for an exponential distribution. The traffic intensity is $t = 0.90$ and $E(X) = 1$.

Table 11.1.2 Simulated correlation between W_n and W_n^e for various combinations of traffic intensity and shape parameters in a GI|GI|1 queue

The traffic intensity is t and the simulation is based on $m = 300$ replications. Thus the standard error of the correlation estimates is approximately 0.059.

	Shape parameters	
t	$k_S = 3.0$ $k_X = 3.0$	$k_S = 0.5$ $k_X = 0.5$
0.9	−0.425	−0.499
0.75	−0.271	−0.440
0.5	−0.082	−0.352

11.1.3, either U_1 or $1 - U_1$ is in the range $(0, \frac{1}{2})$. Say that it is U_1. Then $X_1 = F_X^{-1}(U_1)$ is generated randomly in an interval $(-\infty, F_X^{-1}(\frac{1}{2}))$. But given the value of X_1 (or U_1), the value of X_1^a in $(F_X^{-1}(\frac{1}{2}), \infty)$ is determined. The same holds for $X_2, \ldots, X_{m/2}$. Systematic sampling carries this idea further, as follows.

Divide the unit interval into n equal subintervals. Generate U_1 and add $i/(n)$ to U_1 $n - 1$ times, modulo 1, to create $U_1^{(1)}, \ldots, U_1^{(n)}$. Repeat for each of $U_2, \ldots, U_m$. Then each of the samples $U_1^{(l)}, U_2^{(l)}, \ldots, U_m^{(l)}, l = 1, 2, \ldots, n$ is used to generate

$$\bar{g}(X^{(l)}) = \sum_{j=1}^{m} \frac{g(X_j^{(l)})}{m}$$

and these averages are averaged over l. For details, and generalizations, see J. R. Wilson (1984) or Fishman and Huang (1983). In particular, the division of the unit interval into equal subintervals is not necessary or optimal.

Another extended antithetic or systematic sampling scheme called *rotational sampling* has been used effectively by Fishman and Huang (1983), who show that this scheme gives, in some circumstances, variances that, in contrast to the usual case, are decreasing faster than $1/m$, where m is the number of replications in the simulation.

11.1.3   Summary of Antithetic Sampling

In summary, the use of antithetic variates can be an effective way to reduce variance in simulation studies. Points to note are the following.

• Quite a lot of skill and experience is required to devise an antithetic variable.

• The antithetic variate must be programmed and computed. In particular, in a straightforward simulation, standard, packaged schemes for generating random variates may be used. In using antithetic variates, a program must be available to compute the inverse distribution functions needed, unless one is dealing with symmetric random variables. Consequently, this variance reduction scheme should be contrasted with the alternate scheme of doubling or tripling the number of realizations used in the original crude simulation.

• A disadvantage is that a good antithetic for one quantification of the variate under study (e.g., its mean value) may not help in studying another quantification of the variate (e.g., its quantiles).

• An advantage is that no analytical results are required for the antithetic variate. It must, of course, have the same mean as the variate under study and be negatively correlated with it. Actually, no application of antithetic variates is known in which the variate and the antithetic variate did not have the same distribution.

11.2   Control Variables

The second method of variance reduction listed in the introduction, the use of control variables, uses common random numbers in a much more fundamental way than antithetic variates. It is also usually a broader technique than antithetic variates. This is because of its simplicity and because, with minor changes, it can be used to estimate *any* quantification of a random variable. Thus, a given control variable could be used to estimate the mean, variance, quantiles, and so on, of the random variable. *However, unlike the case of antithetics, application of the technique requires analytic knowledge of the expected value of the control variable.*

The use of control variables in variance reduction in simulation goes back to a very old idea, or rather a question and answer, in statistics.

Suppose we have a bivariate random variable (C, Y) and a sample of size m from this population, namely $(C_1, Y_1), (C_2, Y_2), \ldots, (C_m, Y_m)$, and we wish to estimate some attribute of the marginal distribution of Y: for example, $E(Y)$. Moreover, suppose we know something about C (e.g., $E(C)$). Can we use this information and the C sample to improve the estimate we would obtain about the attribute of Y based solely on the Y sample?

The answer is clearly but vaguely Yes if the random variables are dependent and the problem is dealt with in statistics by the body of methodology called *correlation theory*. The application of this theory in variance reduction is as follows.

11.2.1   Simple Control Variables

To see how to achieve more precise estimation in simulations by using control variables, assume we want to estimate $E(Y)$, and assume that we know $E(C) = \mu_C$. Let $\sigma_C^2 = \text{var}(C)$, $\sigma_Y^2 = \text{var}(Y)$, and $\rho_{CY} = \text{corr}(C, Y)$. These higher order and joint moments may be known, but more often they must be estimated from the sample. This modification will be considered in Section 11.2.3, especially for the case of the estimate to be controlled being a sample average; *however $E(C)$ must be known,* exactly or approximately. If it is known only approximately, this approximate value must have been obtained independently of the simulation in which it is used. The approximation will induce bias in the control variable estimator. For the purposes of exposition we assume, unless indicated, that $E(C)$ is known exactly.

One simple way to bring in the C sample information is to form the sample average $\bar{C}$ of the C sample and the sample average $\bar{Y}$ of the Y sample and combine them to give an estimate that is a simple linear function of $\bar{Y}$, $\bar{C}$, and μ_C:

$$\bar{Y}' = \bar{Y} - (\bar{C} - \mu_C) = \frac{1}{m}\sum_{i=1}^{m}(Y_i - C_i + \mu_C) = \frac{1}{m}\sum_{i=1}^{m}Y_i' \tag{11.2.1}$$

which, like $\bar{Y}$, is unbiased, since

$$E(\bar{Y}') = E(\bar{Y}) - E(\bar{C} - \mu_C) = E(Y) - 0 = E(Y)$$

and *may* have smaller variance than $\bar{Y}$. Note, too, that $\bar{Y}'$ is the sample average of the iid random variables Y_i' defined in Equation 11.2.1. The negative sign is used in front of $(\bar{C} - \mu_C)$ in Equation 11.2.1 on the assumption that C and Y, and therefore $\bar{C}$ and $\bar{Y}$, are *positively correlated*. Then in cases where $\bar{Y}$ is large (i.e., well above its unknown mean μ_Y), $\bar{C}$ will tend to be large, $\bar{C} - \mu_C$ will be positive, and $\bar{Y}$ will be reduced by $\bar{C} - \mu_C$ to give $\bar{Y}'$. The analogous adjustment occurs when $\bar{Y}$ (on a different replication) is small (i.e., well below its unknown mean); the control variable $(\bar{C} - \mu_C)$ pulls it up. Thus we would expect the variance of $\bar{Y}'$ to be smaller than the variance of $\bar{Y}$ if $\bar{C}$ and $\bar{Y}$ are sufficiently positively correlated.

We now make this variance reduction precise. First note that $E(\bar{Y}') = E(\bar{Y}) = E(Y)$. Then we have, using Equation 10.1.22 for the variance of the sum of two random variables,

$$\sigma_{\bar{Y}'}^2 = \text{var}(\bar{Y}') = \text{var}(\bar{Y}) + \text{var}(\bar{C} - \mu_C) - 2\,\text{cov}\{(\bar{C} - \mu_C), \bar{Y}\}$$

For simplicity, consider the case of $\sigma_C^2 = \sigma_Y^2 = \sigma^2$ and let $\rho_{CY} = \text{corr}(C, Y)$.

Then, using Equations 10.1.14 and 10.1.17, we have:

$$\sigma_{\bar{Y}'}^2 = \frac{2\sigma^2(1 - \rho_{CY})}{m} = \text{var}(\bar{Y})\{2(1 - \rho_{CY})\}$$

Therefore, if $2(1 - \rho_{CY}) < 1$, *or equivalently* $\rho_{CY} > \frac{1}{2}$, *the use of a control variable reduces the variance over the case where just* $\bar{Y}$ *is used to estimate* μ_Y. Note the assumption behind this result, that $\sigma_C^2 = \sigma_Y^2$. It is left as an exercise to explore the case of $\sigma_C^2 \neq \sigma_Y^2$.

EXAMPLE **11.2.1 The Trimmed *t* Statistic**

In the example of Section 3.4, we discussed the idea of a trimmed t statistic in which the usual t statistic T_n, from a sample $X_1, \ldots, X_n$ of size n, is *trimmed*. This simple form of trimming is done by removing the largest and smallest values in the X sample, namely $X_{(1)}$ and $X_{(n)}$, and computing $T_{n,1}$ from the trimmed sample, as at Equation 3.4.9.

Now the properties of T_n in Normal samples are known analytically and tabulated (see, e.g., Conover, 1980, Table A25, or any statistics textbook). However, the properties of $T_{n,1}$ are unobtainable analytically and must be estimated by simulation. Moreover, if we compute T_n and $T_{n,1}$ *from the same sample* (i.e., *with common random numbers*), we expect them to be a highly correlated bivariate pair. Thus we may want to use $C = T_n$ as a control in estimating, for example, $E(T_{n,1})$ in Normal samples, since we know that $E(C) = E(T_n) = 0$. (Estimation of other

Table 11.2.1 Simulation of $m = 20$ samples of the t statistic T_n and the trimmed t statistic $T_{n,1}$ from common Normal(0, 1) populations of size $n = 30$; estimated corr$(C, Y) = \rho_{CY} = 0.995$

j	$Y_j = T_{n,1}(j)$	$C_j = T_n(j)$	rank(C_j)	$Y_j' = Y_j - C_j$
OBSERVATION				
1	2.52	2.25	20.00	.27
2	2.42	1.95	19.00	.46
3	1.58	1.42	18.00	.16
4	1.44	1.15	16.00	.29
5	1.38	1.29	17.00	.09
6	1.23	1.06	15.00	.17
7	.83	.55	12.00	.28
8	.69	.65	14.00	.04
9	.44	.25	10.00	.19
10	.44	.56	13.00	−.12
11	.41	.24	11.00	−.01
12 ▶	.31	.24	9.00	.07
13	.11	.13	8.00	−.02
14	−.14	.06	7.00	−.20
15	−.23	−.27	6.00	.04
16	−.40	−.40	5.00	.00
17	−.67	−.76	4.00	.09
18	−.95	−.95	3.00	.00
19	−1.02	−.95	2.00	−.07
20	−1.43	−1.32	1.00	−.11

$\bar{Y} = 0.45$	$\bar{C} = 0.36$	$\bar{Y}' = 0.085$
$S_Y = 1.10$	$S_C = 0.98$	$S_{Y'} = 0.155$
$S_{\bar{Y}} = 0.25$	$S_{\bar{C}} = 0.22$	$S_{\bar{Y}'} = 0.035$

quantifications of $T_{n,1}$, such as quantiles, is of more importance but will be discussed later.)

Table 11.2.1 is a boxplotted table that gives a sample of size $m = 20$ values of T_n and $T_{n,1}$, where each pair is computed from the same iid Normal sample of size $n = 30$. For illustration the $T_{n,1}$ values are given in descending order (i.e., as order statistics). Note, again, that $E(C) = 0$, so that no subtraction of the mean appears in the equation for Y_j' in the table.

The boxplots dividing the columns show that the variables $T_{n,1}$ and T_n are similarly distributed but that Y_j' is much more tightly dispersed than either $T_{n,1}$ or T_n, indicating that the control is very effective in reducing the variance of the new estimate of $E(T_{n,1})$. Coincidentally, another surprising fact exhibited in the table is that the statistic $T_{n,1}$ is apparently more dispersed than T_n. The maximum and minimum of $T_{n,1}$ are 2.52 and -1.43, respectively, while for T_n they are 2.25 and -1.32.

It is also clear from the column marked "rank(C_j)" that ordering the Y_j's (trimmed t statistic $T_{n,1}$) almost orders the C_j's, so that the T_n's and $T_{n,1}$'s are highly correlated. In fact, the Pearson product moment correlation coefficient is estimated to be $\rho_{CY} = 0.995$. Thus the variance of Y should be reduced by a factor of $2(1 - \rho_{CY}) = 2(0.005) = 0.01$ and its standard deviation by $(0.01)^{1/2} = 0.1$.

The empirical values given at the bottom of Table 11.2.1 are $S_{\bar{Y}} = 0.25$ and $S_{\bar{Y}'} = 0.035$, a ratio of $0.035/0.25 = 0.14$, which is in close agreement with the value 0.1. Another way to look at the variance reduction is to note that the use of a control gives the same effect as increasing the sample size by a factor of $(0.14)^{-1/2} \doteq 51$.

Note that since $\bar{Y}'$ is a sample average, we have used the fact that its standard deviation can be estimated from the sample in the usual way to give $S_{\bar{Y}'} = 0.035$. Estimating the precision of an estimate obtained in a simulation is, as always, important.

11.2.2   Regression-Adjusted Control

We subtracted $(\bar{C} - \mu_C)$ from $\bar{Y}$ in Equation 11.2.1 on the basis of C and Y being positively correlated. If we thought that the correlation between C and Y was negative, we would add $(\bar{C} - \mu_C)$ to $\bar{Y}$. More generally, we could use a *regression-adjusted estimator*

$$\bar{Y}^* = \bar{Y} + \beta(\bar{C} - \mu_C) \tag{11.2.2}$$

which is still an unbiased estimator of $E(Y)$ if β is a constant. Thus Y' given at Equation 11.2.1 is the special case of $\beta = -1$. This refinement serves to take the guesswork out of the variance reduction procedure and also to maximize the variance reduction that can be obtained with the control variable. In particular, we hope to obviate the fact demonstrated above that if ρ_{CY} is not large enough, we may get a *variance inflation*. We also hope to eliminate the fact that the simple control can give no variance reduction if σ_Y is much greater than σ_C and variance inflation if σ_Y is much less than σ_C. (See Exercise 11.5 for details.)

We proceed to find the value of β that minimizes the variance of $\bar{Y}^*$. We have

$$\text{var}(\bar{Y}^*) = \text{var}(\bar{Y}) + \beta^2 \text{var}(\bar{C}) + 2\beta \text{cov}(\bar{Y}, \bar{C}) \tag{11.2.3}$$

By differentiating Equation 11.2.3 with respect to β and setting the resulting equation equal to zero, we find that $\text{var}(\bar{Y}^*)$ is minimum if

$$-2\beta \text{var}(\bar{C}) = 2 \text{cov}(\bar{Y}, \bar{C})$$

or

$$\beta = \frac{-\text{cov}(\bar{Y}, \bar{C})}{\text{var}(\bar{C})} = \frac{-\text{cov}(Y, C)}{\text{var}(C)} = -\rho_{CY} \frac{\sigma_Y}{\sigma_C} = \beta_{C \cdot Y} \tag{11.2.4}$$

We have used the fact, from Equation 10.1.17, that $\text{cov}(\bar{Y}, \bar{C}) = (1/m)\text{cov}(Y, C)$ and $\text{var}(\bar{C}) = (1/m)\text{var}(C)$. The notation $\beta_{C \cdot Y}$ is standard in correlation theory.

If $\sigma_Y = \sigma_C = \sigma$, then $\beta = -\rho_{CY}$. However it is not necessary to assume that the standard deviations are equal, since the optimal value of β compensates for differences in σ_Y and σ_C. Thus inserting the value for β given at Equation 11.2.4 into the expression 11.2.3, we find the minimum variance to be

$$\text{var}(\bar{Y}^*) = \frac{1}{m} \sigma_Y^2 \{1 - \rho_{CY}^2\} \tag{11.2.5}$$

Thus

$$\frac{\text{var}(\bar{Y}^*)}{\text{var}(\bar{Y})} = 1 - \rho_{CY}^2 \tag{11.2.6}$$

The reciprocal of the square root of this function is plotted in Figure 11.2.1. It is the ratio of the sample size that would be required to achieve a given standard deviation without using this variance reduction technique to the sample size required to achieve the given standard deviation with variance reduction.

From this result we see that when σ_Y, σ_C, and ρ_{CY} are known, by using $\bar{Y}^*$, the regression-adjusted control variate estimate, we *always get variance reduction*, unless $\rho_{CY} = 0$. Also, the variance of $\bar{Y}^*$ is always less than the variance of the straightforward control estimate $\bar{Y}'$ which, if ρ_{CY} is less than $\frac{1}{2}$, may be larger than the variance of $\bar{Y}$. In the special case of $\sigma_Y = \sigma_C$, we can compute

$$\text{var}(\bar{Y}^*) = \frac{1}{m} \sigma_Y^2 (1 - \rho_{CY}^2) = \frac{1}{m} \sigma_Y^2 (1 - \rho_{CY})(1 + \rho_{CY})$$

$$\leqslant \frac{2\sigma_Y^2}{m} (1 - \rho_{CY}) = \text{var}(\bar{Y}')$$

Note, too, that $(1 - \rho_{CY}^2)^{1/2}$ decreases rather slowly as $|\rho|$ increases; in fact, to divide the standard deviation of $\bar{Y}$ by one-half we need the correlation between C and Y to be about 0.87.

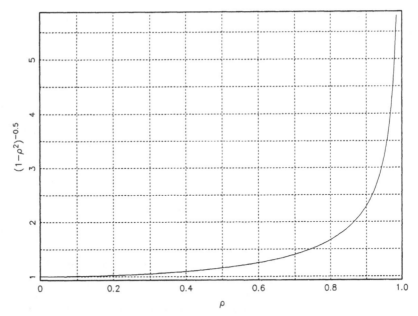

Figure 11.2.1 Plot of the reciprocal of the standard deviation reduction, s.d.$(\bar{Y}^*)$/s.d.$(\bar{Y}) = (1 - \rho_{CY}^2)^{1/2}$, obtained with a regression control. It can be considered to be the ratio of the sample size that would be required to achieve a given standard deviation without using this variance reduction technique to the sample size required to achieve the given standard deviation with variance reduction.

The application of these results in simulation is that, whereas the question posed in the beginning of this section may be unrealistic in real-life situations, in simulation one can very often find, when simulating a function $Y = g(X_1, X_2, \ldots, X_n)$, another function of the same random variables, say $C = h(X_1, \ldots, X_n)$, which will be highly correlated with $g(\cdot)$ and for which some information (e.g., $E\{h(\cdot)\}$), can be obtained analytically either exactly or approximately.

Note that in deriving Equation 11.2.4 we have minimized the variance of $\bar{Y}^*$. If μ_C is not known precisely, and in particular has a bias that goes to zero as $n \to \infty$, we might want to minimize the mean square error of $\bar{Y}^*$ rather than the variance of $\bar{Y}^*$.

11.2.3   Estimation of the Control Parameters and the Variance

If ρ_{CY} and σ_C and σ_Y are not given, they must be estimated from the sample $(C_1, Y_1), \ldots, (C_m, Y_m)$. This introduces small sample bias into $\bar{Y}^*$ and may reduce the savings predicted by Equation 11.2.5. [Jackknifing, from Chapter 9, can also be used to reduce the bias, which is $O(1/m)$. This is only useful if m is small; another scheme is discussed in Exercise 11.10 for larger samples.]

More particularly, estimation of the variance of $\bar{Y}^*$ is no longer straightforward. If σ_Y, σ_C, and ρ_{CY} are *known*, then $\bar{Y}^*$ is a sample average and the variance of $\bar{Y}^*$ is estimated by the sample variance:

$$\hat{\sigma}(\bar{Y}^*) = \left[\frac{1}{(m-1)m} \sum_{i=1}^{m} (Y_i^* - \bar{Y}^*)^2 \right]^{1/2} \tag{11.2.7}$$

where $Y_i^* = Y_i + \beta(C_i - \mu_C)$. Furthermore, if the Y_i^*'s are Normally distributed, then a $100(1 - \alpha)\%$ confidence interval for μ_Y is given by the t statistic (Equation 6.1.40):

$$(\bar{Y}^* - t_{\alpha/2,(m-1)} \times \hat{\sigma}(\bar{Y}^*), \quad \bar{Y}^* + t_{\alpha/2,(m-1)} \times \hat{\sigma}(\bar{Y}^*)) \tag{11.2.8}$$

If the data are not Normally distributed, the Y_i^*'s can be sectioned, as in Chapter 9, to get a confidence interval based on fewer than $m - 1$ degrees of freedom.

Of course, σ_Y, σ_C, and ρ_{CY} will generally not be known, so they must be estimated from the Y_i and C_i samples as at Equations 6.1.10 and 10.2.2, respectively. Then the Y_i^*'s are no longer iid random variables, since they are coupled by the estimated parameters.

Equation 11.2.7 can still be used to estimate $\hat{\sigma}(\bar{Y}^*)$, but it will be biased; the t confidence interval given at Equation 11.2.8 is certainly no longer valid. Instead, batching (sectioning) as discussed in Chapter 9 should be used. Thus, we divide the m values (Y_i, C_i) into k sections and estimate a $\bar{Y}_l^*$, $l = 1, \ldots, m/k$ from each section, with σ_Y, σ_C, and ρ_{CY} estimated separately in each section. Then approximate confidence intervals, as at Equation 11.2.8, can be formed with $(m/k) - 1$ degrees of freedom. All the problems of sectioning discussed in Chapter 9, including bias, pertain. Note, however, that the *point estimator* will still be the same as the original value corresponding to $k = 1$.

Lavenberg and Welch (1981) have studied the problem of adjusting degrees of freedom in Equation 11.2.8 when batching (sectioning) is not used and the data and controls have a multivariate Normal distribution. In that case, the degrees of freedom in Equation 11.2.8 are decreased by 1 from $m - 1$ to $m - 2$. As we would expect, this implies that $\bar{Y}^*$ is not as precise an estimate of $E(Y)$ as when σ_Y, σ_C, and ρ_{CY} are known. The estimate will also be biased, since the parameters are estimated from the data. A different splitting scheme is given in Exercise 11.10, but although there is no bias in the estimate with this scheme, there is some loss of precision.

EXAMPLE   **11.2.2 Simulation of Quantiles: The Trimmed t Statistic Again**
Using control variables to produce variance reduction in simulations of quantiles of a statistic introduces new problems, which we consider here.

  • A quantile is no longer a sample average, so that estimation of the correlation between the Y sample quantile estimate and the control is not simple. Nor is it simple to estimate the variance of the quantile estimate.

  • More fundamentally, consider the ith-order statistic $Y_{(i)}$ of the Y sample. Should we control it with $\bar{C}$, or with a quantile of the C sample? This involves the following questions. We saw in Chapter 10 that $\rho_{YC} = \rho_{\bar{Y}\bar{C}}$, but if Y and C are highly correlated, will $Y_{(i)}$ and $C_{(i)}$ be highly correlated? Which order statistic of C will be the best control for $Y_{(i)}$? It does not necessarily have to be $C_{(i)}$.

• It may well be that we only know true values of moments of C and, thus, cannot use $C_{(i)}$ as a control for $Y_{(i)}$. Will $\bar{C}$, for example, be highly correlated with $Y_{(i)}$ if Y and C are highly correlated?

We pursue these questions throughout this section, starting with the current example, which concerns simulation of quantiles of $T_{n,1}$.

The ordinary t statistic T_n will almost certainly be highly correlated with the trimmed t statistic $T_{n,1}$ computed from the whole sample, as was shown in Example 11.2.1. Moreover, complete distributional results are known for T_n. To extend Example 11.2.1, consider simulation of the quantiles of $T_{n,1}$. We will focus first on estimating the value of the 0.95 quantile of $T_{n,1}$, using as a control the estimated 0.95 quantile of T_n, where the T_n's and $T_{n,1}$'s are generated using common random numbers. That is, T_n and $T_{n,1}$ are generated in each replication *from the same sample* $X_1, \ldots, X_n$. Here, of course, the control is, generally speaking, T_n, while $T_{n,1}$ corresponds to Y. Since high correlations between certain quantizations of T_n and $T_{n,1}$ are attainable, raw simulation of $T_{n,1}$ without control would be criminal. However, we must assume that the 0.95 quantile estimates will also be highly correlated.

Following Chapter 6, Equation 6.3.2, we estimate the 0.95 quantiles of T_{30} and $T_{30,1}$ by the nineteenth-order statistics of the samples, since the sample size is $m = 20$. These order statistics are found from Table 11.2.1 to have values 1.95 and 2.42, respectively. Thus, the *simple* control estimate (i.e., using $\beta = -1$) is

$$Y' = 2.42 - (1.95 - 2.09) = 2.56$$

The value 2.09 is an approximation to the expected value of the nineteenth-order statistic in a sample of 20 of the T_{30} variables; we have used the expected value of the nineteenth-order statistic in a sample of size 20 of $N(0,1)$ random variables, since T_n converges rapidly to a Normal$(0,1)$ random variable as $n \to \infty$. The expected values of normal order statistics are called *normal scores* and are tabulated, for instance, in Pearson and Hartley (1966, Table 28, p. 190). This is a case where the mean of the control variable is not known exactly but where the approximation would become more and more exact as n increases.

Since the controlled quantile estimate is *not a sample average*, we cannot estimate its variance in the straightforward way given in Chapter 4. For small sample sizes m, one could either jackknife, as described in Chapter 9, or generate a much larger sample of size m and section the sample. The sectioning would allow estimation of the correlations and standard deviations of the two estimates so that a regression-adjusted control could be used. When compared to using the whole sample, there is the drawback that one might induce (small-sample) bias in the estimate, since quantile estimates are biased (see Equation 6.3.12). Another point to note is that the high correlation achieved between the mean of T_{30} and the mean of $T_{30,1}$, as in Table 11.2.1, may not be achievable for extreme quantiles.

To demonstrate this idea of sectioning, 10 replications of samples of size $m = 20$ of T_{30} and $T_{30,1}$ were generated: that is, Table 11.2.1 was replicated 10 times. In Table 11.2.2, we show the average and standard deviation of each of the 20

Table 11.2.2   Sample properties of order statistics of $T_{30}(j)$ and $T_{30,1}(j)$, $j = 1, m = 20$ computed from 10 replications

j	av$\{T_{30}(j)\}$	s.d.$\{T_{30}(j)\}$	av$\{T_{30,1}(j)\}$	s.d.$\{T_{30,1}(j)\}$	corr$\{T_{30}(j), T_{30,1}(j)\}$
1	−2.11	0.53	−2.42	0.63	0.96
2	−1.50	0.23	−1.64	0.23	0.94
3	−1.24	0.31	−1.35	0.26	0.97
4	−1.04	0.36	−1.14	0.35	0.98
5	−0.80	0.36	−0.97	0.36	0.95
6	−0.67	0.30	−0.74	0.38	0.97
7	−0.44	0.26	−0.55	0.29	0.97
8	−0.28	0.20	−0.38	0.19	0.81
9	−0.14	0.13	−0.18	0.14	0.79
10	−0.01	0.14	−0.04	0.15	0.81
11	0.07	0.12	−0.09	0.16	0.79
12	0.24	0.20	0.25	0.14	0.90
13	0.30	0.21	0.36	0.20	0.92
14	0.39	0.23	0.45	0.20	0.94
15	0.53	0.26	0.58	0.24	0.95
16	0.66	0.22	0.76	0.25	0.87
17	0.92	0.26	1.02	0.27	0.95
18	1.20	0.22	1.25	0.25	0.92
19	1.51	0.41	1.64	0.42	0.99
20	2.13	0.53	2.42	0.62	0.90

order statistics of both the T_{30} and $T_{30,1}$ samples. In addition, the correlation between the order statistics is shown. If one concentrates on the nineteenth-order statistic to estimate the 0.95 quantiles of T_{30} and $T_{30,1}$, one can use the estimated standard deviations and correlations, s.d.$\{T_{30}(19)\} = 0.41$, s.d.$\{T_{30,1}(19)\} = 0.42$ and $r = 0.99$, respectively, to get $\tilde{\beta} = -0.99(0.42/0.41) = -1.01$ and form 10 regression-controlled estimates of the 0.95 quantile of $t_{20,1}$: for example,

$$t^*_{30,1}(0.95; k) = \tilde{t}_{30,1}(0.95; k) + \tilde{\beta}\{\tilde{t}_{30}(0.95; k) - 2.09\} \qquad k = 1, \ldots, 10$$

The generally high estimated correlation between the order statistics $T_{30}(j)$ and $T_{30,1}(j)$ in Table 11.2.2 indicates an affirmative answer to the question in the third problem mentioned above. The correlations actually increase toward both the high- and low-order statistics. The value at $j = 11$ is low, 0.79, and reflects two things:

1. This simulation of the correlations is a small one; with the 10 replications used in the example, the standard deviation of the estimated correlations $\tilde{\rho}$ is, from Equation 10.2.3, approximately $9^{-1/2} = 0.333$.

2. The sampling distribution of correlation estimates is highly skewed toward zero when the true value of ρ is near one. For details see Equations 10.2.4, 10.2.5, and 10.2.6. Equation 10.2.5 shows that if the sample is Normally distributed and the true value of ρ is 0.9, then the skewness $\gamma_1(r)$ of the sampling distribution of r is -1.7.

The average of these k estimates, $t_{30,1}^*(0.95; k)$, $k = 1, \ldots, 10$, is 2.23, and the standard deviation of this average is approximately $0.063/(10)^{1/2} = 0.020$, which compares with the estimated standard deviation of $0.42/(10)^{1/2} = 0.13$ from the crude estimate, the average of the $\tilde{t}_{30,1}(0.95; k)$. Thus a large variance reduction has been achieved. Note, however, that the standard deviation estimate of $t_{30,1}^*(0.95; k)$ is, as discussed above, not from an iid sample.

EXAMPLE   **11.2.3 Waiting Times in a GI|GI|1 Queue**
Consider, again, the case of simulating the waiting time W_n of the nth customer in a GI|GI|1 queue—that is, a first-come, first-served queue with a single server, service time distribution $F_S(x)$, and interarrival time distribution $F_X(x)$. If we replicate W_n m times, then for each replication, we could generate an MI|MI|1 queue using the same uniform random numbers as are used in the GI|GI|1 queue and in which the mean service and interarrival times are the same as in the GI|GI|1 queue. In other words, the traffic intensities in the two queues are the same. Call the waiting time for the MI|MI|1 queue WE_n. Then we estimate $E(W_n)$ for the GI|GI|1 queue with a regression-adjusted control as follows:

$$\overline{W_n}{}^* = \overline{W}_n + \hat{\beta}\{\overline{WE}_n - E(WE_n)\}$$

where $\overline{W}_n$ is the average of m independent replications of W_n for the GI|GI|1 queue and $\overline{WE}_n$ is the average for the MI|MI|1 queue, and $\hat{\beta} = -(\hat{\rho}\hat{\sigma}_{W_n}/\hat{\sigma}_{WE_n})$; $\hat{\rho}$, $\hat{\sigma}_{W_n}$, and $\hat{\sigma}_{WE_n}$ are the usual estimates of these quantities. Again, $E(WE_n)$ is known and equals $E(X)t/(1 - t)$, with $t = E(S)/E(X)$ the traffic intensity; since σ_{WE_n} is also known, it could be used in place of the estimate. If the service and interarrival times in the GI|GI|1 queue are not too different from those in the MI|MI|1 queue, correlations as high or higher than 0.9 are possible. Note that we are assuming that n is large enough so that $E(WE_n)$ has almost the stationary value for the MI|MI|1 queue.

EXAMPLE   **11.2.4 Sums of Weibull Variables**
This problem comes up in reliability studies. Assume we want to know the quantiles of the sum of j Weibull(λ, k) variates. A Weibull(λ, k) variate is a power law transformation of an exponential variate E with parameter λ (i.e., $E^{1/k}$). Its properties are given in Table 6.1.2. Thus we want quantiles of the sum

$$S = E_1^{1/k} + \cdots + E_j^{1/k}$$

This can be set up as a numerical integration, but the procedure is messy. One can simulate to obtain an approximate answer and improve the precision of the approximation by using as control the quantiles of the sum of the exponentials $S^+ = E_1 + \cdots + E_j$, which has a known Gamma(λ, j) distribution. Although there is no closed form for quantiles of the Gamma(λ, j) distribution, the quantiles are tabulated. Note that in these examples we have used a *known relationship* between the exponential and Weibull distributions; there is no need to compute the inverse of the distribution of the sum of j Weibull-distributed random variables. This distribution is a complicated function.

There are several points to be noted about these examples and about the regression-adjusted control variable method of variance reduction. Unlike antithetics, a control variable can be used for *any attribute* of the random variable Y that is being simulated (e.g., mean, variance, percentiles, quantiles), although different functions of the control variable C (e.g., C^2, C^3) could be more appropriate if the mean of these functions of C can also be obtained. In many cases only $E(C)$ is known, and not, say, $E(C^2)$ or $E(C^3)$, so we are limited to $\bar{C}$ to control *any attribute* of Y. The knowledge of the approximate mean of the quantity, which is used as the control, is of course, essential to the method.

11.2.4   Nonlinear Control

The straightforward estimator of Equation 11.2.1 is called simply a *control*, while Equation 11.2.2 is called a *regression-adjusted control*. However, the use of the estimate $\bar{Y}^*$, which is a *linear* function of $\bar{C}$ and $\bar{Y}$, is based on Normal theory, for which $\bar{Y}^*$, with the value $\beta = -\rho_{CY}\sigma_Y/\sigma_C$, is the minimum variance estimate of $E(Y)$.

When C and Y do not have a bivariate Normal distribution, one might want to use a nonlinear estimate for $E(Y)$, in particular, a quadratic function if $E(\overline{C^2}) = E(C^2)$ is known:

$$Y^{**} = \bar{Y} + \beta_1\{\bar{C} - E(C)\} + \beta_2\{\overline{C^2} - E(C^2)\} \tag{11.2.9}$$

with $\overline{C^2} = (1/m)\sum_{i=1}^{m} C_i^2$.

Except when Y_i and C_i have a bivariate Normal distribution, for which linear regression is optimal, this quadratic control will generally give better results. However, $E(C^2)$ as well as $E(C)$ must be known, and both β_1 and β_2 must be estimated. Also note that the function is linear in the parameters β_1 and β_2 but quadratic in the observations C_i, since the estimate Y^{**} can be written as follows:

$$Y^{**} = \frac{1}{m}\Sigma[Y_i + \beta_1\{C_i - E(C)\} + \beta_2\{C_i^2 - E(C^2)\}] \tag{11.2.10}$$

that is, the average of a quadratic function of C_i, plus Y_i.

11.2.5   Multiple Regression-Adjusted Controls

The estimate Y^{**} is a special case of *multiple regression-adjusted controls*, in that one might have several control variables $C_1, C_2, C_3, \ldots$, with which one wants to control Y. Generally $C_1, C_2, C_3, \ldots$, are also correlated with each other. Consider first the case of two controls, so that we have just C_1 and C_2; then we have:

$$\hat{Y} = \bar{Y} + \beta_1\{\bar{C}_1 - E(C_1)\} + \beta_2\{\bar{C}_2 - E(C_2)\} \tag{11.2.11}$$

In the nonlinear control, Y^{**}, given at Equation 11.2.9, we had $C_2 = C_1^2$. Again, $E(C_1)$, $E(C_2)$ must be known, exactly or approximately.

The values of β_1 and β_2 that minimize the variance of $\hat{Y}$ are worked out in multiple correlation theory, and for the case of two controls, involve the joint and higher moments ρ_{YC_1}, ρ_{YC_2}, $\rho_{C_1C_2}$, σ_{C_1}, σ_{C_2}, and σ_Y.

We have

$$\mathrm{var}(\hat{Y}) = \mathrm{var}(\bar{Y}) + \beta_1^2\,\mathrm{var}(\bar{C}_1) + \beta_2^2\,\mathrm{var}(\bar{C}_2) + 2\beta_1\,\mathrm{cov}(\bar{Y},\bar{C}_1)$$
$$+ 2\beta_2\,\mathrm{cov}(\bar{Y},\bar{C}_2) + 2\beta_1\beta_2\,\mathrm{cov}(\bar{C}_1,\bar{C}_2)$$

and the values of β_1 and β_2 that minimize this variance are obtained by differentiating with respect to β_1 and β_2 and setting the two resulting equations equal to zero. After some simplification, we get the two equations

$$\beta_1\,\mathrm{var}(C_1) + \beta_2\,\mathrm{cov}(C_1, C_2) = -\mathrm{cov}(Y, C_1) \tag{11.2.12}$$
$$\beta_1\,\mathrm{cov}(C_1, C_2) + \beta_2\,\mathrm{var}(C_2) = -\mathrm{cov}(Y, C_2) \tag{11.2.13}$$

where we have dropped the $\bar{C}_1$ for C_1, and so on, since the averages have the same variances and covariances as the random variables, except for a common factor of $1/m$.

Now if $|\rho_{C_1C_2}| \neq 1$, we can solve these equations for β_1 and β_2 and get

$$\beta_1 = -\frac{\sigma_Y}{\sigma_{C_1}}\frac{\rho_{YC_1} - \rho_{C_1C_2}\rho_{YC_2}}{1 - \rho_{C_1C_2}^2} = -\frac{\sigma_Y}{\sigma_{C_1}}\frac{(1 - \rho_{YC_2}^2)^{1/2}}{(1 - \rho_{C_1C_2}^2)^{1/2}}\,\rho_{YC_1 \cdot C_2} \tag{11.2.14a}$$

$$\beta_2 = -\frac{\sigma_Y}{\sigma_{C_2}}\frac{\rho_{YC_2} - \rho_{C_1C_2}\rho_{YC_1}}{1 - \rho_{C_1C_2}^2} = -\frac{\sigma_Y}{\sigma_{C_2}}\frac{(1 - \rho_{YC_1}^2)^{1/2}}{(1 - \rho_{C_1C_2}^2)^{1/2}}\,\rho_{YC_2 \cdot C_1} \tag{11.2.14b}$$

The quantity $\rho_{YC_1 \cdot C_2}$ is called the *partial correlation of Y and C_1 given C_2*. For Normal random variables this is exactly the *conditional correlation of Y and C_1* for given C_2, and it is independent of the fixed value of C_2, say $C_2 = c_2$. Thus adding the second control will help only if the correlation between Y and C_2, *given C_1*, is large in absolute value.

Note that a solution to Equations 11.2.12 and 11.2.13 requires that $|\rho_{C_1C_2}| \neq 1$, which is simply saying that we have some new information in the added control, namely, that the two controls are not perfectly correlated.

It is left as an exercise to put the values β_1 and β_2 back into the expression for the variance and show that the minimum relative variance is, for the case of two controls,

$$\frac{\mathrm{var}(\hat{Y})}{\mathrm{var}(\bar{Y})} = 1 - \frac{\rho_{YC_1}^2 + \rho_{YC_2}^2 - 2\rho_{YC_1}\rho_{YC_2}\rho_{C_1C_2}}{1 - \rho_{C_1C_2}^2} \tag{11.2.15}$$

$$= 1 - \frac{(1 - \rho_{YC_2}^2)^{1/2}}{(1 - \rho_{C_1C_2}^2)^{1/2}}\,\rho_{YC_1}\rho_{YC_1 \cdot C_2} - \frac{(1 - \rho_{YC_1}^2)^{1/2}}{(1 - \rho_{C_1C_2}^2)^{1/2}}\,\rho_{YC_2}\rho_{YC_2 \cdot C_1} \tag{11.2.16}$$

In particular, if the two controls are independent, then $\rho_{C_1C_2} = 0$. Therefore, Equation 11.2.15 reduces to

$$\frac{\text{var}(\hat{Y})}{\text{var}(\bar{Y})} = 1 - \rho_{YC_1}^2 - \rho_{YC_2}^2 \tag{11.2.17}$$

Comparing this to Equation 11.2.6, there is clearly, in this rather ideal case, a gain in variance reduction from adding the second control. In deriving Equations 11.2.15 and 11.2.17, we have assumed that β_1 and β_2 are *known* constants; generally they must be estimated from the simulation data, and consequently the variance reduction obtained will not be as great as predicted by, for example, Equation 11.2.17.

Extensions of these results to the case of more than two controls are given in Exercise 11.11.

11.2.6   Internal Controls and Concomitants; General Ideas

It is useful to distinguish between *internal* controls and *external* controls. (Gaver and Thompson, 1973, refer to internal controls as concomitants.) Internal controls are, roughly, random variables with known mean that occur in the computation of the statistic or response that is being investigated. Thus, their use involves no new data computations. External controls on the other hand, involve copies of the statistic or system under different conditions or configurations and, thus, entail additional computation. All the examples above are external controls, involving different functions of the common random variables. For example, $T_{n,1}$ and T_n are the same random variables under different sets of parameter values. The same comment applies to W_n and WE_n.

The distinction between internal and external controls is subtle and not airtight. For example, the waiting time W_n in a GI|GI|1 queue is a function of the service times $S_i, i = 0, \ldots, n - 1$, and the interarrival times $X_i, i = 1, \ldots, n$, and we might use direct functions of these random variables as an internal control or concomitant. In particular, the waiting time W_n is probably highly correlated with the differences between the sum of the service times and the sum of the interarrival times

$$C = (S_0 + \cdots + S_{n-1}) - (X_1 + \cdots + X_n) \tag{11.2.18}$$

Here $E(C) = nE(S) - nE(X) = nE(X)(1 - t)$, where t is the traffic intensity. The waiting time WE_n for the corresponding MI|MI|1 queue is an external control that requires more computing than the control at Equation 11.2.18, although it may work much better than that control.

More generally, we might use the two quantities $(S_0 + \cdots + S_{n-1})$ and $(X_1 + \cdots + X_n)$, which are just unrestricted random walks, as the two controls in the multiple regression-adjusted control Equation 11.2.9. Actually, experience has shown it to be better to use only the average of recent S's and X's (i.e., $S_{n-1}, S_{n-2},$ S_{n-3} and X_n, X_{n-1}, X_{n-2}). Using separate controls based respectively on the S's and the X's is a fairly ideal application of multiple regression-adjusted control because the two controls are, by construction of the model, independent, so that Equation 11.2.17 holds. Moreover, the average of the S's are positively correlated with W_n

while the average of the X's are negatively correlated with W_n. This is readily verified intuitively.

EXAMPLE   **11.2.5 Internal Controls for the Busy Period of a GI|GI|1 Queue**
Internal controls are particularly useful when simulating the busy period in a queue, since *external* controls are very difficult to find.

Thus, suppose we want to know something about the number of customers served in a busy period in a queueing system. To understand the idea of a busy period, consider the GI|GI|1 queue of Section 3.2. By definition, the queue starts empty at time zero with a customer ready for service. The server will continue serving customers who arrive after this first customer, but will eventually finish a service and find that there are no more customers waiting. The server will then experience an idle period (the time to the next arrival), followed by another busy period. Since service times are, by definition, iid, as are interarrival times, it is not hard to see that the numbers of customers served in each busy period are iid random variables. The queueing system is said to *regenerate* at the beginning of each busy period. Notice that within each busy period, the length of the busy period and the number of customers served are highly dependent.

The random number of customers served in a busy period, denoted by α, takes on discrete values 1, 2, ... For an MI|GI|1 queue, it is known that the random variable α has mean

$$E(\alpha) = \frac{1}{1 - t} \tag{11.2.19}$$

where, as before, $t = E(\text{service time})/E(\text{interarrival time})$.

An attempt to run a GI|GI|1 queue for m busy periods (the numbers α_1, α_2, ..., α_m in each busy period are iid) and also run an MI|GI|1 queue with common random numbers will result in a different number of cycles (not m) in the MI|GI|1 queue. Hence it is difficult to match up cycles in such a way that they contain common random numbers. Even if one starts a busy period for an MI|GI|1 queue every time a GI|GI|1 busy period starts, there will be different numbers of customers in the two cycles unless the queues are only marginally different. Programming complexity is also great.

To find an internal control for this problem, consider the random number of service times $S_0, \ldots, S_{\alpha-1}$ and the random number of interarrival times $X_1, \ldots, X_\alpha$ in a generic busy period. It is particularly easy to see that in any GI|GI|1 queue $\alpha = 1$ if and only if the time to arrival of the first customer is longer than S_0, the service time of the customer who starts service at time 0. Thus, assuming that the distribution functions $F_X(x)$ and $F_S(x)$ of the interarrival and service times are absolutely continuous, we can compute

$$p_1 = P(\alpha = 1) = P(X_1 \geqslant S_0) = \int_0^\infty \{1 - F_X(y)\} f_S(y) \, dy$$

$$= \int_0^\infty F_S(y) f_X(y) \, dy \tag{11.2.20}$$

For an MI|MI|1 queue, Equation 11.2.20 is $p_1 = 1/(1 + t)$. Thus, as a control for α we could use the binary random variable

$$
\begin{aligned}
C_1 &= 0 && \text{if } \alpha = 1 \\
C_1 &= 1 && \text{if } \alpha > 1
\end{aligned}
\tag{11.2.21}
$$

where we know that $E(C_1) = 1 - p_1$.

Another possibility is to use a control, say C_2, which is equal to the (positive) difference between S_0 and X_1 when α is greater than 1. This might have greater correlation with α than does C_1. The idea here is that if $S_0 - X_1$ is very large, the waiting time is likely to be large. Thus, repeating that all random variables come from within the same busy period, we have:

$$
\begin{aligned}
C_2 &= 0 && \text{when } \alpha = 1 && \text{(probability } p_1) \\
C_2 &= S_0 - X_1 && \text{when } \alpha > 1 && \text{(probability } p_2 = 1 - p_1)
\end{aligned}
\tag{11.2.22}
$$

In fact, C_2 is just the waiting time of the second customer in the cycle, since the waiting time W_0 of the first customer is zero:

$$
C_2 = W_1 = [W_0 + S_0 - X_1]^+
$$

For an MI|GI|1 queue, given that C_2 is positive, C_2 will be, because of the memoryless property of exponential random variables, just exponentially distributed with the same parameter μ as the service time: that is, $\mu = 1/\{\text{mean service time}\}$.

Thus, for an MI|GI|1 queue

$$
E(C_2) = (p_1 \times 0) + \left[(1 - p_1) \frac{1}{\mu} \right]
\tag{11.2.23}
$$

For an MI|MI|1 queue, this is $t/\{(1 + t)\mu\}$.

More elaborate internal controls can be obtained (Iglehart and Lewis, 1979), but the analytical computations become fierce [try $P(\alpha = 2)$].

Note that C_1 and C_2 will work well when t is small and α is almost always 1 or 2. When t becomes large, the correlation between α and C_1 will decrease because α is very skewed and occasionally takes on very large values. In fact for the MI|MI|1 queue, $\text{var}(\alpha) = t(1 + t)/(1 - t)^3$, which becomes very large as t approaches 1.

EXAMPLE **11.2.6 Internal Controls for the Trimmed t Statistic**

It is possible to devise a simple internal control for simulating, say, quantiles of the trimmed t statistic for Normal populations. This eliminates the computing time necessary to calculate the regular t statistic T_n.

One proposed internal control for the trimmed t statistic in a sample of size m is the mean $\bar{T}_{n,1}$ the usual sample average. Our ability to use the mean derives from the fact that for symmetric random variables, $X_1, \ldots, X_n$, the trimmed t statistic

Table 11.2.3 Estimated correlation between the
sample mean and the order statistics from a sample
of trimmed t statistics $T_{20,1}$ of size $m = 20$ for 100
simulation replications

Order statistic index	Estimated correlation
1	0.51
2	0.66
3	0.69
4	0.71
5	0.74
6	0.78
7	0.79
8	0.79
9	0.77
10	0.76
11	0.73
12	0.71
13	0.75
14	0.75
15	0.68
16	0.65
17	0.67
18	0.65
19	0.55
20	0.46

defined at Equation 3.4.9 is a symmetric random variable. Consequently, $T_{n,1}$ has mean equal to zero and $E(\bar{T}_{n,1}) = E(C) = 0$ is known. Intuitively, if the sample mean is high, one would expect the higher order sample statistics or the sample quantiles to be greater than their respective means.

Table 11.2.3 shows the estimated correlation between the mean and 20 order statistics for $T_{20,1}$. To obtain Table 11.2.3, we have effectively sectioned a sample of 2000 $T_{20,1}$'s into 100 samples (replications) of size 20 each. It is clear that the correlations are fairly high, falling off at the extreme order statistics.

Since a quantile estimate, say $\tilde{t}_{n,1}(\alpha)$, is not an average, the correlation between $\bar{T}_{n,1}$ and $\tilde{t}_{n,1}(\alpha)$ cannot be estimated without sectioning. Thus, with no sectioning, a simple linear control (Equation 11.2.1) would have to be used. This is not a problem with correlations of the order of 0.98, but inequality of variances of the estimate and the control could make the control ineffective, especially with correlations of the order of 0.75. To avoid this, the sample would have to be sectioned. The resulting regression-controlled estimates would then be combined by averaging. The penalty, as is usual with sectioning, is the introduction of small-sample bias into the quantile estimate.

Because in many cases the mean of a statistic is known, but not any other property, the idea of using the mean as an internal control can be very useful in practice. Notice that the median of $T_{n,1}$ could also have been used as a control, since the distribution of $T_{n,1}$ is symmetric about the known value 0; another internal control is suggested in Exercise 11.13.

11.2.7   Combinations of Antithetic and Control Variables

It is completely possible to apply different techniques of variance reduction to the same problem to obtain greater variance reduction. Several individual techniques and combinations are explored in Example 11.2.7.

EXAMPLE   **11.2.7 Summary of a Simulation of the Waiting Times in a GI|GI|1 Queue**
In Example 11.2.3, we considered the use of the waiting times in an MI|MI|1 queue, generated with common random numbers, to control the waiting time in a GI|GI|1 queue. Antithetics for this case when the service and interarrival times had Weibull distributions were considered in Example 11.2.1. In all cases the simulated queues were started empty. This example explores combining these two approaches to

Table 11.2.4   Results on a simulation of the expected waiting time of the 1500th customer in a GI|GI|1 queue

$E(X) = 1$	Weibull interarrival times $k_X = 3.00$
$t = 0.90$	Weibull service times $k_S = 3.00$
$m = 1200$	

	Estimated mean	Estimated s.d. mean	Variance reduction factor
1. Program validation through antithetic estimation of the MI|MI|1 queue mean: $WE_{1500}^A = (WE_{1500} + WE_{1500}^a)/2$	8.220	0.15	—
2. Crude simulation estimate using U_i: W_{1500}	39.12	1.16	—
3. Crude simulation estimate using $1 - U_i$: W_{1500}^a	36.40	1.07	—
4. Antithetic estimate: $W_{1500}^A = (W_{1500} + W_{1500}^a)/2$	37.76	0.72	2.6
5. Exchange estimate: $W_{1500}^E = (W_{1500} + W_{1500}^e)/2$	37.85	0.60	3.8
6. Combination antithetic and exchange estimate: $(W_{1500} + W_{1500}^a + W_{1500}^e)/3$	37.36	0.53	4.7
7. Estimate controlled internally by service and waiting times: $W_{1500} - \sum_1^{1500} S_i + \sum_1^{1500} X_i$	39.12	1.16	1.0
8. Combination antithetic estimate controlled by antithetic MI|MI|1 estimate: W_{1500}^A controlled by WE_{1500}^A	38.54	0.64	3.3
9. Combination antithetic estimate regression controlled by antithetic MI|MI|1 estimate: W_{1500}^A regression controlled by WE_{1500}^A	37.10	0.59	4.3
10. Combination antithetic estimate, multiple regression controlled by MI|MI|1 estimate and antithetic MI|MI|1 estimate: W_{1500}^A regression controlled by WE_{1500} and WE_{1500}^a	37.09	0.59	4.3

variance reduction, as well as adding and combining several other ideas introduced in this section.

Results for one particular set of parameter values are given in Table 11.2.4. The number of replications m is 1200 and the waiting time W_n investigated is that of the 1500th customer ($n = 1500$). This number was chosen to try to ensure that the simulation is of the stationary waiting time distribution. [Methods for determining when this stationarity is attained, and the use of averaging along single sample paths to estimate $E(W_N)$ are considered in Volume II.] This case uses a traffic intensity of $t = 0.9$, so that the queue is close to saturation. Regular skewed Weibull-distributed interarrival and service times are generated with parameters $k_X = 3.0$ and $k_S = 3.0$, respectively. Because we have set $E(X) = 1$ and $t = 0.9$, we have $E(S) = 1.111$.

In equilibrium, the waiting time WE_n in an MI|MI|1 queue $E(WE_n) = E(X)t/(1 - t)$, so that WE_{1500} can be used as a control for simulating the waiting time in our GI|GI|1 queue.

To validate our simulation program, the first line of Table 11.2.4 shows the simulated estimate of $E(WE_n)$ to be 8.220, as opposed to the true value of 9.0 when $t = 0.9$. The estimator used in this first line is actually the antithetic estimator WE_{1500}^A, the average of the MI|MI|1 waiting times, WE_{1500}, and their antithetics, WE_{1500}^a. Since the standard deviation of the mean is 0.15, there is some indication that either the queue has not reached steady state at $n = 1500$ (problems of convergence in queues will be discussed in Volume II) or that the waiting times are so skewed that 2 standard deviations does not provide adequate coverage of the true mean. Of course the computer program could be invalid, but in light of the discrepancy this was carefully checked.

Calculations show that the precisions of all the estimates in the table (i.e., the estimated standard deviations of the waiting times divided by their estimated means) are 3.0% or less.

Consider now the waiting time in the GI|GI|1 queue for which simulation results are given in lines 2 through 10 of Table 11.2.4. Looking at the variance reduction factor in the last column, we see that the antithetic and the exchange estimates (lines 4 and 5) achieve variance reductions of 2.6 and 3.8, respectively, although these should be divided by 2 to compensate for the fact that $2n$ output variables (pairwise dependent) go into W_{1500}^A and W_{1500}^E. These amounts of variance reduction are consistent with the correlations given on the first two lines of Table 11.2.5.

On the sixth line of Table 11.2.4, an estimator combining W_{1500}, W_{1500}^a, and W_{1500}^e gives even better precision because W_{1500}^a and W_{1500}^e are, from Table 11.2.5, not highly positively correlated (0.172). Again this variance reduction factor should perhaps be divided by 3 to compensate for the fact that $3n$ output variables go into the estimate.

Line 7 of Table 11.2.4 shows a controlled estimate that is ineffective because the control, the difference of the cumulated waiting and service times, has a correlation with W_{1500} that is too small to be effective without regression adjustment (lines 5 and 6 of Table 11.2.5). On the eighth and ninth lines, we have decided to start with the antithetic estimate W_{1500}^A, since this already has a sharply smaller standard

Table 11.2.5 Correlations between various waiting time estimates in a simulation of a GI|GI|1 queue

Variables	Correlation
W_{1500}, W_{1500}^a	-0.156
W_{1500}, W_{1500}^e	-0.445
W_{1500}^a, W_{1500}^e	0.172
W_{1500}, WE_{1500}	0.673
$W_{1500}, \sum_1^{1500} X_i$	-0.408
$W_{1500}, \sum_1^{1500} S_i$	0.462
WE_{1500}, WE_{1500}^a	-0.296

deviation than W_{1500}. We then use as a regular control, and a regression-adjusted control, WE_{1500}^A, the antithetic in the MI|MI|1 case. These produce variance reduction factors of 3.3 (line 8) and 4.3 (line 9), respectively. Splitting WE_{1500}^A into its components WE_{1500}^a and WE_{1500} and using these two variables as a multiple regression control for W_{1500}^A produces, on line 10, only a very marginal improvement over the single regression-adjusted control in line 9.

It is tempting to feel that one might control W_{1500} with W_{1500}^a, WE_{1500}^a, and WE_{1500}; note however that $E(W_{1500}^a)$ is *not known*. This does, however, raise the possibility of throwing all possible controls into a multiple control procedure. The drawback is that all the controls have to be computed, and it might be more efficient to simply compute more values of W_{1500}. Also, if two of the controls have correlations close to one, the computation will be unstable and the marginal gain in variance reduction will be negligible.

Results not presented here for a second case with Weibull inputs with parameters $k_X = 0.5$ and $k_S = 0.5$ show mainly that the "best" method of variance reduction depends on the parameters in the model. In this case, the exchange estimate W_{1500}^E and the antithetic W_{1500}^A are equivalent, while the triple combination does worse. The best methods are the controlled antithetic estimates corresponding to lines 8 and 9 in Table 11.2.4.

Both the cases considered are parametrically far from the MI|MI|1 case, so that more variance reduction than has been found in this example could well occur.

11.2.8   An Extended Statistics Problem: Simulating the Distribution of the Anderson–Darling Statistic Using Control Variables

We give now a slightly more advanced example of an application of variance reduction via control variables to a problem in statistics. *This example may be skipped by readers with minimal preparation in statistics.*

The problem considered is the tabulation of quantiles of the Anderson–Darling goodness-of-fit statistic (Anderson and Darling, 1952).

In Chapter 6, we introduced the Kolmogorov–Smirnov statistic D_n, which is used to test the null hypothesis that an iid sample $X_1, \ldots, X_n$ comes from a completely specified distribution $F_X(x)$. The test statistic D_n is given at Equations 6.2.18 to 6.2.21 and is based on a metric that measures the "distance" between the empirical cdf, $\hat{F}_X(x)$, defined at Equation 6.2.14, and the true distribution $F_X(x)$ as their maximum absolute deviation over all x. Another metric, used in the Cramér–von Mises statistic (see, e.g., Stephens and Maag, 1968), measures the distance between $\hat{F}_X(x)$ and $F_X(x)$ as the integral of the squared distance between the curves (i.e., the area of the squared difference between the two curves). In both cases, the metric converges to zero as the sample size n increases if $F_X(x)$ is the true underlying distribution of the sample.

A drawback to both the Kolmogorov–Smirnov and Cramér–von Mises statistics is that they are not sensitive to departures from the null hypothesis that occur in the tails of the distribution. This is because, for any x, $E\{\hat{F}_X(x) - F_X(x)\} = 0$ when $\hat{F}(x)$ is defined as at Equation 6.2.14; however

$$\text{s.d.}\{\hat{F}_X(x) - F_X(x)\} = \left[\frac{F_X(x)\{1 - F_X(x)\}}{n}\right]^{1/2} \tag{11.2.24}$$

which varies between 0 when $x \to -\infty$, $(1/(4n))^{1/2}$ when $F_X(x) = \frac{1}{2}$, and 0 when $x \to \infty$. To overcome this unequal variability, it has been suggested that one use a weighted distance measure, the weight being the reciprocal of the standard deviation of the difference. This leads to the weighted integral distance measure

$$W_n^2 = n \int_{-\infty}^{\infty} \frac{\{\hat{F}_X(x) - F_X(x)\}^2}{F_X(x)\{1 - F_X(x)\}} \, dF_X(x) \tag{11.2.25}$$

called the Anderson–Darling statistic (Anderson and Darling, 1952). Since $\hat{F}_X(x)$ is a step function, an alternate computable form of Equation 11.2.25 is given (Anderson and Darling, 1952; Lewis, 1961) after some manipulation, as follows:

$$\begin{aligned} W_n^2 = -n - (n^{-1}) \sum_{i=1}^{n} [(2i - 1)\ln F_X(X_{(i)}) \\ + \{2(n - i) + 1\}\ln\{1 - F_X(X_{(i)})\}] \end{aligned} \tag{11.2.26}$$

where the $X_{(i)}$ are the order statistics of the sample of size n. However, since the probability integral transform shows that $F_X(X_{(i)}) = U_{(i)}$, the ith-order statistic from a uniform$(0, 1)$ population of size n, the statistic is, under the null hypothesis, distribution free and may be written as follows:

$$W_n^2 = -n - (n^{-1}) \sum_{i=1}^{n} [(2i - 1)\ln U_{(i)} + \{2(n - i) + 1\}\ln(1 - U_{(i)})] \tag{11.2.27}$$

This expression involves only the order statistics $U_{(i)}, i = 1, \ldots, n$, from a uniform$(0, 1)$ population. Thus, we do not have to tabulate the null distribution of W_n^2 for all distributions, only for the uniform distribution.

A crude simulation of the percentiles and quantiles of W_n^2 was published by Lewis (1961). The indication in that paper was that the distribution of W_n^2 converges very rapidly to its known asymptotic distribution, but the precision of the simulation was rather low.

Is it possible to improve precision of the simulation of the distributional properties of the W_n^2 statistic by using variance reduction methods? Note, first, that antithetics will not work because the statistic, as given at Equation 11.2.27, is unchanged if $U_{(i)}$ is replaced by $1 - U_{(i)}$.

Turning to the method of control variables for variance reduction, two controls suggest themselves:

1. The Kolmogorov–Smirnov statistic also measures "distance" between $\hat{F}_X(x)$ and $F_X(x)$, and should be large when W_n^2 is large if the same $U_{(i)}$'s are used to compute the two statistics (common random numbers).

2. Since W_n^2 is sensitive to departures in the tails of the distribution (e.g., when the lower order statistics are small), use the smallest order statistic $U_{(1)}$ or $\ln U_{(1)}$ as a control. Since an exponential random variable is generated (Section 4.2.5) as $-\ln(1 - U)$, then the quantity $-\ln U_{(1)} = -\ln\{1 - (1 - U_{(1)})\}$ in Equation 11.2.27 will be the largest order statistic from an exponential population, and its mean is known (see, e.g., Cox and Lewis, 1966, p. 26) to be:

$$E\{-\ln U_{(1)}\} = \sum_{j=1}^{n} \frac{1}{n + 1 - j}$$

However, its distribution is not known. Instead, we concentrate on using $-\ln(1 - U_{(1)})$ as the control. The reason for this is that $-\ln U_{(1)}$ is highly (negatively) correlated with $-\ln(1 - U_{(1)})$, which thus should also be an effective control. Moreover, $-\ln(1 - U_{(1)})$ has the same distribution as the *smallest* order statistic from an exponential population, namely an exponential distribution with mean $1/n$. This distribution is easier to work with than the distribution of $-\ln(U_{(1)})$ if, for instance, we wish to use a function of $-\ln(U_{(1)})$ as a control.

Of course, all three controls, D_n, $-\ln U_{(1)}$, and $-\ln(1 - U_{(1)})$, could be used, but for now, we restrict the problem to two controls. Note that the first of these controls, the Kolmogorov–Smirnov statistic, is an *external* control whereas the second $-\ln(1 - U_{(1)})$ is an *internal* control.

To test these two controls, the three statistics D_n, W_n^2, and $-\ln(1 - U_{(1)})$ were generated in the program SMTBPC from the software supplement, using common random numbers, and 50 super-replications; see Chapter 8 for details. An additional program called the AFTERBURNER computes correlations between the three simulated statistics using the 50 super-replications, and also computes the controlled estimates, along with variance estimates. (See the software supplement for details.) This is equivalent to using 50 sections in the simulation, as described in Chapter 9, to obtain variance estimates.

Figures 11.2.2 and 11.2.3 give the correlations between all estimated quantities (means, quantiles, etc.) for the Kolmogorov–Smirnov and Anderson–Darling statistics, and for $-\ln(1 - U_{(1)})$ and the Anderson–Darling statistic, respectively.

SIMTBED AFTER-BURNER   50 SUPER-REPLICATIONS   CORRELATION BETWEEN ESTIMATORS 1-2

SUBSAMPLE SIZE	2	4	6	8	10	12	14	16
MEAN	0.7823	0.8296	0.8377	0.8471	0.8777	0.8703	0.8236	0.8615
STD	0.6504	0.7390	0.6849	0.5140	0.7000	0.6915	0.7365	0.7348
SKEWNESS	0.3002	0.3964	0.2670	0.5284	0.6408	0.7136	0.5617	0.5394
KURTOSIS	0.2355	0.4199	0.3705	0.5225	0.6164	0.7952	0.6358	0.5769
SER.COR.	0.6192	0.6694	0.6765	0.7675	0.7920	0.7004	0.7513	0.6130
QUANTILES								
0.010	0.9214	0.6984	0.4017	0.5400	0.4446	0.5489	0.4273	0.2810
0.025	0.9261	0.5232	0.4672	0.6147	0.6295	0.5895	0.4881	0.4100
0.050	0.8507	0.3750	0.6695	0.6926	0.5273	0.4120	0.4721	0.4674
0.100	0.8706	0.3238	0.7080	0.7040	0.6732	0.6508	0.4366	0.5271
0.250	0.6852	0.7240	0.6508	0.7420	0.7531	0.6853	0.4813	0.5703
0.500	0.5037	0.7752	0.7038	0.5958	0.7008	0.7628	0.6970	0.8046
0.750	0.5972	0.8228	0.7514	0.7154	0.7190	0.6428	0.7381	0.8074
0.900	0.5997	0.7274	0.5534	0.5810	0.6315	0.6869	0.6038	0.6806
0.950	0.6188	0.6985	0.5640	0.6272	0.5671	0.6154	0.5745	0.6501
0.975	0.6808	0.6082	0.5395	0.5205	0.6637	0.6316	0.5328	0.5828
0.990	0.8010	0.4279	0.4400	0.4933	0.5719	0.5432	0.4644	0.3240

ESTIMATORS:
1 - CANONICAL KS-TEST OF ORDERED UNIFORMS
2 - CANONICAL ANDERSON-DARLING OF ORDERED UNIFORMS
3 - CONTROL -LOG(SMALLEST UNIFORM)

SAMPLE SIZE (N): 5000   NO. OF REPLICATIONS (M):   5   DEGREE OF REGRESSION (D): 3

Figure 11.2.2 Entries are correlations between different quantifications of the Anderson–Darling statistic and the Kolmogorov–Smirnov statistic, simulated using common random numbers, in SMTBPC. For example, since $N = 5000$ and $M = 5$, at $n = 2$, a total of 5000 values of the mean of the two statistics were estimated and averaged. This was based on 50 super-replications so that 50 means for each statistic were used to obtain the estimated correlation of 0.7823 in the row labeled "MEAN" in the "$n = 2$" column.

The means of the Anderson–Darling and Kolmogorov–Smirnov statistics exhibit correlations ranging from 0.7823 at $n = 2$ to 0.8615 at $n = 14$. Correlations between other quantities (e.g. quantiles) are on the order of 0.5 to 0.80, indicating that use of the statistic D_n as a single regression-adjusted control would halve the sample size needed to achieve a given variance of the estimates.

SIMTBED AFTER-BURNER   50 SUPER-REPLICATIONS   CORRELATION BETWEEN ESTIMATORS 2-3

SUBSAMPLE SIZE	2	4	6	8	10	12	14	16
MEAN	0.3625	0.3797	0.4233E-01	0.2074	0.1946	0.1399	0.3085	-0.1079
STD	0.6519	0.5325	0.3573	0.3240	0.1564	-0.1977E-01	0.2811	0.7318E-01
SKEWNESS	0.4630	0.4667	0.2155	0.3389	0.7449E-01	0.3660	0.1776	0.2283
KURTOSIS	0.5026	0.4506	0.2996	0.3620	0.1293	0.4196	0.1199	0.4411
SER.COR.	0.1166	-0.9915E-01	0.2847	0.3332E-01	-0.1078	0.8555E-01	0.2477	0.2464E-01
QUANTILES								
0.010	-0.2458	-0.4871E-01	0.1642	-0.2057	-0.1729	-0.1225	-0.9012E-01	-0.3482
0.025	-0.3971	-0.3530	-0.1373	0.3422E-01	-0.2513	-0.2350E-01	0.8843E-01	-0.2027
0.050	-0.1747	-0.2088	-0.2208	-0.8957E-01	-0.1401	-0.3171E-01	0.5349E-01	-0.1608
0.100	-0.3007	-0.1933	-0.2203	-0.4144E-01	-0.2002	-0.1506	-0.7513E-01	-0.2284
0.250	-0.3661	0.1993	-0.2638	-0.3313E-01	-0.8515E-01	-0.1342	-0.2302	-0.8181E-01
0.500	-0.7361E-01	0.7877E-01	-0.1768	0.1185	0.6102E-02	0.1993	0.2924E-01	-0.8443E-01
0.750	0.2058	0.3196	-0.1985	0.4658E-01	0.8512E-01	0.1955	0.3372	-0.1955
0.900	0.2241	0.3751	0.1919	0.9692E-01	0.1750E-01	0.5656E-01	0.2446	-0.6958E-01
0.950	0.3517	0.2126	0.1053	0.4436	0.1351	-0.1731E-01	0.2125	0.9227E-01
0.975	0.3827	0.4045	0.2466	0.2071	-0.7531E-02	0.9503E-01	0.7661E-01	0.2063
0.990	0.4787	0.3830	0.1949	0.1266	-0.3416E-02	-0.2264E-01	0.1490	0.5419E-01

ESTIMATORS:

1 - CANONICAL KS-TEST OF ORDERED UNIFORMS
2 - CANONICAL ANDERSON-DARLING OF ORDERED UNIFORMS
3 - CONTROL -LOG(SMALLEST UNIFORM)

SAMPLE SIZE (N): 5000   NO. OF REPLICATIONS (M):   5   DEGREE OF REGRESSION (D): 3

Figure 11.2.3   Entries are correlations between different quantifications of the Anderson–Darling statistic and the control $-\ln(1 - U_{(1)})$, simulated using common random numbers, in SMTBPC. For example, since $N = 5000$ and $M = 5$, at $n = 2$, a total of 5000 values of the mean of the two statistics were estimated and averaged. This was based on 50 super-replications so that 50 means for each statistic were used to obtain the estimated correlation of 0.3625 in the row labeled "MEAN" in the "$n = 2$" column.

A mitigating factor, though, is that the *external* control D_n is fairly time-consuming to compute, and although its quantiles are tabulated for small samples (Conover, 1980, Table A14), means, variances, and other moments are not known.

The statistic $-\ln(1 - U_{(1)})$ is an *internal* control that is computed when W_n^2 is computed; however, Figure 11.2.3 shows that its correlation with W_n^2 for various quantiles and sample sizes is too low to be useful as a control. Moreover, since it is positively correlated with D_n (figure not shown), its use as a second (multiple) control is not warranted.

An alternative scheme for obtaining variance reduction is the following.

Since the variables $-\ln(1 - U_{(i)})$ are the order statistics from a unit exponential population, let $E_{(i)} = -\ln(1 - U_{(i)})$. Then we rewrite W_n^2 from Equation 11.2.27 as follows:

$$
\begin{aligned}
W_n^2 = -n + (n^{-1}) \sum_{i=1}^{n} & [-(2i - 1)\ln\{1 - \exp(-E_{(i)})\} \\
& + \{2(n - i) + 1\}E_{(i)}]
\end{aligned}
\tag{11.2.28}
$$

Now the order statistics from an exponential(1) population can be written (see, e.g., Cox and Lewis, 1966, p. 26) in terms of iid exponential(1) random variables E_j' as follows:

$$
E_{(i)} = \sum_{j=1}^{i} \frac{E_j'}{n - j + 1}
\tag{11.2.29}
$$

Equations 11.2.28 and 11.2.29 together give W_n^2 as a function of *independent exponential random variables*, all of which can be used to control W_n^2 in a multiple control situation, for the following reasons.

1. The computing of the E_j' is part of the computation of W_n^2 and, therefore, involves no extra computing time if used as internal controls.
2. The E_j''s are independent, the ideal situation for multiple controls, so that they contribute independently to the control of W_n^2.
3. The right-hand part of the sum in Equation 11.2.28 is a linear function of the E_j'''s.
4. The left-hand part of the sum is completely determined by the right-hand part, but is not necessarily completely correlated with it. (Recall that the correlation between an exponential random variable and its completely determined antithetic is -0.64.)

The case $n = 2$ is studied here; other values of n are left as an exercise in multiple controls.

Assume that we are interested in determining by simulation the 0.95 quantile of W_2^2. The variables E_1' and E_2' can be used as controls, and the correlations of their simulated 0.95 quantiles with the estimated 0.95 quantile of W_2^2 are found experimentally to be 0.4008 and 0.2882, respectively. These are low because Points 3 and 4 above are deceptive: W_2^2 is actually a very nonlinear function of the independent random variables E_1' and E_2'. One can also use $(E_1')^2$ and $(E_2')^2$ as controls; the correlations between the simulated 0.95 quantiles of W_2^2 and the 0.95 quantiles of those two variables are estimated to be 0.4011 and 0.2894. This similarity to the values 0.4008 and 0.2882 above reflects the fact that E_1' and $(E_1')^2$ have (theoretical) correlation of about 0.9.

Another control derives from the fact that Equations 11.2.28 and 11.2.29 can be used to show that $E(W_n^2) = 1$ for all n.

Thus, in a simulation to obtain an estimate of the 0.95 quantile of W_n^2 at $n = 2$, one could use the following multiple controls:

1. $\overline{W_2^2}$ with mean value 1
2. $\overline{E}_1'$ and $\overline{E}_2'$, each with mean value 1
3. $(\overline{E}_1')^2$ and $(\overline{E}_2')^2$, each with mean value 2
4. If necessary, the mean or other quantifications of the external control D_2, the Kolmogorov–Smirnov statistic computed with the same E_1' and E_2' as are used to compute $\overline{W_2^2}$

In practice, the 0.95 quantile of the D_n statistic is used as a control because its expected value is known. The sample mean of D_n is easier to compute, but its expected value is not known.

The computations proceed as in Exercise 11.11. This is left as an exercise; the variance reduction obtained, however, is disappointing because the correlations between the statistic W_n^2 and the controls are relatively low. For instance, from Figure 11.2.2, the correlation between the sample 0.95 quantiles of W_n^2 and D_n is on the order of 0.6 at all sample sizes. Moreover, correlations between the controls suggested above are high, so that their inclusion as multiple controls does not give much more variance reduction.

11.2.9   Summary of the Use of Control Variables

Control variables, both internal and external, have been shown to be effective in reducing variance in some simple examples. A criticism of these examples is that in simulating more complex systems or statistics, such as the Anderson–Darling statistic discussed in Section 11.2.8, effective controls are hard to find. Several examples of these more complex situations can be found in Lavenberg, Moeller, and Welch (1982), Gaver and Shedler (1971), and Venkatraman and Wilson (1986), while Swain and Schmeiser (1987) have given an example in statistics in which control variables give large variance reductions. A recent renewal of interest in using concomitants is reflected in the work of J. R. Wilson and Pritsker (1984a, b). Use of these techniques in simulating stochastic systems and processes will be discussed further in Volume II.

One final point needs to be made. Validation of simulation programs is absolutely essential, and in this sense, computation of a control variable is almost mandatory. Thus, having results on the MI|MI|1 queue *computed with the same program* is extraordinarily helpful in validating the program. In fact, any known quantities should be computed, even if they are not used for variance reduction.

11.3   Conditional Sampling

11.3.1   Introduction

We discuss now a technique from the second class of variance reduction methods discussed in Section 11.0. This class exploits *conditioning arguments, conditional expectations, and substructure of the random variable or system response.* It is related in concept to stratified sampling—which is considered in Section 11.5. However, it is not an attempt to change the amount of sampling in different subpopulations, but rather an attempt to use *known analytical results*. It is very useful in improving results of statistical simulations; its most systematic use, under the name "Monte Carlo swindles," has been in studies of robustness (see, e.g., Relles, 1970; Andrews et al., 1972; Gross, 1973; Simon, 1976; and Johnstone and Velleman, 1985). The first four references become more understandable once you know that they are using conditional sampling. Variance reductions up to 1000 are claimed, although these involve a lot of work on the part of the simulator and exploit, in particular, special properties of the Normal distribution and, in general, special properties of symmetric random variables.

Our goal is to use simulation to find properties of a random variable Y that is a function of other random variables. We consider first the case of conditioning on a univariate random variable. The *basic assumption is that when a particular random variable in this function (call it X) is fixed, we know the distribution or moments of Y.* That is, we know $E_Y(Y|X = x)$, the conditional expectation of Y with respect to Y when X is given. More generally, we assume $E_Y(g(Y)|X = x)$ is known in order to include other quantizations of the random variable Y. For example, *for percentiles, the function $g(Y)$ of Y is the zero–one function $I(Y; y)$* with parameter y, defined at Equation 11.3.11 below. Thus, $E_Y(I(Y; y)|X = x) = P(Y \leqslant y|X = x)$.

We illustrate these preliminary ideas by example.

1. Ratios are a particular example in which fixing one random variable allows one to obtain known results; thus let $Y = Z/X$, where Z and X may or may not be independent.

In the independence case, one has trivially that (e.g., for X positive),

$$P(Y \leqslant y|X = x) = P\left\{\frac{Z}{x} \leqslant y\right\} = P(Z \leqslant yx)$$

and if the distribution of Z is known, we have the distribution of Y, given $X = x$. The unconditional distribution of Y is then obtained as $E_X\{P(Z \leqslant yX)\}$.

If the random variables are dependent, it is not as simple to get the conditional distribution of Y, given $X = x$. The outstanding example of the possibility of obtaining conditional results occurs when Z and X have a bivariate Normal distribution. Then, using the results given at Equations 10.1.24 and 10.1.25, the conditional distribution of Y, given $X = x$, is Normal with mean $\mu_Z/x + (\rho\sigma_Z/\sigma_X)(1 - \mu_X/x)$ and variance $\sigma_Z^2(1 - \rho^2)/x^2$, provided $x \neq 0$.

The same argument is true for sums of random variables; however, in that case if we just want moments, then $E(Z + X) = E(Z) + E(X)$ even if Z and X are dependent.

2. Another common example of conditioning that can lead to known results occurs with random sums of random variables:

$$Y = S_X = Z_1 + \cdots + Z_X$$

where the Z_i's may be iid and independent of the discrete valued variable X. The case where X is dependent on the Z_i's arises in renewal process counts (see, e.g., Cox, 1962). Consider the time to the last event before time t in the renewal process; thus let $S_{N(t)} = Z_1 + \cdots + Z_{N(t)}$, where the Z_i's are the successive times between events in the renewal process, $N(t)$ is the number of events in $(0, t)$ in the renewal process, and $S_{N(t)}$ is the time to the last renewal before t. Clearly $N(t)$ and the Z_i's are dependent, since, for example, $N(t) = 0$ if and only if $Z_1 > t$. (For a discussion of renewal theory see, e.g., Cox, 1962.)

The use of conditional sampling to get variance reduction in simulations is based on the formula

$$E(Y) = E_X\{E_Y(Y|X)\} \tag{11.3.1}$$

where, when densities exist, we have:

$$E_Y(Y|X) = \int_{-\infty}^{\infty} y\, f_{Y|X}(y|X)\, dy = \int_{-\infty}^{\infty} y\, \frac{f_{XY}(X; y)}{f_X(X)}\, dy$$

Note that $h(X) = E_Y(Y|X)$ is a *random variable*, and Equation 11.3.1 follows by integrating to get $E(Y) = \int h(x) f_X(x)\, dx$.

If X is discrete, with $p_x = P(X = x)$, we have

$$E(Y) = \sum_{\forall x} h(x) p_x = \sum_{\forall x} E_Y(Y|X = x) p_x$$

11.3.2   The Conditional Sampling Algorithm

The idea in using these results for variance reduction in simulation is to do one of the following:

Crude Simulation   Take m realizations of Y, namely $Y_1, \ldots, Y_m$ and obtain the usual estimate of $E(Y)$,

$$\bar{Y} = \frac{1}{m} \sum_{i=1}^{m} Y_i$$

for which $\operatorname{var}(\bar{Y}) = (1/m)\operatorname{var}(Y)$.

Conditional Sampling   Sampling on X—that is, $X_1, \ldots, X_m$, estimate $E(Y)$ as the X average in Equation 11.3.1 using the *known* $E_Y(Y|X)$ to get the following:

$$Y^* = \frac{1}{m} \sum_{i=1}^{m} E_Y(Y|X_i) \tag{11.3.2}$$

This is again an unbiased estimator of $E(Y)$, with theoretical variance:

$$\operatorname{var}(Y^*) = \frac{1}{m} \operatorname{var}_X\{E_Y(Y|X)\} \tag{11.3.2a}$$

To see that one achieves variance reduction with this conditional sampling scheme, we must evaluate $\text{var}_X\{E_Y(Y|X)\}$. Now $\text{var}_X\{E_Y(Y|X)\}$ is obtained by conditioning arguments, using Equation 11.3.1, as follows:

$$
\begin{aligned}
\text{var}(Y) &= E(Y^2) - E^2(Y) \\
&= E_X\{E_Y(Y^2|X)\} - [E_X\{E_Y(Y|X)\}]^2 \\
&= E_X\{E_Y(Y^2|X)\} - E_X\{E_Y^2(Y|X)\} + E_X\{E_Y^2(Y|X)\} - [E_X\{E_Y(Y|X)\}]^2
\end{aligned}
$$

where in the last line we have simply added and subtracted the middle two terms. Therefore, collecting terms, we have

$$
\text{var}(Y) = E_X\{\text{var}_Y(Y|X)\} + \text{var}_X\{E_Y(Y|X)\} \tag{11.3.3}
$$

or

$$
\text{var}_X\{E_Y(Y|X)\} = \text{var}(Y) - E_X\{\text{var}_Y(Y|X)\} \tag{11.3.4}
$$

But since $\text{var}_Y(Y|X)$ is a positive random variable with positive expectation, Equation 11.3.4 gives the result:

$$
\text{var}_X\{E_Y(Y|X)\} \leqslant \text{var}(Y) \tag{11.3.5}
$$

Returning to the estimator Y^* given at Equation 11.3.2, we then have:

$$
\text{var}(Y^*) = \frac{1}{m}\text{var}_X\{E_Y(Y|X)\} = \frac{1}{m}[\text{var}(Y) - E_X\{\text{var}_Y(Y|X)\}] \tag{11.3.6}
$$

$$
\leqslant \frac{1}{m}\text{var}(Y) = \text{var}(\bar{Y}) \tag{11.3.7}
$$

That is, $\text{var}(Y^*) \leqslant \text{var}(\bar{Y})$.

This is the basic result; we will always get variance reduction with this conditioning technique. The larger $E_X\{\text{var}_Y(Y|X)\}$, the greater the variance reduction.

EXAMPLE   **11.3.1 Sums of Random Variables**
As an absolutely trivial but illustrative case of the results of Equations 11.3.4 and 11.3.5, consider $Y = Z + X$, with Z and X independent and $E(X)$ known. Then $\text{var}_Y(Y|X = x) = \text{var}(Z + x) = \text{var}(Z)$, so that $E_X\{\text{var}_Y(Y|X)\} = \text{var}(Z)$. Also, by independence, $\text{var}(Y) = \text{var}(Z) + \text{var}(X)$.

Then using Equation 11.3.4 we have

$$
\text{var}_X\{E_Y(Y|X)\} = \text{var}(Z) + \text{var}(X) - \text{var}(Z) = \text{var}(X)
$$

This result implies the intuitive result that if X had variance much greater than variance Z, we would get much more variability if we estimated $E(Y)$ by using, for example,

$$\bar{Y} = \frac{1}{m} \sum_{i=1}^{m} (X_i + Z_i) = \frac{1}{m} \sum_{i=1}^{m} Y_i$$

than if we used the known value of $E(X)$ and estimated $E(Y)$ by $\bar{Z} + E(X)$.

EXERCISE   In Example 11.3.1, what happens if Z and X are correlated and $E(X)$ is known? ∎

The exact computational algorithm for simulation using *conditional sampling* is as follows:

1. Generate a sample of size m of the X_i's, say $X_1, \ldots, X_m$.
2. Compute the *known* values

$$W_i = E_Y(Y|X_i = x_i) \qquad (11.3.8)$$

3. Average to get

$$Y^* = \frac{1}{m} \sum_{i=1}^{m} W_i$$

4. Since Y^* is an average of iid random variables W_i, estimate var(Y^*) as the sample variance of the W_i's

$$\text{vâr}(Y^*) = \frac{1}{m(m-1)} \sum_{i=1}^{m} (W_i - Y^*)^2 \qquad (11.3.9)$$

Note that the computation of $E_Y(Y|X_i = x_i)$ in Step 2 may be quite onerous and this may decrease the utility of the conditional sampling technique. This is particularly true if X is a continuous-valued random variable, since *it might be* time-consuming to compute $E_Y(Y|X = x)$ for all m different values of X. If X is a discrete-valued random variable, the procedure can be simplified. Assume that X takes on only k values and the number of times that $X = x_i$, $i = 1, \ldots, k$, in the sample of size m is m_i, where $m_1 + m_2 + \cdots + m_k = m$. Then we have:

$$Y^* = \sum_{i=1}^{k} \frac{m_i}{m} E_Y(Y|X = x_i) \qquad (11.3.10)$$

11.3.3   Illustrative Examples: Ratios

EXAMPLE   **11.3.2 Normalized Moments**
Ratios again! This is a very common situation in which analytical results are hard to obtain and simulation is needed. For example, ratios occur in normalized moments such as the squared coefficient of variation

$$C^2(X) = \frac{\text{var}(X)}{\{E(X)\}^2} \qquad E(X) \neq 0$$

with estimate, from Equation 6.1.19, of $\tilde{C}^2(X) = S_X^2/(\bar{X})^2$, and in the correlation coefficient

$$\rho_{XY} = \frac{\text{cov}(X, Y)}{\sigma(X)\sigma(Y)}$$

with estimate $r = \hat{\rho}_{XY} = \hat{\text{cov}}/(S_X S_Y)$ given at Equations 10.2.1 and 10.2.2.

EXAMPLE   **11.3.3 Ratios of Independent Random Variables**

Taking generally $Y = Z/X$, with Z and X independent, if we know something about Z, we sample on X (or vice versa, if we know something about $1/X$, sample on Z).

Then, if $g(Y) = Y$,

$$E(Y) = E_X\{E_Y(Y|X)\} = E_X\left(\frac{1}{X}\right)E(Z)$$

because of independence. Of course $E(1/X)$ must exist, and this would not be true, for example, if X were exponentially distributed.

Also, if

$$g(Y) = Y^2 \qquad E(Y^2) = E_X\{E_Y(Y^2|X)\} = E_X\left(\frac{1}{X^2}\right)E(Z^2)$$

again using independence, and $E(1/X^2)$ must exist.

If

$$g(Y; y) = I(Y; y) = 1 \qquad \text{if} \quad Y = \frac{Z}{X} \leqslant y$$

$$= 0 \qquad \text{if} \quad Y = \frac{Z}{X} > y \qquad\qquad (11.3.11)$$

then

$$\begin{aligned} E\{g(Y; y)\} &= E_X[E_Z\{I(Z; yX)\}] \\ &= E_X\{P(Z \leqslant yX)\} \\ &= E_X\{F_Z(yX)\} \end{aligned} \qquad\qquad (11.3.12)$$

where we have assumed for this percentile problem that X is a positive random variable.

EXERCISE   Write down the expectation for a random variable X taking values on the whole real line.   ∎

A simple application of these ideas is to the simulation of the percentile, $p_t = P\{T_n \leqslant t\}$, of the T_n random variable with n degrees of freedom. The straight-forward way is, for each replication, to generate n Normal$(0, 1)$ random variables and form $(n)^{1/2}\bar{X}/S_X$. However, recall that in Normal samples X and S_X^2 are independent. Thus, it is more efficient to generate one chi-square random variable with $n - 1$ degrees of freedom, divide by $(n - 1)$ to get S_i, and compute the conditional (Normal) probability that $\bar{X}_i$, an independent Normal$(0, 1)$ random variable, is less than tS_i in order to estimate the p_tth percentile.

EXAMPLE **11.3.4 Ratios of Dependent Random Variables**
As an example of ratios of *non-independent* random variables, assume that Z and X are a pair of bivariate Normal random variables. Then, using well-known results for conditioning in Normal, bivariate populations (see Chapter 10: Equations 10.1.16 and 10.1.17), we have:

$$
\begin{aligned}
E(Y) = E_X\{E_Y(Y|X)\} &= E_X\left\{E\left(\frac{Z}{X}\bigg|X\right)\right\} \\
&= E_X\left\{\frac{1}{X}E(Z|X)\right\} \\
&= E_X\left[\frac{1}{X}\left\{\mu_Z + \frac{\sigma_Z}{\sigma_X}\rho_{XZ}(X - \mu_X)\right\}\right] \\
&= E_X\left\{\frac{\mu_Z}{X} + \frac{\sigma_Z}{\sigma_X}\rho_{XZ}\left(1 - \frac{\mu_X}{X}\right)\right\}
\end{aligned}
\tag{11.3.13}
$$

As usual, special properties of Normal distributions make everything easier. The exact conditional sampling procedure is then as follows:

1. Generate a sample of m Normal random variables $X_i, \ldots, X_m$ with mean μ_X and σ_X.
2. Compute

$$
W_i = \frac{\mu_Z}{X_i} + \frac{\sigma_Z\rho_{XZ}}{\sigma_X} - \frac{\sigma_Z\mu_X\rho_{XZ}}{\sigma_X X_i} \qquad \text{for } i = 1, \ldots, m
$$

3. Average over these m quantities to obtain the estimate Y^* of $E(Z/X)$ and use the sample variance of the W_i to estimate var(Y^*).

Note that it is not necessary to generate a bivariate Normal sample, only the marginal (Normal) random variable X. If we use the straightforward method, we must first generate the m bivariate variables (Z_i, X_i), form $Z_i/X_i = Y_i$, and average to get:

$$
\bar{Y} = \frac{1}{m}\sum_{i=1}^{m} Y_i
$$

One expects that the distribution of the Y_i's is going to be very long tailed, simply because the ratio of two independent Normal random variables has a Cauchy distribution for which the mean does not exist. In fact, ratios of correlated

Normal random variables always have infinite mean (Hinkley, 1969). Thus, the simulation is invalid if one is considering moments of the ratios. It does illustrate the need for careful validation of analytic assumptions in simulations; they are very difficult to validate by simulation. In particular, the ratio $Y = Z/X$ has, when μ_X is far, relative to σ_X, from 0 a distribution that is close to a Normal distribution, except in the tails. The mean is still undefined.

The procedure defined above, though invalid for simulating the mean of Y, still has utility in simulating the *distribution* of $Y = Z/X$. This is done using an extension of Equation 11.3.12 and the known Normal distribution (see Chapter 10) of Y given $X = x$.

The results in Equations 11.3.1 and 11.3.2 and the estimator Y^* are easily extended to cases in which X is vector valued. A particular case would be that of functions defined on a continuous-time Markov chain. By fixing the n states in a simulation up to the nth transition, it is easy to calculate $E_Y(Y|X)$ for any set of states $\mathbf{X}$ obtained in the simulation of the Markov chain. See Hordijk, Iglehart, and Schassberger (1976) for details. In this case, the X_i's are not independent and variance reduction is not automatic.

We examine now two important multivariate cases in which the X_i's are iid.

11.3.4   Minima of Subexponential Gamma Variates: The Beta–Gamma Transformation

In Section 3.5, we discussed the statistic

$$M_n = \min(X_1, \ldots, X_n) = X_{(1)} \tag{11.3.14}$$

where the X_i's are iid, and we showed that

$$F_{M_n}(x) = P(M_n \leqslant x) = 1 - \{1 - F_X(x)\}^n \tag{11.3.15}$$

This equation can be solved numerically for most distributions, and it has a closed-form solution for extreme value distributions, of which the exponential and Weibull distributions are special cases. In particular, if the X_i's are exponential$(1/\beta)$ then $F_{M_n}(x) = 1 - e^{-nx/\beta}$: that is, M_n is exponential with parameter (n/β).

In the general case, moments of minima are much harder to find than percentiles. For example, if X is a positive valued random variable, we have:

$$E(M_n) = \int_0^\infty P(M_n > x)\,dx = \int_0^\infty \{1 - F_X(x)\}^n\,dx$$

and this usually cannot be directly evaluated. Quantiles may also be difficult to evaluate, since they involve $F_X^{-1}(x)$, which is generally harder to compute than $F_X(x)$.

Now consider the case where X has a Gamma(β, k) distribution with $k \leqslant 1$. The inverse distribution is difficult to compute (see, e.g., Best and Roberts, 1975). However, a Gamma(β, k) random variable can be expressed (Lewis, 1983) as

$B(k, 1 - k)G(\beta, 1)$, where $B(k, 1 - k)$ is a Beta random variable (Table 6.1.2) distributed independently of the $G(\beta, 1)$ random variable, which is just an exponential variable. Since β is a scale parameter, we set it equal to 1 in what follows.

Now if the X_i are expressed as $B_i(k, 1 - k)E_i$ in Equation 11.3.14 and we condition on the B_i's (i.e., $B_i = b_i, i = 1, \ldots, n$), then M_n becomes the minimum of n independent exponential random variables that are *not identically distributed*, since $\beta_i = b_i$. However, the (conditional) minimum is still exponentially distributed:

$$P(M_n > x | \mathbf{B} = \mathbf{b}) = \prod_{i=1}^{n} e^{-x/b_i} = e^{-(\sum(1/b_i))x} \tag{11.3.16}$$

This suggests the following schemes for determining distributional properties of the minimum of n independent Gamma variables, where the shape parameter k is less than 1.

i. Crude Simulation

1. Generate n $G(\beta, k)$ random variables $X_1, \ldots, X_n$.
2. Find the minimum of these n random variables as $X_{(1)} = M_n$.
3. Repeat m times to obtain a sample of size m of $X_{(1)}$'s. This is the basis of the usual estimates of moments (Section 6.1), percentiles (Section 6.2), and quantiles (Section 6.3) of the distribution of $X_{(1)} = M_n$.

Besides the need to generate mn Gamma variates, a bottleneck in this procedure is finding the minimum of n random variables, which requires n comparisons, though not an ordering of the whole sample.

ii. Crude Simulation and Direct Generation of $X_{(1)}$

It is possible to generate the minimum $X_{(1)}$ of a sample of n random variables $X_1, \ldots, X_n$ sampled from $F_X(x)$ directly using the inverse probability integral transform. Thus from Equation 11.3.1:

$$F_{M_n}^{-1}(p) = F_X^{-1}\{1 - (1 - p)^{1/n}\}$$

so that one can generate $X_{(1)}$ as follows:

$$X_{(1)} = F_X^{-1}\{1 - (1 - U)^{1/n}\}$$

No sample of n Gamma variates is needed; the drawback is the *need to compute* $F_X^{-1}(\cdot)$, which is difficult for the Gamma distribution with $k < 1$. Once, however, m $X_{(1)}$'s have been generated, the simulation proceeds as in subsection i. Of course, U can be used instead of $1 - U$, since both are uniform(0, 1) random variables, and from now on we will assume that this change is implemented.

iii. Control Variables Using Direct Generation of $X_{(1)}$

If $X_{(1)}$ is generated as in subsection ii, then from the same uniform variate U one generates an exponential minimum:

$$C_{(1)} = G_X^{-1}(1 - U^{1/n}) = -\frac{1}{\beta}\ln(U^{1/n})$$

This variable $C_{(1)}$ should be highly positively correlated with the corresponding $X_{(1)}$, suggesting its use as a regression-adjusted control variable for any moment, percentile, or quantile of $X_{(1)}$ that is to be obtained in the simulation.

iv. Conditional Sampling Based on the Beta–Gamma Transform

Using the representation of X_i as $B_i(k, 1 - k)E_i$ given above, and the result given at Equation 11.3.16, proceed as follows, assuming for convenience that the quantity of interest is $E(M_n)$.

- Generate n Beta$(k, 1 - k)$ variables B_i with observed values b_i.
- Since $E(M_n|\mathbf{B} = \mathbf{b}) = 1/(\beta\sum 1/b_i)$, form this conditional mean from the n b_i's.
- Repeat m times and average to estimate $E(M_n)$.
- Since this estimate of $E(M_n)$ is an average, form the usual variance estimate from the sample variance.

No inverses are required in this scheme. Its utility depends on the kind of variance reductions that can be obtained. With this in mind, a simulation was performed using the SMTBPC program.

Results are shown for $k = 0.96$ for eight different sample sizes n in Figures 11.3.1 and 11.3.2. Figure 11.3.1 shows the straightforward simulation of subsection i and Figure 11.3.2 shows the conditional estimate given in subsection iv. Actually, the quantity estimated is $n \times M_n$, which would have expected value 1 for all n for the exponential case $(k = 1)$. Vertical scales in the two figures are identical.

At $n = 5$ the standard deviation of the mean is 0.0152 for the straightforward estimator (Figure 11.3.1) and 0.0026 for the conditional estimator (Figure 11.3.1) a factor of approximately 6, giving a variance reduction of 36. At $n = 40$, the ratio of standard deviations is quite similar, $0.0419/0.0080 \approx 5.3$. Note, however, how different the distributions of the estimators are. The straightforward estimator is almost exponential, indicated by the positive skewness in the boxplots, while the conditional estimate is highly negatively skewed. This negative skewness reflects the fact that Beta variables with parameters close to zero or one produce values close to zero and one.

In Figures 11.3.3 (regular estimator) and 11.3.4 (conditional estimator) the case of $k = 0.60$ is illustrated. This is a better and more realistic test of the method, since it is further removed than the case $k = 0.96$ from the known, conditioning case, of $k = 1.0$. Again the vertical scales in the two figures are identical. One's intuition is that here, fairly far from the exponential case, the conditioning would not be as effective in producing variance reduction. However, the ratio of standard deviations of the means is $0.00806/0.0406 \approx 2$ at $n = 5$. Note, particularly, the different distributions obtained from the two estimators. The distribution obtained for the conditional estimator is much less positively skewed than that for the regular estimator.

The results for a number of values of k are summarized in Table 11.3.1. The *percentage variance reduction factor* there (last column) is another commonly used

SAMPLE SIZE (N): 2000   NO. OF REPLICATIONS (M): 10   DEGREE OF REGRESSION (D): 3

```
2.8  -   201                                      36                              24
     |     0                                       0                               0
     |     0          103                          0         43                    0
     |     0           0          72               0          0                    0
2.5  -     0           0           0               0          0         44    29
     |     0           0           0               0          0          0     0
     |     0           0           0               0          0          0     0    0
     |     0           0           0               0          0          0     0    0
     |     +           0           0               0          0          0     0    0
2.2  -     |           0           0               +          +          0     0    +
     |     |           +           +                                     0     0
     |     |                                                             0     0
     |     |                                                             0     0
     |     |                                                             +
2.0  -     |           |           |               |          |               +
     |     |           |           |               |          |
     |     |           |           |               |          |
1.7  -     |           |           |               |          |
     |     |           |           |               |          |
     |     |           |           |               |          |
1.4  -     |           |           |               |          |
     |    ---         ---         ---             ---        ---             ---   ---         ---
1.1  - -*|:|........|*|.........|*|..........|*|.......|*|........|*|...|*|.....|*|...........|*|....
0.80 -
     |    |+|         |+|         |+|             |+|                      |+|   |+|         |+|
0.52 -     |           |           |               |                       |     |           |
     |     |           |           |               |                       |     |           |
     |    ---         ---         ---             ---                     ---   ---         ---
0.23 -     |           |           |               |                       |     |           |
     |     +           +           +               +                       +     +           +
     +---|-----------|-----------|-----------|----------|----------|-----------|----------|
          5          10          15          20         25         30          35         40
```

SUBSAMPLE SIZE	5	10	15	20	25	30	35	40
MEAN	0.9184	0.8807	0.8923	0.8631	0.9181	0.8774	0.8719	0.8815
STD	0.9601	0.9494	0.9233	0.8590	0.9531	0.9724	0.9358	0.9372
STD MEAN	0.1518E-01	0.2123E-01	0.2532E-01	0.2716E-01	0.3370E-01	0.3785E-01	0.3920E-01	0.4191E-01
SKEWNESS	2.1715	2.2078	2.1505	1.7574	1.9183	2.2634	2.1049	1.7901
KURTOSIS	7.3848	7.1138	6.9885	3.9075	4.6107	7.0855	5.8990	3.6116

MEAN OF REGRESSION ON AVERAGES - COEFFICIENTS:	0.855488	1.31905	-16.3641	56.7171
VARIANCE OF REGRESSION - COEFFICIENTS:	0.485088E-02	12.1945	1791.79	17821.9
STD DEV OF REGRESSION - COEFFICIENTS:	0.696482E-01	3.49206	42.3295	133.499

REGRESSION ON VARIANCE - COEFFICIENTS: 168.985   -1368.53   3912.71   -3744.19

ESTIMATOR: MEAN-STRAIGHT MINIMUM FOR GAMMA;k=.96
VERTICAL SCALE: YMIN =   0.0001
                 YMAX =   2.8133

Figure 11.3.1 Investigation, using SMTBPC, of the sampling properties of the straightforward estimation of $n \times M_n$, the minimum of n Gamma(β, k) random variables, for $k < 1$; the true k is 0.96.

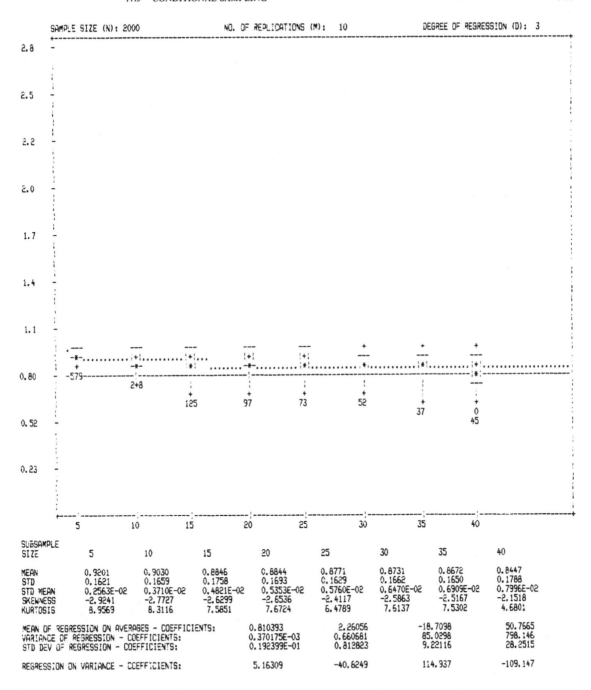

Figure 11.3.2 Investigation, using SMTBPC, of the sampling properties of the conditional estimator of $n \times M_n$, given at Equation 11.3.16; the true k is 0.96. For this simulation, it is necessary to generate n Beta$(k, 1 - k)$ variates.

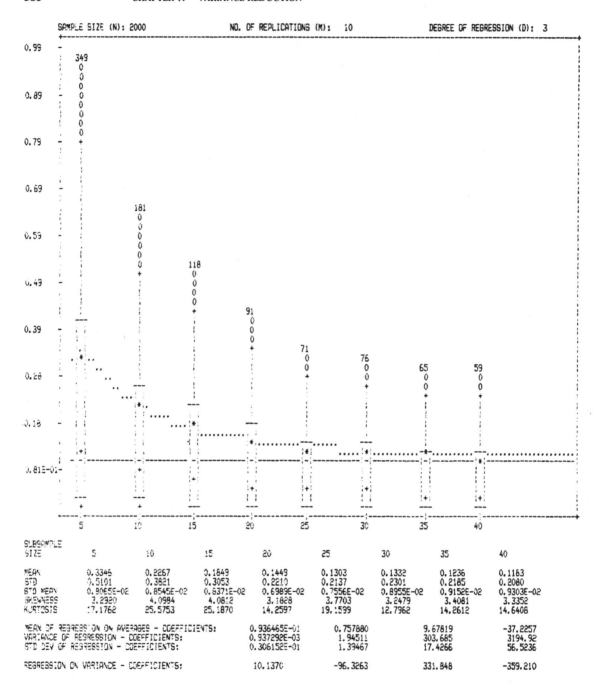

Figure 11.3.3 Investigation, using SMTBPC, of the sampling properties of the straightforward estimation of $n \times M_n$, the minimum of n Gamma(β, k) random variables, for $k < 1$; the true k is 0.60.

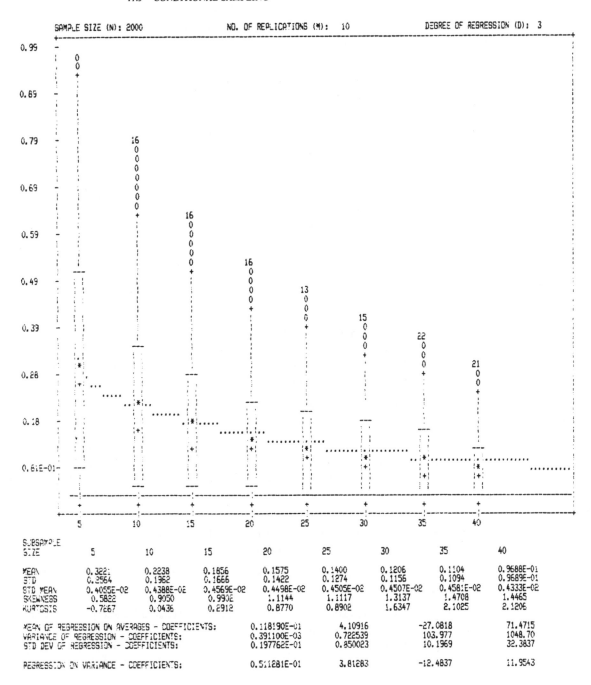

Figure 11.3.4 Investigation, using SMTBPC, of the sampling properties of the conditional estimation of $n \times M_n$, given at Equation 11.3.16; the true k is 0.60. For this simulation, it is necessary to generate n Beta(k, $1 - k$) variates.

Table 11.3.1 Comparison of crude and conditional simulations for the minimum of n Gamma(β, k) random variables with various values of $k < 1$

k	n	Straightforward estimate of the mean			Conditional estimate using Beta–Gamma transform			
		Mean	s.d.	s.d. mean	Mean	s.d.	s.d. mean	VR[a]
0.40	5	0.1135	0.2465	0.0039	0.1161	0.1663	0.0026	54%
	40	0.0102	0.0291	0.0013	0.00876	0.01773	0.00079	63%
0.60	5	0.3346	0.5101	0.0081	0.3221	0.2564	0.0041	74%
	40	0.1183	0.2080	0.0093	0.0969	0.0969	0.0043	78%
0.80	5	0.6326	0.7672	0.0121	0.6246	0.2820	0.0045	86%
	40	0.4452	0.5951	0.0266	0.3961	0.2162	0.0097	86%
0.90	5	0.9184	0.9601	0.0152	0.9201	0.1621	0.0026	97%
	40	0.8815	0.9372	0.0419	0.8447	0.1788	0.0080	96%

[a] VR = Percentage variance reduction factor. This is the absolute difference between the variances divided by the straightforward variance, times 100. It is different from the measurement used in, for example, Table 11.2.4, where the ratio of the straightforward variance to that obtained with variance reduction is used.

measure. It is the absolute difference between the variances with and without variance reduction, divided by the variance without variance reduction, times 100.

Several other points about these methods of simulating the minimum of n Gamma(k, β) variates are in order.

• The conditional estimator is adaptable to estimates of variances and percentiles of the distribution of M_n, since the conditional distribution is exponential.

• The conditional estimator based on Equation 11.3.16 can be generalized to the case of nonidentically Gamma-distributed random variables (see Exercise 11.17). In this case, it will be clearly superior to the method of subsection ii since, for that method, n inverse distributions $F_X^{-1}(\cdot)$ would have to be computed.

11.3.5   Linear Functions of Random Variables Expressible as Ratios of Normals to Other Random Variables

In Section 11.3.4 we exploited the following facts to achieve variance reduction using conditional sampling:

• A Gamma(β, k) variate, for $k < 1$, can be expressed as a Beta$(k, 1 - k)$ variate times an exponential variate.
• Conditioning on the Beta$(k, 1 - k)$ variate leaves a scaled exponential variate.
• Minima of independent, but not necessarily identically distributed, exponential random variables have an exponential distribution.

The methodology does not extend to, say, linear functions of Gamma variates, because sums of Gamma variates are Gamma distributed only if the variates have the *same* rate parameter β. Otherwise, one obtains essentially noncentral chi-square distributions, whose distributional properties are difficult to evaluate. They are, of course, easy to generate if these properties are to be evaluated by simulation.

   In statistics, a large role is played by statistics that are *linear functions* of iid random variables and, in particular, of two-sided symmetric random variables, say $X_1, \ldots, X_i, \ldots, X_n$. Normal and Laplace variables (Table 6.1.2) are examples. Moments are easily obtainable, but not distributions, unless the X_i's are Normally distributed. Then the linear function is Normally distributed.

   Can we exploit this latter fact to estimate, by conditional sampling and simulation, the distribution of a linear function of the X_i's such as that described in Equation 11.3.17?

$$Y = h(\mathbf{X}) = \sum_{i=1}^{n} a_i X_i \tag{11.3.17}$$

The answer is clearly Yes if the X_i can be expressed as ratios, $X_i = N_i/Z_i$, or equivalently products, $X_i = N_i W_i$, of Normal random variables, $N_i \sim N(\mu_i, \sigma_i)$, and independent, positive-valued random variables Z_i and W_i. Then

$$h(\mathbf{X}|\mathbf{W} = \mathbf{w}) = \sum_{i=1}^{n} a_i w_i N_i \tag{11.3.18}$$

is Normally distributed and percentiles can be computed for each set of w_i's. Replicating m times and averaging gives the final estimate.

   Note, too, that if $H_l = h(\mathbf{X}_l|\mathbf{W}_l = \mathbf{w}_l)$ denotes the lth realization $h(\mathbf{X}|\mathbf{W} = \mathbf{w})$ in m replications, then the range $H_{(m)} - H_{(1)}$ of the m variables has known mean and can be used as a control variable for the percentile estimate.

   The breadth of this scheme depends on the extent to which random variables can be expressed as N/Z or NW. Two cases have to be distinguished:

- A random variable X is given. Is there a Z or W?
- If one is exploring alternatives to Normality in linear statistics (i.e., in robustness studies), *why not use random variables* X that are expressible as N/Z or NW? A much used case occurs when Z is uniform$(0, 1)$ (i.e., the slash distribution).

   Some symmetrical random variables that are expressible as N/Z or NW are given in subsections i through iv, where N is always Normal$(0, 1)$ and independent of Z and W.

   i. *t-distributed random variables*   Here Z is well known, for a t-distributed random variable with $n - 1$ degrees of freedom, to be $(C/(n - 1))^{1/2}$, where C is a chi-square variable with $n - 1$ degrees of freedom.
   ii. *Contaminated Normal random variables*   Let $Z = 1$ with probability $1 - p$, and $Z = 1/k$, for $k \neq 0$, with probability p. Then X is $N(0, 1)$ with probability $1 - p$, and $N(0, k)$ with probability p. This distribution is used in robustness studies.
   iii. *Cauchy random variables*   Let Z be the absolute value of an $N(0, 1)$ random variable; then $X = N/Z$ is Cauchy.
   iv. *Laplace random variables*   If Z has the density function (Hsu, 1979)

$$F_Z(z) = z \exp\left\{-\frac{z^2}{2}\right\} \qquad 0 < z$$

or alternatively (Simon, 1976), the reciprocal of Z, namely W, has the density function

$$f_W(w) = w^{-3} \exp\left\{ \left(-\frac{w^2}{2} \right)^{-1} \right\} \qquad 0 < w$$

then X is a standard Laplace-distributed variable (Table 6.1.2) with $\mu = 0$ and $\beta = 1$. The Laplace-distributed random variable is symmetric and longer tailed than the Normal.

v. *l-Laplace random variables*   The Laplace-distributed random variable is a special case of the *l*-Laplace-distributed random variables, where $l > 0$ and $l = 1$ gives the Laplace distribution. If W_v is the square root of a Beta$(l, 1 - l)$ random variable and W_{iv} is the multiplier distribution given in (iv), then

$$W_v W_{iv} N$$

is *l*-Laplace. This gives, for $l < 1$, an even longer tailed class of random variables than the Laplace random variable. For details see Dewald, Lewis, and McKenzie (1988).

This scheme for using conditional sampling with random samples X_i, $i = 1$, ..., n that are expressible as $X_i = N_i/W_i$ is broader than just linear functions. However, beyond linear functions, results are fragile. For instance, it will not work in general with quadratic functions of the X_i. Further details may be found in Arnold et al. (1956), Trotter and Tukey (1956), Relles (1970), Gross (1973), and Simon (1976). Recent work by Johnstone and Velleman (1985) is also relevant. Symmetry is also exploited in these papers, although it is not essential to the idea of expressing a random variable as a ratio of a Normal random variable to an independent random variable.

The breadth of the representation of a random variable X as the ratio N/W has been investigated by Efron and Olshen (1978).

11.4   Importance Sampling

One of the original uses of simulation, or Monte Carlo as it is called in this context, was in the evaluation of definite integrals (see Hammersley and Handscomb, 1964). For instance, consider the following simple example.

EXAMPLE   **11.4.1 Evaluation of a Definite Integral**

We wish to find the volume in a unit cube in four space of a region determined by

$$x_1 \times x_3 \leqslant 0.5$$
$$x_1 + \sin(x_4) \leqslant 0.8$$
$$x_1 + x_2^2 + x_3^3 + x_4^4 \leqslant 0.7$$

In principle, if this region is denoted by R, we have

$$V = \iiint\limits_{R} \int dx_1 dx_2 dx_3 dx_4$$

and the integration is straightforward. However, defining the region of integration R is difficult. Thus, we write

$$V = \int_0^1 \int_0^1 \int_0^1 \int_0^1 I(x_1, x_2, x_3, x_4) \, dx_1 dx_2 dx_3 dx_4$$
$$= E\{I(x_1, x_2, x_3, x_4)\}$$

where $I(x_1, x_2, x_3, x_4)$ has value one if its argument is in R and is zero otherwise, and the expectation in the second line holds if it is with respect to four independent uniform$(0, 1)$ random variables. The expectation suggests generating four uniform$(0, 1)$ random numbers and, if they are in R, scoring the outcome as 1. Otherwise it is scored as zero. Repeating this simple procedure m times, we have a binomial estimate with variance $V(1 - V)/n$. A simulation shows this volume estimate to be approximately 0.132. Since the m was 100, the estimated standard deviation of this estimate is $[0.132(1 - 0.132)/100]^{1/2}$.

Much of simulation and Monte Carlo can, in fact, be thought of as the evaluation of integrals, since an expected value can be written as an integral.

Continuing, and for simplicity considering a two-dimensional problem, let the integral to be evaluated be

$$I = \int_{-\infty}^{\infty} \int_{-\infty}^{\infty} z(x, y) \, dx \, dy \tag{11.4.1}$$

where $z(x, y)$ is some integrable function of x and y. Now let $f(x, y)$ be a two-dimensional (bivariate) density function (see Equation 10.1.4) and rewrite Equation 11.4.1 as follows:

$$I = \int_{-\infty}^{\infty} \int_{-\infty}^{\infty} \frac{z(x, y)}{f(x, y)} f(x, y) \, dx \, dy \tag{11.4.2}$$
$$= \int_{-\infty}^{\infty} \int_{-\infty}^{\infty} z^*(x, y) f(x, y) \, dx \, dy$$
$$= E\{z^*(X, Y)\} \tag{11.4.3}$$

where it is assumed that $f(x, y)$ is nonzero unless $z(x, y)$ is zero, in which case the ratio is assumed to be zero. Also, the expectation is with respect to the bivariate random variables (X, Y) with density $f(x, y)$.

The result (Equation 11.4.3) suggests that to estimate the integral I by simulation, we proceed as follows:

- Generate a sample of size m from the bivariate density $f(x, y)$, namely $(X_1, Y_1), \ldots, (X_m, Y_m)$.

- Compute the sample average:

$$\bar{I} = \frac{1}{m} \sum_{i=1}^{m} z^*(X_i, Y_i)$$

$$= \frac{1}{m} \sum_{i=1}^{m} \left[\frac{z(X_i, Y_i)}{f(X_i, Y_i)} \right] \tag{11.4.4}$$

- Estimate the variance of $\bar{I}$, since it is a sample average, as follows:

$$\hat{var}(\bar{I}) = \frac{1}{m} \left[\frac{1}{m-1} \sum \{z^*(X_i, Y_i) - \bar{I}\}^2 \right] \tag{11.4.5}$$

As at Equation 4.2.1, the actual, theoretical variance of $\bar{I}$ is:

$$var(\bar{I}) = \frac{1}{m} var\{z^*(X, Y)\} = \frac{1}{m} \int_{-\infty}^{\infty} \int_{-\infty}^{\infty} \{z^*(x, y) - I\}^2 f(x, y) \, dx \, dy$$

$$= \frac{1}{m} \left[\int_{-\infty}^{\infty} \int_{-\infty}^{\infty} \{z^*(x, y)\}^2 f(x, y) \, dx \, dy - I^2 \right] \tag{11.4.6}$$

Now the bivariate density function in Equation 11.4.2 is specified quite arbitrarily, and we want to know whether we can choose $f(x, y)$ to minimize the variance of $\bar{I}$ for given m and, possibly, to simplify sampling of the random variables (X, Y). Thus one might want to make (X, Y) a bivariate Normal pair, since we have seen in Chapter 10 that sampling from that distribution is simple. We then choose the bivariate Normal parameters μ_X, μ_Y, σ_X, σ_Y, and ρ, so as to minimize the variance of $\bar{I}$.

Before investigating the choice of $f(x, y)$, let us broaden the question to the case of an initial problem that is to actually evaluate an expectation $E\{z^*(X, Y)\}$, where X and Y have the given bivariate density $f(x, y)$. *Do we have to sample from $f(x, y)$?* The answer is No, since we can do a weighted sampling. Thus, let $f^+(x, y)$ be another bivariate density function and write

$$I = \int_{-\infty}^{\infty} \int_{-\infty}^{\infty} z^*(x, y) \frac{f(x, y)}{f^+(x, y)} f^+(x, y) \, dx \, dy$$

$$= \int_{-\infty}^{\infty} \int_{-\infty}^{\infty} z^+(x, y) f^+(x, y) \, dx \, dy$$

The problem is still the same; if we sample from $f^+(x, y)$ and form the average $\bar{I}^+$, then $E(\bar{I}^+) = I$. However, by sampling from $f^+(x, y)$ instead of $f(x, y)$, $var(\bar{I}^+)$ will be different from $var(\bar{I})$. *Importance sampling is the variance reduction technique in which the sampling distribution [i.e., $f^+(x, y)$] is chosen to reduce, and possibly minimize, $var(\bar{I}^+)$.*

How do we obtain $f^+(x, y)$ to minimize the variance of $\bar{I}^+$, and what is the minimum value of the variance so obtained? As at Equation 11.4.6, we have:

$$\text{var}(\bar{I}^+) = \frac{1}{m}\left[\int_{-\infty}^{\infty}\int_{-\infty}^{\infty}\left\{\left[\frac{z^*(x,y)f(x,y)}{f^+(x,y)}\right] - I\right\}^2 f^+(x,y)\,dx\,dy\right] \tag{11.4.7}$$

and the calculus of variations shows that this is a minimum if:

$$f^+(x,y) = \frac{|z^*(x,y)|f(x,y)}{\displaystyle\int_{-\infty}^{\infty}\int_{-\infty}^{\infty}|z^*(x,y)|f(x,y)\,dx\,dy} \tag{11.4.8}$$

When $z^*(x,y)$ is nonnegative, this optimal bivariate density is:

$$f^+(x,y) = \frac{z^*(x,y)f(x,y)}{I} \tag{11.4.9}$$

Inspection of Equation 11.4.9 shows roughly that $f^+(x,y)$ gives more weight than does $f(x,y)$ to the points where $z^*(x,y)$ is greater: that is, the sample values X_i^+, Y_i^+, $i = 1, \ldots, m$, generated from the density $f^+(x,y)$ tend to occur where $z^*(x,y)$ is greater.

More importantly, when $z^*(x,y)$ is nonnegative and consequently Equation 11.4.9 holds, the variance of I, from Equation 11.4.7, can be shown to actually be zero!

Of course, attaining this minimum variance depends on knowing I, the quantity we are trying to estimate, so that we seem to have come full circle. The utility of the result of Equation 11.4.9 is that there is no lower bound to the variance of $\bar{I}^+$, so that with a judicious choice of $f^+(x,y)$, one might be able to obtain large variance reductions. Guidance in choosing $f^+(x,y)$ comes from the observation that if the integrand $z^*(x,y)f(x,y)$ and the proposed density $f^+(x,y)$ are "similar," in the sense that their ratio is nearly constant, then by Equation 11.4.9, this constant must be I and by Equation 11.4.7, the variance will be small. Pilot studies may help in choosing a suitable $f^+(x,y)$.

Another alternative is to minimize the variance for a parametric family of $f^+(x,y)$'s, say $f^+(x,y;\theta)$, from which it is presumably easier to sample. Thus, one might choose the bivariate Normal density discussed above. One can also do importance sampling on X or Y alone. For details see, for example, Kahn (1956).

To summarize, importance sampling tells us to draw samples from a distribution that is different from the one suggested in the initial formulation of the problem. We then carry along an appropriate weighting factor in the averaging, which corrects for having used the "wrong" distribution in the sampling.

EXAMPLE   **11.4.2   Importance Sampling in Simulation Percentiles**
The x_0 percentile $p(x_0)$ of a distribution $F_X(x)$ was defined generally at Equation 6.2.1, and for an exponential(λ) distribution becomes

$$p(x_0) = P\{X \leqslant x_0\} = \int_0^{x_0} \lambda e^{-\lambda u}\, du$$

$$= 1 - \int_{x_0}^{\infty} \lambda e^{-\lambda u}\, du$$

If we consider estimating $1 - p(x_0)$ by simulation, then

$$1 - p(x_0) = \int_0^{\infty} [1 - I(u; x_0)]\lambda e^{-\lambda u}\, du$$

with $I(u; x_0)$ defined as at Equation 11.3.11, and the method of crude simulation suggests that we generate m exponential(λ) random variables $X_1, \ldots, X_n$, count the proportion of these that are greater than x_0 and, as at Equation 6.2.8, use this as an estimator, $1 - \hat{p}(x_0)$, of $1 - p(x_0)$. Its variance, as at Equation 6.2.7, is $[\{1 - p(x_0)\}p(x_0)]/m$.

However, it is easy to show, and is left as an exercise, that in this one-dimensional problem $f^+(x) = \lambda e^{-(x-x_0)}$, $x \geqslant 0$, and $f^+(x) = 0$ if $x < 0$. Then $z^+(x) = e^{-\lambda x_0}$. Thus importance sampling says to generate a sample X_1 from the shifted exponential density $f^+(x)$ and estimate $p(x_0)$ as $z^+(X_1) = e^{-\lambda x_0}$, which is independent of X_1. In fact $e^{-\lambda x_0}$ is $p(x_0)$ and the variance of this estimate is 0, as predicted above.

Of course this problem is again tautologous; if one knows the answer, one can find $f^+(x, y)$!! However, Example 11.4.2 illustrates that direct sampling can be improved upon by importance sampling.

More realistic cases in which importance sampling would be important are as follows.

1. In estimating the probability that a bivariate Normal random variable falls in a region $P(A)$, where the region is far from the center of the distribution, importance sampling can improve the precision of the crude estimate of $P(A)$. This problem arises in multivariate statistical analysis, in search theory in operations research, and in other areas of military operations research. For attempts to use variance reduction for this problem, see, for example, Moran (1984).

2. Stable laws are a two-parameter class of random variables that include the Normal ($\alpha = 2$, $\beta = 0$) and Cauchy ($\alpha = 1$, $\beta = 0$). In general, however, only their characteristic function is known analytically, not their distribution. However, a method due to Chambers, Mallows, and Stuck (1976) allows one to generate these random variables. Since for $\beta = 0$, these are symmetric random variables, it is possible that importance sampling could be useful in estimating tail probabilities for these random variables.

The next example illustrates some of the ideas of importance sampling in a problem from physics. The random variable involved takes values in a discrete, finite space so that the expectation is a summation rather than an integral. However, all the ideas of importance sampling given above apply to the discrete case. For other expositions and discussions of computing problems that arise with the Ising model, see Bak (1983), Binder (1984), or Kirkpatrick and Swendsen (1985).

EXAMPLE   **11.4.3 Importance Sampling in the Ising Model**

The Ising model from statistical thermodynamics provides a conceptually simple model for a common physical system, a magnet. The magnet can be described as a two-dimensional lattice whose points represent atomic magnetic moments or "spins." If we denote the spin in the ijth position as S_{ij}, then the spin can be either "up" ($S_{ij} = +1$) or "down" ($S_{ij} = -1$). The configuration of the entire lattice, assumed to be of size n by n with $n^2 = N$, is defined by the matrix of spins $S = \{S_{ij}\}$. Since 2^N different configurations of the S_{ij} are possible, we need to index the configurations through an index r. In this way, we can speak of the system in state S_r for $r = 1, 2, \ldots, 2^N$.

The configuration of the spins determines many of the properties of the system, including its energy $E(S_r)$ (do not confuse this standard mnemonic notation from physics with the expected value notation) and its magnetism $M(S_r)$. A simple equation for the energy associated with a configuration S_r is given by

$$E(S_r) = -\sum J_{ij,kl} S_{ij} S_{kl} - H \sum_{(ij)} S_{ij} \qquad (11.4.10)$$

where the first summation is over all combinations (ij), (kl) such that $(ij) \neq (kl)$ and the S_{ij} and S_{kl} are the individual spins within configuration S_r. The $J_{ij,kl}$ are *given*, positive constants that create negative (positive) energy for parallel (antiparallel) spins and behave so that $J_{ij,kl} \to 0$ as either $|i - k|$ or $|j - l|$ gets large. The values of $J_{ij,kl}$ will change depending on the physical substance whose magnetism is being modeled. Similarly, H is a *given* constant that adjusts the energy to reflect whether the lattice as a whole points up or down (relative to an external magnetic field).

The magnetism is simpler to define, being given as follows:

$$M(S_r) = \sum_{(ij)} \frac{S_{ij}}{N} \qquad (11.4.11)$$

thereby reflecting whether more spins are up or down.

Although any configuration of spins is possible in nature, some are more likely than others, and the probability of randomly observing a particular configuration S_r is given in the Ising model by:

$$P(S = S_r) = \frac{\exp\{-E(S_r)/kT\}}{\sum\limits_{l=1}^{2^N} \exp\{-E(S_l)/kT\}} \qquad (11.4.12)$$

where T is the temperature of the system, k is Boltzmann's constant, and $E(S_r)$ is given at Equation 11.4.10.

Notice that the probability of any one configuration decreases exponentially with the ratio of the energy of that configuration to the temperature. The dependence on temperature is such that at high temperatures $\exp\{-E(S_r)/kT\}$ is usually on the order of 1, and spins are relatively independent of each other and as likely to be up as down. Conversely, at low temperature $\exp\{-E(S_r)/kT\}$ is very sensitive to variations in $E(S_r)$ and spins are more likely to be in the same state.

The problem we wish to investigate is the calculation, for a given $J_{ij,kl}$ function, of the expected magnetism of a randomly observed lattice at temperature T. Since only a finite number of configurations are possible, this expectation can be written explicitly as follows:

$$
\begin{aligned}
\text{expectation}\{M(S)\} &= \sum_{l=1}^{2^N} M(S_l)P(S = S_l) \\
&= \frac{\sum_{l=1}^{2^N} M(S_l)\exp\{-E(S_l)/kT\}}{\sum_{l=1}^{2^N} \exp\{-E(S_l)/kT\}}
\end{aligned}
\tag{11.4.13}
$$

and computed for any T. Unfortunately, the summations extend over 2^N terms and become infeasible in terms of computing time even for a small lattice. K. G. Wilson (1979) claims that the limit of practical direct computation of Equation 11.4.13 is not much larger than a 6×6 lattice, involving approximately 7×10^{10} configurations.

Instead, we can try to use crude simulation to estimate the expected magnetism by selecting randomly from the 2^N possible configurations. If we let $Y^{(j)}$, $j = 1, \ldots, m$ be a set of identically distributed random variables representing such a sampling scheme, then we demand that for each j,

$$
P(Y^{(j)} = S_r) = \frac{1}{2^N} \qquad \text{for } r = 1, 2, \ldots, 2^N
\tag{11.4.14}
$$

An estimator for expectation$\{M(S)\}$ is given by

$$
\hat{M} = \frac{\sum_{j=1}^{m} M\{Y^{(j)}\}\exp\{-E(Y^{(j)})/kT\}}{\sum_{j=1}^{m} \exp\{-E(Y^{(j)})/kT\}}
\tag{11.4.15}
$$

where Exercise 11.22 asks you to verify that the numerator of $\hat{M}$ is an unbiased estimate of $(2^N/m)\sum_{l=1}^{2^N} M(S_l)\exp\{-E(S_l)/kT\}$. Two drawbacks to this approach are the following:

1. Generation of the $Y^{(j)}$ can be very time-consuming if not done carefully. One suggested algorithm is:

    a. Generate $Y^{(1)}$ by determining each of the 2^N individual spins through independent Bernoulli($\frac{1}{2}$) variates that define up or down spins. This is the initialization step, and it is very time-consuming.

    b. Instead of repeating Step a to obtain the next realization, choose one of the 2^N spins at random, say S_{ij}, and flip it to the opposite direction with probability $\frac{1}{2}$. Let this new configuration be $Y^{(2)}$. (Note that either $Y^{(2)} = Y^{(1)}$ or, with probability $\frac{1}{2}$, $Y^{(2)}$ and $Y^{(1)}$ differ only in spin S_{ij}.)

    c. Continue as in Step b, choosing a random spin $S_{i_k j_k}$ and flipping it with probability $\frac{1}{2}$. The resulting configuration is denoted $Y^{(k)}$.

Exercise 11.23 establishes that $Y^{(j)}$ defined in this way has the desired marginal distribution. It should be clear, however, that the $Y^{(j)}$ are dependent and that this will inflate the variance of the resulting estimate. One way to reduce this problem is to use, say, only every sixth $Y^{(j)}$ so that the time-consuming evaluation of $M(Y^{(j)}) \exp\{-E(Y^{(j)})/kT\}$ is performed on relatively independent observations.

2. Many of the randomly chosen configurations $Y^{(j)}$ will be improbable in nature and thereby will contribute negligibly to the average: that is, they will contribute only $M(Y^{(j)})\exp\{-E(Y^{(j)})/kT\}$, where $\exp\{-E(Y^{(j)})/kT\}$ is relatively small. The extent of inefficiency introduced in this way cannot be quantified explicitly, and we must rely on the general principle of *importance sampling* that sampling weights should be incorporated to make the estimand as constant as possible.

Incorporating *importance sampling* into this example is conceptually straightforward, since nonuniform weights,

$$\frac{\exp\{-E(S_r)/kT\}}{\sum_{l=1}^{2^N} \exp\{-E(S_l)/kT\}}$$

already appear in the function (Equation 11.4.13) to be estimated. Hence, instead of generating the $Y^{(j)}$'s, we need to generate an identically distributed sequence $X^{(j)}$, $j = 1, 2, \ldots, l$, such that for each j,

$$P(X^{(j)} = S_r) = \frac{\exp\{-E(S_r)/kT\}}{\sum_{l=1}^{2^N} \exp\{-E(S_l)/kT\}} \qquad \text{for } r = 1, 2, \ldots, 2^N$$

Our estimator then becomes

$$\tilde{M} = \sum_{j=1}^{m} \frac{M(X^{(j)})}{m} \tag{11.4.16}$$

where Exercise 11.24 demonstrates that expectation$(\tilde{M})$ = expectation$\{M(S)\}$. The main problem with this approach is that the sampling probabilities

$$\frac{\exp\{-E(S_r)/kT\}}{\sum_{l=1}^{2^N} \exp\{-E(S_l)/kT\}}$$

are not known, since computing the normalizing factor $\sum_{l=1}^{2^N} \exp\{-E(S_l)/kT\}$ is as hard a problem as computing expected$\{M(S)\}$. Instead, the following algorithm is usually used:

a. Generate $X^{(1)}$ by determining each of the 2^N individual spins through independent Bernoulli$(\frac{1}{2})$ variates that define up or down spins.

b. Choose one of the 2^N spins at random, say $S_{i_1 j_1}$ and flip it to the opposite state with probability p, where

$$p = \frac{\exp\{-E(X^{(1)} \oplus S_{i_1 j_1})/kT\}}{\exp\{-E(X^{(1)})/kT\}} \tag{11.4.17}$$

if the expression on the right-hand side of the equation is less than one, or $p = 1$ otherwise. In Equation 11.4.17 $X^{(1)} \oplus S_{i_1 j_1}$ represents configuration $X^{(1)}$ with spin $S_{i_1 j_1}$ flipped. The result of this flip is a new configuration denoted by $X^{(2)}$.

c. Repeat Step b, each time selecting a spin $S_{i_l j_l}$ at random and flipping it with probability

$$\frac{\exp\{-E(X^{(l)} \oplus S_{i_l j_l})/kT\}}{\exp\{-E(X^{(l)})/kT\}}$$

The resulting configuration is denoted $X^{(l+1)}$.

This algorithm does not give precisely the sampling scheme required. In fact, the first configuration, $X^{(1)}$, is chosen equiprobably from all possible S_r, (as was $Y^{(1)}$). However, with each iteration of Step c, we tend to flip toward the more likely state, either $X^{(l)} \oplus S_{i_l j_l}$ or $X^{(l)}$, so that in the long run, the probability of a configuration S_{r_0} relative to a configuration S_r follows the correct ratio

$$\frac{\exp\{-E(S_{r_0})/kT\}}{\exp\{-E(S_r)/kT\}}$$

Hence, when the sampling algorithm reaches "equilibrium," we will be sampling from the space of possible configurations with the correct weights. To help ensure that we are in equilibrium, the first few $X^{(l)}$ are often discarded and not used in computing $\tilde{M}$. As with $Y^{(j)}$ the dependence in the $X^{(l)}$ sequence is broken by choosing, say, every sixth value.

EXAMPLE   **11.4.4 Computations for the Ising Model with $n = 4$**
A comparison of the three possible approaches for computing the magnetism in an Ising model—exact calculation, crude simulation, and simulation with importance sampling—was carried out for a small 4×4 lattice. The simulations were done using the algorithms described above, keeping every third configuration, and discarding the first 100 configurations from the importance sampling approach.

The energy function chosen allows only adjacent spins (including diagonal separations) to have influence through the $J_{ij,kl}$ function. For $(i, j) \neq (k, l)$ let

$$J_{ij,kl} = 1 \quad \text{if } |i - k| = 1 \quad \text{and} \quad |j - l| = 1$$
$$J_{ij,kl} = 0 \quad \text{otherwise}$$

For simplicity, we choose $H = 1$.

Table 11.4.1 Expectation of the magnetism in a 4 × 4 lattice

The expectation of the magnetism was calculated in closed form and also estimated through crude simulation and importance sampling. Each estimate represents an average over 100 replications of the respective algorithm.

Estimation method	Average estimate: (s.d.(av))			Timing (minutes)
	$kT = 25$	$kT = 15$	$kT = 10$	
Closed numerical calculation	0.06620	0.16915	0.42357	50.0
Crude simulation[a]				
$m = 150$	0.049(0.016)	0.087(0.021)	0.123(0.027)	5.2
$m = 300$	0.062(0.014)	0.111(0.021)	0.164(0.029)	9.5
$m = 900$	0.070(0.007)	0.150(0.013)	0.264(0.020)	28.0
$m = 3600$	0.065(0.004)	0.153(0.009)	0.314(0.014)	112.0
Importance sampling[a]				
$m = 150$	0.067(0.015)	0.178(0.022)	0.381(0.034)	17.0
$m = 300$	0.053(0.011)	0.164(0.014)	0.391(0.027)	35.0
$m = 900$	0.074(0.006)	0.177(0.011)	0.443(0.017)	150.0

[a] Although m gives the number of random configurations generated, only one-third of these were used in each replication in estimating the expected magnetism.

Table 11.4.1 shows the estimates and timings for the different schemes implemented on an accelerated COMPAQ Portable 286. To calculate variance estimates for the simulations, 100 replications were run for each case. Hence the estimated expectation of the magnetism represents, in each case, an average over 100 replications of the respective algorithms above. Similarly, timings reflect all 100 replications.

Comparisons of respective standard deviations and timings in Table 11.4.1 do not tell the whole story. The importance sampling approach leads often, but not always, to a slight variance reduction, at the cost of a threefold increase in computing time. Note, however, that the crude simulation produces a biased estimate (recall the general discussion of bias in ratio estimators in Chapter 6), making it unusable for small values of kT and m, the size of the simulation. In addition, in larger, more realistic lattices (i.e., for $n > 4$), the computation required for importance sampling will not grow proportionally to the size of the lattice; by clever programming, it can be limited according to the influence of the $J_{ij,kl}$ function. The work for the crude simulation will, however, increase with lattice size.

11.5 Stratified Sampling

We discuss here, in a more formal setting than we have previously used, the variance reduction technique called *stratified sampling*. This setting is due to J. R. Wilson (1984) and Cheng (1984). For other approaches to stratified sampling, see, for example, Kahn (1956).

Thus, assume that we are estimating $\theta = E(Y)$, where

$$Y = \psi(\mathbf{U})$$

and $\mathbf{U}$ is the finite set of pseudo-random numbers that constitute the input process. Let the size of this set be k, and the uniform density over the unit cube in k dimensions I^k be denoted $f_U(\mathbf{u})$. Then we can write

$$\theta = E(Y) = \int_{I^k} \psi(\mathbf{u})\, d\mathbf{u}$$

Cheng (1984) introduced the idea of an *intermediary variable*, $X = \beta(\mathbf{U})$, such that the following obtain:

- The distribution $F_X(x) = P(X \leqslant x)$, $-\infty < x < \infty$, is known.
- Y and X are (hopefully) correlated.
- Given $X = x_0$, one can sample U according to the conditional density $f_U\{\mathbf{u}\,|\,\beta(\mathbf{u}) = x_0\}$.

The intermediary variable may be multivariate, but we consider only the univariate case here.

In stratified sampling, the space of outcomes of X is partitioned (stratified) into L strata $\{S_l\colon 1 \leqslant l \leqslant L\}$, where the probability of X falling in each stratum S_l is *known* and denoted by p_l.

Choice of the strata is the first problem in this method. For the present, assume that the stratification is given.

Now we sample randomly in each stratum a *fixed* number n_l of pairs $\{Y_{lj}, X_{lj}\colon 1 \leqslant j \leqslant n_l\}$ and calculate a mean response in each stratum:

$$\bar{Y}_l = \frac{1}{n_l} \sum_{j=1}^{n_l} Y_{lj} \qquad 1 \leqslant l \leqslant L \tag{11.5.1}$$

Note that $E(\bar{Y}_l) = E\{Y_{lj}\,|\,X \in S_l\}$.

Let $n = n_1 + n_2 + \cdots + n_L$. The stratified estimator of θ is then

$$\hat{\theta}_n = \sum_{l=1}^{L} p_l \bar{Y}_l \tag{11.5.2}$$

which is *unbiased*.

Note that a *fixed number* of observations n_l is being obtained in each stratum, rather than the *random* number that would be obtained by complete unstratified sampling. This does not affect the expectation of $\hat{\theta}_n$, but does affect its distribution. In fact, we hope the variance of $\hat{\theta}_n$ will be very different from $\mathrm{var}(Y)/m$. Also, the n_l's, $l = 1, \ldots, L$, may be *very* different from the expected values $p_l n$ we observe from stratum S_l in random, unstratified sampling.

Now let

$$\mu_{Yl} = E(Y\,|\,X \in S_l) \tag{11.5.3}$$

$$\sigma_{Yl}^2 = E(Y^2\,|\,X \in S_l) - \mu_{Yl}^2 \tag{11.5.4}$$

$$\bar{\sigma}_Y = \sum_{l=1}^{L} p_l \sigma_{Yl} \tag{11.5.5}$$

The first two quantities are the mean and variance of Y in the lth stratum. Since we are simulating Y to find $E(Y)$, these will clearly be unknown.

It can be shown, for the given stratification, that the variance of $\hat{\theta}_n$ is minimized with the *optimal sampling allocation* $n_n^*(l)$, where

$$n_n^*(l) = \frac{np_l\sigma_{Yl}}{\bar{\sigma}_Y} \qquad 1 \leqslant l \leqslant L \tag{11.5.6}$$

Note two things about this result.

With random sampling, n_l would be random and would have expectation np_l, which could be very different from the *fixed* value $n_l = n_n^*(l)$.

Intuitively, Equation 11.5.6 says that one should sample more in the stratum where Y is most variable. This will be the main guide in picking n_l, since the quantities in (11.5.6) are unknown.

Also, note that the problem of picking the "best" stratification is still one layer back.

The variance of $\hat{\theta}_n$ under the *optimal sampling allocation* (call it $\hat{\theta}_n^*$) is:

$$\text{var}(\hat{\theta}_n^*) = \frac{1}{n}\left[\text{var}(Y) - \sum_{l=1}^{L} p_l\{(\mu_{Yl} - \theta)^2 + (\sigma_{Yl} - \bar{\sigma}_Y)^2\}\right]$$

One stratified sampling scheme that is often used is *proportional allocation*, in which $n_l = np_l$, which is the expected number of samples in stratum l under random sampling. The variance reduction for estimating the mean of Y comes from the fact that fixed numbers of observations are taken in each stratum, thereby eliminating a source of variability. The disadvantage is that one must use different estimators for each quantification of Y of interest. Thus the formulation is broad enough to cover the estimation of, say, the mean, variance, or percentiles of a response, but is specific to that quantification. Variance estimates of the estimate have to be obtained by sectioning of the data.

The scheme described above is actually *prestratified sampling*. It is possible and more practical to use an initial sampling to *poststratify*. Details are not given here.

A very simple example of stratified sampling is the following.

Assume that we have a GI|GI|k queueing system (Section 3.3) in which the number of servers k, while scheduled to be 4, can be from 1 to 4 on different days, with probabilities p_1, p_2, p_3, p_4. It is assumed that the number of servers who show up on different days are independent random variables. The probability of only one server showing up on any given day is very small, as is the probability of only two showing up. If one is interested in a customer waiting at least 5 minutes for service, this will most likely occur on days on which only one server is present. Therefore, one should sample, when doing a simulation consisting of n days, the case where $k = 1$ much more often than np_1. Even sampling the different values of k the fixed number of times $p_1 n, p_2 n, p_3 n, p_4 n$ is a help. This as denoted above is the case of *proportional allocation*.

Exercises   **11.1.** List random variables (e.g., Cauchy's, binomials) for which it seems simple or efficient to generate antithetic pairs X and X^a. What is the role of symmetry in this process? Where the fixed relationship between X and X^a, as given in Section 11.1.1, is not obvious, plot the relationship between X and X^a. Do this either by generating a sample from $F_X(x)$ and plotting the values of the sample against the corresponding X^a's or by using a fixed grid of x values.

11.2. In the GI|GI|1 queue in Example 11.1.1, assume that you want to estimate $p_0 = P(W_n = 0)$. Describe how to do this and how you would get an estimate of the variance of your estimate.

11.3. Describe how you would use the variables $W_n(j)$, $j = 1, \ldots, m$ and $W_n^a(j)$, $j = 1, \ldots, m$ to estimate the variance of W_n. Note that the sample of size $2m$ is not an independent sample. What effect does this have on the variance of the variance estimate?

^T**11.4.** It is desired to study the correlations in Table 11.1.1 in more detail. Can you think of a way of obtaining variance reduction in this simulation? In particular, refer back to methods discussed in Chapter 8.

11.5. Derive a condition for variance reduction with the straightforward control $\bar{Y}'$ given at Equation 11.2.1 for the case where $\sigma_C^2 \neq \sigma_Y^2$. Is it in fact possible that in this case there exist conditions under which $\text{var}(\bar{Y}')$ is always greater than $\text{var}(Y)$ no matter what the value of ρ_{CY}? Study the cases $\sigma_C < \sigma_Y$ and $\sigma_C > \sigma_Y$ separately.

11.6. Assume that you know the mean value $E(X)$ of a sample of random variables $X_1, \ldots, X_m$, and wish to estimate a quantile of the distribution of X. Can you use $\bar{X}$ as a control in the estimation of the quantile? Discuss how you would do this, with particular reference to the cases of m small, say 100, and m very large, say 10,000.

^T**11.7.** In using the regression-adjusted control, it is assumed that $E(C)$ is known. We have seen that in the simple GI|GI|1 queue where the control is the waiting time WE_n in the MI|MI|1 queue, run with common random numbers, that $E(WE_n) \rightarrow t/(1 - t)$ as $n \rightarrow \infty$, where $t = E(X)/E(S)$ is the traffic intensity. If we choose n too small, then $E(WE_n) \neq t/(1 - t)$. Discuss the effect of an incorrect value of $E(C)$ in the simple and multiple control situations.

^T**11.8.** As an internal control for a GI|GI|1 queue when simulating the lengths of the busy period, it is suggested to use

$$C_1 = 0 \quad \text{with probability } p_1 \quad\quad \text{if } \alpha = 1$$
$$C_1 = 1 \quad \text{with probability } p_2 \quad\quad \text{if } \alpha = 2$$
$$C_1 = 2 \quad \text{with probability } 1 - p_1 - p_2 \quad \text{if } \alpha > 2$$

Compute p_1, p_2 when the queue has exponential service and interarrival times.

^T**11.9.** Repeat Exercise 11.8 when the interarrival times are no longer exponentially distributed. Can you suggest any other controls?

^T**11.10.** In an application of control with regression, one does not usually know ρ_{CY} or σ_C and σ_Y. Thus we may have to estimate these quantities from the m observations on C and Y to estimate $\beta_{C \cdot Y}$ given at Equation 11.2.4. Then it is no longer true that $E(Y^*) = E(Y)$. A way of getting around this bias is to split the observations into two equal groups and determine $\beta_{C \cdot Y}$ for each group separately; each of these estimates of $\beta_{C \cdot Y}$ is used with the other half of the sample to form an estimate Y^* from a sample of size $m/2$. The two estimates Y^* are then averaged to form the complete estimate.

Explain why this complete estimate will be unbiased. Are there any drawbacks to this scheme? In particular, how would the variance of this new estimator compare with the variance of the original estimator?

^{T}**11.11** The use of control variables can be generalized to the case of controlling p responses, $Y_1, \ldots, Y_p$ with k controls, as follows (J. R. Wilson, 1984; Rubinstein and Marcus, 1985):

Denote the response as a p-dimensional column vector

$$\mathbf{Y} = \{Y_1, \ldots, Y_p\}'$$

and the controls $X_1, \ldots, X_k$ as the k-dimensional column vector

$$\mathbf{X} = \{X_1, \ldots, X_k\}'$$

The column vector of means of the controls is $\boldsymbol{\mu_X}$ and the quantity to be estimated is $E(\mathbf{Y}) = \mathbf{0}$.

Denote the linearly controlled responses as

$$\mathbf{Y(b)} = \mathbf{Y} - \mathbf{b}(\mathbf{X} - \boldsymbol{\mu_X})$$

where $\mathbf{b}$ is a $p \times k$ matrix of control coefficients. Let $\boldsymbol{\Sigma_Y}$ and $\boldsymbol{\Sigma_X}$ be the covariance matrices of $\mathbf{Y}$ and $\mathbf{X}$, respectively, and $\boldsymbol{\Sigma_{YX}}$ be the covariance between $\mathbf{Y}$ and $\mathbf{X}$.

^{T}**a.** Show that

$$\boldsymbol{\Sigma_{Y(b)}} = \text{cov}\{\mathbf{Y(b)}\} = \boldsymbol{\Sigma_Y} - \mathbf{b}\boldsymbol{\Sigma_{YX}'} - \boldsymbol{\Sigma_{YX}}\mathbf{b}' + \mathbf{b}\boldsymbol{\Sigma_X}\mathbf{b}' \qquad *$$

and that the *generalized variance* of $\mathbf{Y(b)}$ (i.e., $\det[\boldsymbol{\Sigma_{Y(b)}}]$) is minimized if $\mathbf{b}$ is given by

$$\boldsymbol{\beta} = \boldsymbol{\Sigma_{YX}}\boldsymbol{\Sigma_X^{-1}} \qquad\qquad\qquad \dagger$$

^{T}**b.** Find the resulting minimum generalized variance when $\mathbf{b} = \boldsymbol{\beta}$.

^{T}**c.** Consider the simulation of the 0.95 quantile of the Anderson–Darling statistic W_n^2, given at Equation 11.2.26, so that $p = 1$, and use as the $k = n$ controls $\mathbf{X}$, the iid exponentials E_i', $i = 1, \ldots, n$ in Equation 11.2.29. Work out * and $\dagger$ for this particular case, noting that the correlations between the E_i''s are zero, and the means and variances are known. In fact, for all i, E_i has mean and variance equal to one.

^{C}**d.** Perform a simulation using the results of part c and determine, for $n > 2$, what kind of variance reduction is obtained over the crude simulation. Note that the covariances between W_n^2 and the E_i''s will have to be estimated.

11.12. Suggest one or more control variables for simulating the distribution of the Kolmogorov–Smirnov statistic. Recall that their mean values must be known, exactly or approximately, and it is helpful if the standard deviations of the controls and the correlations between the controls are known.

^{T}**11.13.** As an internal control for the quantiles of the trimmed t statistic, it is suggested that one use $T_{n,1}(n - i) + T_{n,1}(i)$, $i = 1, \ldots, n/2$ to control $T_{n,1}(n - i)$ in a quantile estimate. Comment on this in general and in relationship to the use to $\bar{T}_{n,1}$ as an internal control as in Example 11.2.2 and in Exercise 11.6.

^T**11.14.** One wishes to calculate the probability of obtaining a total of 3 when one tosses two six-faced dice, where the dice are labeled from one to six, and each face has equal probability of appearing on a throw (Kahn, 1956).

 a. Compute the probability of a total of 3.

 b. In straight sampling to estimate p_3, one tosses the pair of dice m times and if there are m_3 occurrences of a total of 3, then

$$\hat{p}_3 = \frac{m_3}{m}$$

What is the variance of $\hat{p}_3$, and how is it estimated?

 c. Let X be the outcome of the first die and X the outcome of the second, and $p = P(X_1 + X_2 = 3 | X_1)$. What is the range of values of the random variable p and its distribution on those values?

 d. Use the result in part c to estimate p by conditional sampling (i.e., sampling only X_1).

 e. How do you estimate the variance of this estimate?

^T**11.15.** Repeat Exercise 11.14 when the two dice have probabilities 6/100, 12/100, 15/100, 18/100, 20/100, and 29/100 of outcomes 1, 2, 3, 4, 5, and 6, respectively.

^T**11.16.** Describe how you would use the conditional distribution of M_n given in Equation 11.3.16 to estimate by simulation the $x = 2$ percentile of the distribution of nM_n for $k < 1$. Describe how you would use the same distribution of M_n given at the same equation to estimate by simulation the variance, skewness, and kurtosis of M_n.

^T**11.17.** Generalize the result of Equation 11.3.16 so that one can estimate, using conditional sampling, properties of the minimum of n Gamma(k_i, β_i) random variables $X_1, \ldots, X_i, \ldots, X_n$. Note that the k_i's and β_i's may be different, although the k_i's must be less than one.

^T**11.18.** Use the answer to Exercise 11.16 to suggest a method of estimating the quantiles of M_n.

^T**11.19.** Show that when $z(x, y)$ is nonnegative, the density $f^+(x, y)$ makes the variance of $\bar{I}^+$ zero.

^T**11.20.** Examine the effect of choice of $f^+(x, y)$ on the estimates of kth-order moments of $\bar{I}^+$ [i.e., on $E((\bar{I}^+)^k)$ and var$((\bar{I}^+)^k)$].

^T**11.21.** Show that if $\rho_{YC_1} = \rho$, $\rho_{YC_2} = \rho^2$, and $\rho_{C_1 C_2} = \rho$, then $\rho_{YC_2 \cdot C_1}$ is zero, and using C_2 as a second control in addition to using C_1 as a control will give no variance reduction. Can you generate a trivariate Normal random variable having this correlation structure from three iid Normal random variables?

^T**11.22.** Show that the numerator of $\hat{M}$ in Equation 11.4.15 is an unbiased estimate of $(2^N/m)\sum_{i=1}^{2^N} M(S_i)\exp\{-E(S_i)/kT\}$.

^T**11.23.** Show that the first scheme for generating $Y^{(j)}$ given under Equation 11.4.15 has the desired marginal distribution.

^T**11.24.** Show that the estimator $\tilde{M}$ given by Equation 11.4.16 has the correct expected value, expectation($\tilde{M}$) = expectation$\{M(S)\}$.

^T**11.25.** Suggest an *importance sampling* scheme for the simple example of evaluation of a definite integral given in Example 11.4.1.

^T**11.26.** Suggest a *stratified sampling* scheme for the simple example of evaluation of a definite integral given in Example 11.4.1.

References

Anderson, T. W., and Darling, D. A. (1952). Asymptotic Theory of Certain "Goodness-of-fit" Criteria Based on Stochastic Processes. *Annals of Mathematical Statistics, 23,* 193–212.

Andrews, D. F., Bickel, P. J., Hample, F. R., Huber, P. J., Rogers, W. H., and Tukey, J. W. (1972). *Robust Estimates of Location: Survey and Advances.* Princeton University Press: Princeton, NJ.

Arnold, H. J., Bucher, B. D., Trotter, H. F., and Tukey, J. W. (1956). Monte Carlo Techniques in a Complex Problem about Normal Samples. In *Symposium on Monte Carlo Methods,* H. A. Meyer, Ed. Wiley: New York, pp. 80–83.

Bak, P. (1983). Doing Physics with Microcomputers. *Physics Today, 25,* 20–26. (December).

Best, D. J., and Roberts, D. E. (1975). AS91 The Percentage Points of the χ^2 Distribution. *Applied Statistics* (Ser. C), *24*(3), 385–388.

Binder, K., Ed. (1984). *Applications of the Monte Carlo Method in Statistical Physics.* Springer: New York.

Chambers, J. M., Mallows, C. L., and Stuck, B. W. (1976). A Method for Simulating Stable Random Variables. *Journal of the American Statistical Association, 71,* 340–344.

Cheng, R. C. H. (1981). The Use of Antithetic Variates in Computer Simulations. *Proceedings of the 1981 Winter Simulation Conference.* IEEE, Atlanta, pp. 313–315.

Cheng, R. C. H. (1982). The Use of Antithetic Variates in Computer Simulations. *Journal of the Operational Research Society, 33,* 229–237.

Cheng, R. C. H. (1984). Antithetic Variate Methods for Simulation of Processes with Peaks and Troughs. *European Journal of Operations Research, 15,* 227–236.

Conover, W. J. (1980). *Practical Nonparametric Statistics,* 2nd ed. Wiley: New York.

Cox, D. R. (1962). *Renewal Theory.* Methuen: London.

Cox, D. R., and Lewis, P. A. W. (1966). *The Statistical Analysis of Series of Events.* Methuen: London.

Dewald, L. S., Lewis, P. A. W., and McKenzie, E. (1988). *l*-Laplace Processes and the Square-Root Beta-Laplace Transform. In press.

Efron, B., and Olshen, R. A. (1978). How Broad Is the Class of Normal Scale Mixtures? *Annals of Statistics, 6,* 1159–1164.

Fishman, G. S., and Huang, B. D. (1983). Antithetic Variates Revisited. *Communications of the Association for Computing Machinery, 26*(11), 964–1071.

Gaver, D. P., and Shedler, G. S. (1971). Control Variable Methods in the Simulation of Multiprogrammed Computer System. *Naval Research Logistics Quarterly, 18,* 435–450.

Gaver, D. P., and Thompson, G. L. (1973). *Programming and Probability Models in Operations Research.* Wadsworth: Belmont, CA.

Gross, A. M. (1973). A Monte Carlo Swindle for Estimators of Location. *Applied Statistics, 22,* 347–353.

Hammersley, J. M., and Handscomb, D. C. (1964). *Monte Carlo Methods.* Methuen: London.

Handscomb, D. C. (1969). Variance Reduction Techniques. In *The Design of Computer Simulation Experiments,* T. H. Naylor, Ed. Duke University Press: Durham, NC.

Hinkley, D. V. (1969). On the Ratio of Two Correlated Normal Random Variables. *Biometrika, 56*(3), 635–639.

Hordijk, A., Iglehart, D. L., and Schassberger, P. (1976). Discrete Time Methods for Simulating Continuous Time Markov Chains. *Advances in Applied Probability, 8,* 772–788.

Hsu, D. A. (1979). Long-tailed Distributions for Position Errors in Navigation. *Journal of the Royal Statistical Society, C, 28*(1), 62–71.

Iglehart, D. L., and Lewis, P. A. W. (1979). Regenerative Simulation with Internal Controls. *Journal of the Association for Computing Machinery, 26*(2), 271–282.

Johnstone, I. M., and Velleman, P. F. (1985). Efficient Scores, Variance Decompositions, and Monte Carlo Swindles. *Journal of the American Statistical Association, 80*(392), 851–862.

Kahn, H. (1956). Use of Different Monte Carlo Sampling Techniques. In *Symposium on Monte Carlo Methods*, H. A. Meyer, Ed. Wiley: New York, pp. 146–190.

Kirkpatrick, S., and Swendsen, R. H. (1985). Statistical Mechanics and Disordered Systems. *Communications of the Association for Computing Machinery, 28*(4), 363–394.

Kleijnen, J. P. C. (1974). *Statistical Techniques in Simulation*, Part I. Dekker: New York.

Lavenberg, S. S., Moeller, T. L., and Welch, P. D. (1982). Statistical Results on Control Variables with Application to Queueing Network Simulation. *Operations Research, 30*(1), 182–202.

Lavenberg, S. S., and Welch, P. D. (1981). A Perspective on the Use of Control Variables to Increase the Efficiency of Monte Carlo Simulations. *Management Science, 27*(3), 322–335.

Lewis, P. A. W. (1961). Distribution of the Anderson–Darling Statistic. *Annals of Mathematical Statistics, 32*, 1118–1124.

Lewis, P. A. W. (1983). Generating Negatively Correlated Gamma Variates Using the Beta–Gamma Transformation. *Proceedings of the 1983 Winter Simulation Conference*, S. Roberts, J. Banks, and B. Schmeiser, Eds. IEEE Press: New York, pp. 175–176.

Moran, P. A. P. (1967). Testing for Correlation Between Nonnegative Variates. *Biometrika, 54*, 395–401.

Moran, P. A. P. (1984). The Monte Carlo Evaluation of Orthant Probabilities for Multivariate Normal Distributions. *Australian Journal of Statistics, 26*(1), 39–44.

Nelson, B. L., and Schmeiser, B. W. (1986). Decomposition of Some Well-known Variance Reduction Techniques. *Journal of Statistical Computation and Simulation, 23*, 183–209.

Page, E. S. (1965). On Monte Carlo Methods in Congestion Problems. II. Simulation of Queueing Systems. *Operations Research, 13*, 300–305.

Pearson, E. S., and Hartley, H. O. (1966). *Biometrika Tables for Statisticians*, Vol. 1. The University Press: Cambridge.

Relles, D. A. (1970). Variance Reduction Techniques for Monte Carlo Sampling from Student Distributions. *Technometrics, 12*, 499–515.

Rubinstein, R. Y., and Marcus, R. (1985). Efficiency of Multivariate Control Variates in Monte Carlo Simulation. *Operations Research, 33*(3), 661–667.

Simon, G. (1976). Computer Simulation Swindles, with Applications to Estimates of Location and Dispersion. *Applied Statistics, 25*(3), 266–274.

Stephens, M. A., and Maag, V. R. (1968). Further Percentage Points for W_n^2. *Biometrika, 55*, 428–430.

Swain, J. J., and Schmeiser, B. W. (1987). Monte Carlo Estimation of the Sampling Distribution of Nonlinear Model Parameter Estimators. *Annals of Operations Research, 8*, 245–256.

Trotter, H. F., and Tukey, J. W. (1956). Conditional Monte Carlo from Normal Samples. In *Symposium on Monte Carlo Methods*, H. A. Meyer, Ed. Wiley: New York, pp. 64–79.

Venkatraman, S., and Wilson, J. R. (1986). The Efficiency of Control Variates in Multiresponse Simulation. *Operations Research Letters, 5*(1), 37–42.

Wilson, J. R. (1984). Variance Reduction Techniques for Digital Simulation. *American Journal of Mathematical and Management Sciences, 4*(3 and 4), 277–312.

Wilson, J. R., and Pritsker, A. A. B. (1984a). Variance Reduction in Queueing Simulation Using Generalized Concomitant Variables. *Journal of Statistical Computation and Simulation, 19*, 129–153.

Wilson, J. R., and Pritsker, A. A. B. (1984b). Experimental Evaluation of Variance Reduction Techniques for Queueing Simulation Using Generalized Concomitant Variables. *Management Science, 30*(12), 1459–1472.

Wilson, K. G. (1979). Problems in Physics with Many Scales of Length. *Scientific American, 242*, 158–179 (August).

Author Index

Subject Index

T - #0622 - 101024 - C0 - 234/185/23 - PB - 9781138561878 - Gloss Lamination